From Biped to Strider

The Emergence of Modern Human Walking, Running, and Resource Transport

From Biped to Strider

The Emergence of Modern Human Walking, Running, and Resource Transport

Edited by

D. Jeffrey Meldrum

Idaho State University
Pocatello, Idaho

and

Charles E. Hilton

Western Michigan University
Kalamazoo, Michigan

Kluwer Academic/Plenum Publishers
New York, Boston, Dordrecht, London, Moscow

Library of Congress Cataloging-in-Publication Data

From biped to strider : the emergence of modern human walking, running, and resource transport / edited by D. Jeffrey Meldrum and Charles E. Hilton.
p. cm.
Based on a symposium which was held as a session of the 69th annual meeting of the AAPA held in San Antonio, Texas, in 2000.
Includes bibliographical references and index.
ISBN 0-306-47999-0 – ISBN 0-306-48000-X (pbk.)
1. Human locomotion–Congresses. I. Meldrum, D. Jeffrey, 1958- II. Hilton, Charles E. III. American Association of Physical Anthropologists. Meeting (69th : 2000 : San Antonio, Tex.)

QP303.F685 2004
612.7′6—dc22 2003060287

ISBN 0-306-47999-0 (hardbound)
0-306-48000-X (paperback)

233 Spring Street, New York, New York 10013

http://www.kluweronline.com

10 9 8 7 6 5 4 3 2 1

A C.I.P. record for this book is available from the Library of Congress.

Printed in the United States of America

Participants in the symposium, *From Biped to Strider: The Emergence of Modern Human Walking*, held at the 69th annual meeting of the AAPA held in San Antonio, Texas, April 15, 2000. Front row (l to r): Roshna Wunderlich, Laura Gruss, Marsha Ogilvie, Karen Rosenberg, David Begun, Jeff Meldrum. Back row (l to r): Dennis Bramble, Chris Ruff, Chuck Hilton, David Carrier [not pictured: Ron Clarke, Yvette Deloison, Peter Schmid, Dan Schmitt, Patricia Kramer].

Contributors

David R. Begun, Department of Anthropology, University of Toronto, Toronto, ON M5S 3G3, Canada

Gilles Berillon, CNRS, Paris, France

David Carrier, Department of Biology, University of Utah, Salt Lake City, UT 84112

Yvette Deloison, CNRS, Paris, France

Russell D. Greaves, Department of Anthropology, University of Texas, San Antonio, TX 78249

Laura Tobias Gruss, Department of Biological Anthropology & Anatomy, Duke University, Durham, NC 27708

Charles E. Hilton, Department of Anthropology, Western Michigan University, Kalamazoo, MI 49008

Patricia Ann Kramer, Departments of Anthropology and Orthopaedics, University of Washington, Seattle, WA 98104–2499

Henry M McHenry, Department of Anthropology, University of California, Davis, CA 95616

D. Jeffrey Meldrum, Department of Biological Sciences, Idaho State University, Pocatello, ID 83209–8007

Marsha D. Ogilvie, Department of Anthropology, University of New Mexico, Albuquerque, NM 87131

Peter Schmid, Anthropology Institute, University of Zurich-Irchel, Zurich CH8057, Switzerland

Daniel Schmitt, Department of Biological Anthropology & Anatomy, Duke University, Durham, NC 27708

Preface

The inspiration for this volume of contributed papers stemmed from conversations between the editors in front of Chuck Hilton's poster on the determinants of hominid walking speed, presented at the1998 meetings of the American Association of Physical Anthropologists (AAPA). Earlier at those meetings, Jeff Meldrum (with Roshna Wunderlich) had presented an alternate interpretation of the Laetoli footprints based on evidence of midfoot flexibility. As the discussion ensued we found convergence on a number of ideas about the nature of the evolution of *modern* human walking. From the continuation of that dialogue grew the proposal for a symposium which we called *From Biped to Strider: the Emergence of Modern Human Walking*. The symposium was held as a session of the 69th annual meeting of the AAPA, held in San Antonio, Texas in 2000. It seemed to us that the study of human bipedalism had become overshadowed by the often polarized debates over whether australopithecines were wholly terrestrial in habit, or retained a significant degree of arboreality. Furthermore, there remained considerable difference of opinion as to whether bipedalism as practiced by australopithecines was either equivalent to modern human walking, intermediate in nature, or unique. It seemed to us that preoccupation with these issues had lead to a degree of disregard concerning the fact that modern human locomotion is more than merely habitually walking on two legs. It has become quite evident that of all primates, and of the majority of mammals, modern humans are notably adapted for endurance walking and running. We felt that with the advent of the new millennium it was time to shift attention to the pattern and timing of the emergence of modern human walking, characterized by the development of the longitudinal arch, midfoot stability, and plantar fascia modifications that accompanied the human striding stiff-legged gait. When and in what sequence did these morphological traits appear? What were the changes in the bio-behavioral complex of hominin locomotor evolution? What were the implications for the enhancement and the expansion of hominin mobility?

We invited participants that would represent both paleontological and neotological perspectives on the functional morphology and behavioral ecology of hominin locomotor adaptations. These contributions span from the origins of bipedalism to locomotor contrasts among contemporary human populations. We purposely recruited participants largely from the upcoming generation of anthropologists in order to infuse fresh perspectives and insights into the dialogue, while building upon the pioneering researches and discussions of senior colleagues. Their contributions before, during, and after the symposium were thought-provoking.

This volume reports the proceedings of that symposium. The subtitle was expanded to properly recognize the significant roles of running and resource transport in modern human locomotor adaptations. Regretably, not all the participants were able to contribute to the volume, but we also picked up an additional paper along the way from Gilles Berrilion, and Henry McHenry graciously agreed to contribute a discussion paper. The papers in this volume were reviewed by at least three reviewers and the overall quality benefited greatly from the expertise and professionalism displayed by each of them. We also wish to acknowledge the long-suffering and assistance of our publisher, Teresa Krauss.

In the time between the symposium and the completion of this volume, new fossil discoveries continue to add pieces to the puzzle of human locomotor evolution. At the same time, the application of new investigatory techniques and broader interdisciplinary perspectives provide refined interpretations of the paleontological and neontological evidence. It is our hope that this volume will spawn novel perspectives and penetrating questions directed at these issues, and that it will contribute to the ongoing dialogue between those who are interested in the hominin transition from biped to strider.

Jeff Meldrum
Chuck Hilton

Contents

From Biped to Strider

The Emergence of Modern Human Walking, Running, and Resource Transport

Chapter 1

Striders, Runners, and Transporters

CHARLES E. HILTON AND D. JEFFREY MELDRUM

Functional-morphological analyses related to fossil and contemporary hominin locomotion are the focus of this volume. As locomotion is considered a key element in the overall behavior of living primates, allowing them to fulfill such basic needs as avoiding predators, foraging for food, and finding mates, biological anthropologists have generally agreed that it most likely served similar functions in earlier hominins as well. In primates, differing locomotor behaviors and their impact on other biological complexes have produced a diverse range of behavioral and anatomical configurations. In turn, primate locomotion studies are diverse in their scope. Anthropologists interested in hominin locomotion frequently draw on primate and other animal locomotor studies in efforts to understand the complexities associated with the evolution of hominin locomotion. Through comparative analyses on musculo-skeletal structures, positional behavior, and the kinematic and kinetic components of body motion in settings ranging from dissection rooms, laboratories, and in the field, researchers have developed a wide variety of approaches and techniques for investigating the intricacies of locomotor movement in living contemporary hominins and their closest relatives.

Historically within paleoanthropology, locomotor studies have served a wide range of purposes and such analyses have been an integral component in determining the status of fossils purportedly thought to be human ancestors. Direct and indirect assessment of the locomotor anatomy for evidence of bipedality in fossil remains were important considerations for conferring or negating hominin status when Dubois (1894) announced his discovery of *Pithecanthropus erectus*, when Boule (1911–13) concluded that the La Chapelle Neandertal most likely walked with a bent-knee gait, and when Dart (1925, 1926) proposed that *Australopithecus africanus* had been a biped. While our level of inquiry is more sophisticated than that taken by these earlier researchers, functional morphological analyses of locomotion do give paleoanthropologists one of their few windows of opportunity for framing the behavior of our ancestors. The current state of knowledge indicates that hominin bipedalism probably extends into the Miocene to at least 5.2 mya (White et al., 1994,

From Biped to Strider: The Emergence of Modern Human Walking, Running, and Resource Transport, edited by D. Jeffrey Meldrum and Charles E. Hilton. Kluwer Academic/Plenum Publishers, New York, 2004.

1995; Haile-Salassie, 2001), but recent finds may push our bipedal ancestry to nearly 7 mya (Brunet et al., 2002; Senut et al., 2001). Evidence for obligate bipedalism is considered not only *the* critical defining characteristic for the attainment of hominin status but signals some of the earliest extensive morphological changes to occur relative to other apes.

As this locomotor shift marks our divergence from other Miocene hominoids, functional-morphological studies of hominin locomotion have been given special importance in paleoanthropology. Locomotor reconstructions for earlier hominin and hominoid ancestors allow us to conceptualize how this key behavior may have facilitated or constrained movements across the landscape. From these reconstructions, we can begin to understand how locomotor related morphological changes may be linked to other key behavioral changes that contributed to the success of ancestral hominin lifeways.

Current studies on the evolution of hominin lifeways via locomotion research can be broadly characterized as focusing on several major topics of inquiry. What was the state of pre-hominin locomotion and the selective pressures that led to the "origins of bipedalism?" What was the nature of locomotor behavior among the other genera pre-dating and contemporaneous with the genus *Homo*? Did the appearance of the genus *Homo* represent a dramatically different form of overall hominin locomotor behavior that permitted new patterns of landscape use and mobility in association with foraging? Finally, how did patterns of hominin locomotor behavior change with increased sedentism and the use of domesticates during the Holocene? Answers within each of these broad areas will undoubtedly generate many more interesting questions but will also provide a frame of reference towards assessing the capabilities associated with hominin locomotion.

In the last two decades, locomotor researchers have begun recognizing that the modern hominin locomotor complex operates in a wide range of physically demanding situations. Such demands, depending on their frequency in our evolutionary history, most likely placed additional selective pressures on our bipedal gait. As a result, selection on hominin bipedalism has produced vastly different locomotor characteristics relative to non-human primates and these additional characteristics often can be overlooked. Discerning the critical components of living hominin locomotion has received greater attention so as to provide the baseline comparative background for framing earlier fossil hominin locomotion.

As the title of this volume suggests, striding bipedalism is only one component of contemporary hominin locomotor capabilities. For numerous decades, we have known that maintaining a striding gait is a key feature of modern hominin bipedality (Napier, 1964, 1967). A striding gait contributes to the smooth and efficient transfer of body weight from one lower limb to the other (Carsloo, 1972; Eberhart et al., 1954; Elftmann, 1954; Inman et al., 1981; Saunders et al., 1953). However, hominin bipedality is more than just walking smoothly on two lower limbs. For living hominins, our bipedality is further enhanced by the ability to engage in long and extended bouts of striding bipedalism if necessary. Such endurance walking is frequently seen in people in traditional societies, who, without the benefit of transportation technology, can cover greater distances in a single day than those seen in any other living primate (Kelly, 1983; 1992, 1995; Gould, 1969; Lee, 1979; Rodman and McHenry, 1980; Tonkinson, 1991). In modern hominin foragers, endurance walking and other distinctive locomotor characteristics are often combined to produce a complex bio-behavioral pattern enhancing overall foraging range and mobility. Yet, we do not know how these locomotor characteristics interact with each other. For example, numerous anthropologists have qualitatively documented that in living hunter-gatherers, males and females exhibit differing patterns of locomotion across the landscape (Binford, Greaves, and Kaplan, personal communications), with males often covering greater daily distances than females

(Brown, 1970; Murdock and Provost, 1972). Differences in male and female economic organization related to resource targeting clearly contribute to distinctive male and female mobility patterns (Binford, 1978, 1980, 1990, 2001; Kelly, 1995; Kuhn and Stiner, 2002), but there is a need to understand how such patterns differentially affect the locomotor systems for each sex. Males and females foraging in the same overall environmental zone may employ different walking speeds in association with searching and non-searching components of foraging, traverse different terrains, travel as single individuals and/or travel in a group (Hilton, personal observation). We recognize that immature individuals in a hunting and gathering group may not always walk the distances traveled by adults (Gould, 1969; Hawkes et al., 1994; Kaplan et al., 2000; Lee, 1979; Tonkinson, 1991). It is possible that individuals of different ages within a same-sex adult group may also exhibit unexpected patterns in walking distance and load carrying (e.g., surprisingly, younger adults may often carry the lightest loads). Additionally, living hominin hunter-gatherers not only cover large distances during foraging, but will also cover greater distances when visiting neighboring groups and relatives. These factors are all elements of endurance walking for which we could use more information.

Another under-researched aspect of hominin locomotion in paleoanthropological studies is running and its possible role in foraging. Many biological anthropologists are still reluctant to acknowledge the unique cursorial capabilities of living hominins. The ability of individuals to engage in marathons is seen by many biological anthropologists as a product of extreme athletic conditioning rather than a product of our evolutionary history (Carrier, 1984). While living hominins are neither the fastest nor the most graceful-looking runners, both males and females in reasonable physical shape are capable of running for extended time periods (Bramble, 2000; Bramble and Carrier, 1983; Carrier, 1984; Costill, 1986; McArdle et al., 1991). If the critical measurement for cursoriality is speed then clearly our slow running speeds disqualify us. However, if we consider duration as a critical component of cursoriality, then living hominins possess unique capabilities associated with their running clearly not seen in other living primates (Bramble, personal communication; Carrier, personal communication). As noted by Carrier (1984) nearly two decades ago, unlike mammalian quadrupeds, the operation of our respiratory system is not directly coupled to the movements of the upper limbs. Our ventilation is independent of our gait speed, which allows us to run long distances by minimizing the complications associated with thermal stress and heat exhaustion. This hominin feature, along with anatomical structures acting as important elastic storage mechanisms (e.g., the plantar ligaments, Achilles' tendon, and nuchal ligament) enhance our running capacity (Carrier et al., 1995; McMahon, 1984).

Our knowledge regarding the role of running in extant and prehistoric hominin foragers is still limited in comparison to the numerous paleoanthropological investigations into bipedal walking. Predator avoidance is suggested as one selective mechanism for explaining why humans are capable of running, especially given the shift to more open savanna environments (Fleagle, 1999). However, due to our slow running speeds this mechanism may not have been effective in escaping predators. It is suggested that running would be extremely advantageous in capturing prey. Persistent pursuit of many terrestrial quadrupeds would force them into eventual heat exhaustion particularly during the hottest part of the day when many predators are not active (Carrier, 1984). One criticism is that such a prey capture pattern has not been well documented for living or recent hominin foragers. However, it can be noted that even after long bouts of walking, foragers are still often capable of quick bouts of running if necessary (Kaplan, personal communication). While sprinting after prey is sometimes seen in modern foragers, these are relatively infrequent short duration events (Hilton, personal observation). One possible explanation for the reduced frequency of running as a component of hunting is that modern foragers are aided frequently by sophisticated hunting technology (Greaves; Whittaker, personal communications). Running behavior may have been much more frequent in earlier hominins who did not possess the traps, poisons, ambush techniques, or projectile technology seen in extant hunter-gatherer groups. Without the benefit of such technology, earlier hominins in comparison to recent hunter-gatherers most likely engaged in longer and more frequent bouts of running in their efforts to chase down prey.

Another hominin locomotor characteristic is the ability to carry large burdens in conjunction with endurance walking. Like non-human primates, living hominins will transport offspring, but unlike non-human primates, hominins transport valuable economic resources such as food, water, firewood, and tools for extended distances across the landscape (Binford, 1978, 2001; Foley, 1987; Potts, 1988, 1991). Even without the aide of transportation technology, traditional peoples use a variety of different containers and mechanisms to accomplish such resource transport (Gould, 1969; Heglund et al., 1995; Lee, 1979; Kelly, 1995; Maloiy et al., 1986; Tonkinson, 1991). While non-human primates will move a few resources, this behavior is very limited and is obviously constrained by the use of their upper limbs in locomotion. Images of bipedal chimps and bonobos awkwardly carrying food prompted one researcher to quip "that chimps could use a container" (McGrew, personal communication). Paleoanthropologists have long recognized that bipedalism allows the upper limbs to more effectively carry items (see Fleagle, 1997 for a review), but paleoanthropological investigations into the impact of burden carrying as a selective force on hominin locomotion can be improved.

One problem is that interpreting the evidence for resource transport in the archaeological record is a difficult endeavor. Carrying implements such as bags, burden baskets, or a leather sling (such as a Ju/'Hoansi kaross) do not preserve well. Even a tool as minimal as a string belt can be used to carry items while the hands are full or need to be free for other tasks. While we speculate that earlier hominins must have used some type of carrying implement, the earliest evidence for basketry and weaving are the clay impression fragments dating to 28 Kya at the site of Dolní Věstonice in the Czech Republic (Adavasio et al., 1996; Soffer et al., 2000). These clay impressions, indirectly suggesting that individuals were capable of weaving a burden basket, are a late occurrence in the archaeological record.

The carrying implements described above are used by traditional people who regularly carry resources for extended distances to a central place. At issue is the evolutionary time depth of such load carrying and how loads carried with habitual frequency affected the locomotor anatomy (Heglund et al., 1995; Maloiy et al., 1986). The earliest direct evidence of any type of resource transport by hominins is tied to the use of lithic implements dated to 2.5 Mya at the Ethiopian site of Gona (Semaw et al., 2002). The evidence from stone tools does not indicate movement over great distances (Isaac, 1984; Klein, 1997; Potts, 1984, 1988, 1991, 1994; Rogers, 1994; Schick, 1987), possibly 10–12 km at the most. However, without a fine-grained time scale we have no idea as to the time length to move such implements from their source locations to their final resting spots. Transportation of stone tools probably had little appreciable effect on the locomotor system and it is unlikely to represent the beginnings of resource transport. It seems reasonable that perishable organic items were probably carried sometime before the advent of stone tools. Yet, investigations into the hominin movement of animal carcasses and plant resource loads is problematic due to the implications associated with central place foraging in Plio-Pleistocene hominins and increased planning depth. These factors are clearly viewed by some as foraging strategies that appeared much later in our evolutionary history (Binford, 1990, 2001; Kuhn and Stiner, 2002; Stiner, 1991, 1994).

All of these characteristics, seen to varying degrees in modern hominin foragers, are thought to have had an important influence not only on foraging activities but also on the selective forces that produced our modern anatomy and physiology. In particular, the shift by bipedal apes towards a new and distinctively modern-*Homo*-like body form at or immediately prior to the Plio-Pleistocene boundary appears to have given rise to increased foraging opportunities in more open savanna-like environmental zones. This new hominin body form, as exemplified by WT-15000 (*H. ergaster*), possessed longer lower limbs, shorter upper limbs, and a cylindrically shaped torso with narrow hips relative to non-*Homo* hominins (McHenry and Coffing, 2000; Richmond et al., 2002; Ruff, 1993; Ruff and Walker, 1993). The combination of these postcranial morphological features and the evidence for an externally projecting nose (Franciscus and Trinkaus, 1988) indicates a body form that would have facilitated thermoregulation during intense physical activities in the hotter/drier landscapes of the savanna (Aiello and Wheeler, 1995; Ruff, 1993). The limb proportions and the narrower pelvic (waist) and trunk regions in WT-15000 produced a high surface to volume ratio allowing this body form to dissipate heat more effectively than its closely related non-*Homo* forms (McHenry and Coffing, 2000; Ruff, 1993). This new *Homo* body form signals a dramatic shift in the types of environmental zones exploited, the food resources used, and the methods used for procuring those resources. The anatomical changes facilitating thermoregulation are all possibly linked to the increased demands of strenuous physical activities, such as greater daily ranges, a higher frequency of running duration and distances, and the simultaneous transport of resources, in association with foraging. Longer lower limbs in *Homo* relative to *Australopithecus* would have produced a longer stride and allowed one to either cover the same foraging distance in shorter time or travel greater distances in the same amount of time (McHenry and Coffing, 2000). Shorter upper limbs relative to those seen in the australopiths would have helped to minimize the energetic expenditure of upper limb oscillations (Bramble, 1991). Torso modifications, although possibly related to a decrease in the size of the gastrointestinal tract, would have shifted the position of center of mass during locomotion (Aiello and Wheeler, 1995). *Homo ergaster* most likely engaged in activities incorporating new patterns of locomotor behavior with a possible shift in organizational tactics requiring increased demands in transporting resources. These changes mark a successful adaptive complex that did not change in its overall components until sometime after the advent of anatomically modern humans. Thus, it is not surprising that the body form seen in WT-15000 at 1.65 Mya may also be linked to the earliest evidence, at 1.8 Mya, of hominins outside of Africa at the site of Dmanisi (Gabunia et al., 2001), especially if cranial similarities between the Dmanisi fossils and *H. ergaster* are extended to the postcrania.

It is clear that even within the time frame of modern humans further shifts in economic organization and behavioral strategies have led to increasing reliance on technological adaptations affecting the biological adaptations associated with our locomotor systems. Increasingly complex technological innovations in conjunction with social organization, coupled with increased sedentism and the advent of agriculture produced the dramatically gracile skeletons seen in recent and contemporary modern hominins.

The papers in this current volume offer contributions into the variability associated with hominin locomotion by highlighting the range of selective pressures on hominin locomotor behavior throughout our evolutionary history. The volume is organized into two parts. The first addresses the advent and nature of pre-*Homo* bipedality while the second part focuses on locomotor behavior within the genus *Homo*. David Begun addresses the evidence that hominin bipedalism is derived from a knuckle-walking ancestry through his work on comparative anatomy. Peter Schmid provides a discussion of his work on the Laetoli footprints as well as his recent experimental research on chimp and human gait movements. Yvette Deloison combines her recent work on the STW-573 pedal remains assigned to *Australopithecus africanus* and their implications for understanding the bipedality in the Laetoli footprints. The paper by Jeff Meldrum also examines the Laetoli footprints through a comparative approach utilizing modern human trackways through volcanic ash deposits in Hawaii and presents inferences about hominin pedal morphology through the Pleistocene. Patricia Kramer provides an overview of the importance of locomotion for understanding the behavioral ecology of hominins and highlights how theoretical locomotor models can be used to examine the efficiency of walking and load carrying in *Australopithecus afarensis* and *Homo*. Laura Gruss and Dan Schmitt examine the kinematics of walking associated with the dramatic change in limb proportions seen in early *Homo* as exhibited by WT-15000. Dave Carrier highlights how the advent of stone tools may have relaxed certain selective pressures on biological adaptations associated with male-male competition thus facilitating the evolution of hominin running. The paper by Chuck Hilton and Russell Greaves focuses on the patterns of resource transport and distances walked during foraging by males and females in a modern hominin foraging group. Finally, Marsha Ogilvie examines the extremely important transition from foraging to farming in the prehistoric American Southwest in order to address how patterns of mobility changed in males and females with the advent of agriculture.

Each contributor to this volume offers additional insights towards a greater understanding of hominin locomotion. Our hominin locomotor system did not stop evolving with the advent of a striding bipedal gait but has been under numerous selective pressures up to the present. Identifying and investigating these additional gait variables will generate new perspectives on hominin locomotion that further enhance our understanding of hominin life-ways.

REFERENCES

Adovasio, J., Soffer, O., and Klíma, B., 1996, Paleolithic Fiber Technology: Data from Pavlov I, ca. 26,000 b.p. *Antiquity* 70:526–34.

Aiello, L., and Dean, D., 1990, *An Introduction to Human Evolutionary Anatomy*. Academic Press, New York.

Aiello, L., and Wheeler, P., 1995, The Expensive Tissue Hypothesis: The Brain and the Digestive System in Human and Primate Evolution. *Current Anthropology* 37:199–222.

Binford, L., 1978, *Nunamiut Ethnoarchaeology*. Academic Press, New York.

Binford, L., 1980, Willow Smoke and Dog's Tails. *American Antiquity* 45:4–20.

Binford, L., 1987, American Association of Physical Anthropologists Annual Luncheon Address, April 1986: The Hunting Hypothesis, Archaeological Methods, and the Past. *Yearbook of Physical Anthropology* 30:1–9.

Binford, L., 1990, Isolating the Transition to Cultural Adaptations: An Organizational Approach. In *The Emergence of Modern Humans: Biocultural Adaptations in the Later Pleistocene*, edited by Trinkaus, E., pp. 18–41. Cambridge University Press.

Binford, L., 2001, *Reconstructing Frames of Reference: An Analytical method for Archaeological Theory Building Using Ethnographic and Environmental Data Sets*. University of California Press, Berkeley.

Bramble, D., 2000, Head Stabilization in Human Running: Implications for Hominid Evolution (Abstract). *American Journal of Physical Anthropology* Supplement 30:111.

Bramble, D., 1991, Functional and Evolutionary Significance of the Occipital Torus in *Homo* (Abstract). *American Journal of Physical Anthropology* Supplement 12:52–53.

Bramble, D., and Carrier, D., 1983, Running and Breathing in Mammals. *Science* 219:251–219.

Brown, J., 1972, A Note on the Sexual Division of Labor. *American Anthropologist* 72:1073–78.

Brunet, M., Guy, F., Pilbeam, D., Taisso-Mackaye, H., Likius, A., Ahounta, D., Beauvilain, A., Blondel, C., Bocherens, H., Boisserie, J., De Bonis, L., Coppens, Y., Dejax, J., Dennys, C., Duringer, P., Eisenmann, V., Fanone, G., Fronty, P., Geraads, D., Lehmann, T., Lihoreau, F., Louchart, A., Mahamat, A., Merceron, G., Mouchelin, G., Otero, O., Pelaez-Campomanes, P., Ponce de Leon, M., Rage, J., Sapanet, M., Schuster, M., Sudre, J., Tassy, P., Valentin, X., Vinguad, P., Viriot, L., Zazzo, A., and Zollikofer, C., 2002, A New Hominid from the Upper Miocene of Chad, Central Africa. *Nature* 418:145–151.

Boule, M., 1911–1913, L'homme fossile de la Chapelle-aux-Saints. *Annales de Paleontologie* 6:111–172; 7:21–56, 85–192; 8:1–70.

Carrier, D., 1984, The Energetic Paradox of Human Running and Hominid Evolution. *Current Anthropology* 25:483–495.

Carrier, D.R., Heglund, N., and Earls, K., 1995, Variable Gearing During Locomotion in the Human Musculoskeletal System. *Nature* 265:651–653.

Carsloo, S., 1972, *How Man Moves: Kinesiological Studies and Methods*. Heinemann, London.

Costill, D., 1986, *Inside Running, Basics of Sports Physiology*. Benchmark Press, Inc.

Dart, R., 1925, *Australopithecus africanus:* The Man-Ape of South Africa. *Nature* 115:195–199.

Dart, R., 1926, Taung and Its Significance. *Natural History* 26:315–327.

Dubois, E., 1894, *Pithecanthropus erectus, eine menschenähmliche Übergangsform aus Java*. Batavia: Landersdruckerei.

Eberhart, H., Inman, V., and Bresler, B., 1954, The Principal Elements in Human Locomotion. In *Human Limbs and Their Substitutes*, edited by Klopsteg, P., and Wilson, P., pp. 437–471. McGraw-Hill Book Company, New York.

Elftmann, H., 1954, The Functional Structure of the Lower Limb. In *Human Limbs and Their Substitutes*, edited by Klopsteg, P., and Wilson, P., pp. 157–170. McGraw-Hill Book Company, New York.

Fleagle, J., 1999, *Primate Adaptation and Evolution*. Academic Press, New York.

Foley, R., 1987, *Another Unique Species: Patterns in Human Evolutionary Ecology*. Longman Scientific and Technical, London.

Franciscus, R., and Trinkaus, E., 1988, Nasal Morphology and the Emergence of *Homo erectus*. *American Journal of Physical Anthropology* 75:517–527.

Gabunia, L., Anton, S., Lordkipanidze, D., Vekua, A., Justus, A., and Swisher, C., 2001, Dmanisi and Dispersal. *Evolutionary Anthropology* 10:158–170.

Gould, R., 1969, *Yiwara: Foragers of the Australian Desert*. Charles Scribner's Sons, New York.

Haile-Selassie, Y., 2001, Late Miocene Hominids from the Middle Awash, Ethiopia. *Nature* 412:178–181.

Hawkes, K., O'Connell, J., and Blurton-Jones, N., 1995, Hadza Children's Foraging: Juvenile Dependency, Social Arrangements, and Mobility among Hunter-Gatherers. *Current Anthropology* 36:688–700.

Heglund, N., Willems, P., Penta, M., and Cavagna, G., 1995, Energy-saving Gait Mechanics with Head-supported Loads. *Nature* 375:52–54.

Inman, V. T., Ralston, H., and Todd, F., 1981, *Human Walking*. Williams and Wilkins, Baltimore.

Isaac, G., 1984, Foundation Stones: Early Artifacts as Indicators of Activities and Abilities. In *Stone Age Prehistory*, edited by Bailey, G., and P. Callow, P., pp. 221–242. Cambridge University Press, Cambridge.

Kaplan, H., Hill, K., Lancaster, J., and Hurtado, M., 2000, A Theory of Human Life History Evolution: Diet, Intelligence, and Longevity. *Evolutionary Anthropology* 9:156–185.

Kelly, R., 1983, Hunter-Gatherer Mobility Strategies. *Journal of Anthropological Research* 39:277–306.

Kelly, R., 1992, Mobility/Sedentism: Concepts, Archaeological Measures, and Effects. *Annual Review of Anthropology* 21:43–66.

Kelly, R., 1995, *The Forager Spectrum: Diversity in Hunter-gatherer Lifeways*. Smithsonian Institution Press, Washington, D.C.

Klein, R., 1989, *The Human Career*. University of Chicago Press, Chicago.

Kuhn, S., and Stiner, M., 2002, The Antiquity of Hunter-Gatherers. In *Hunter-Gatherers: An Interdisciplinary Perspective*, edited by Painter-Brick, C., Layton, R., and Rowley-Conwy, P., pp. 99–142. Cambridge University Press, Cambridge.

Lee, R., 1979, *The !Kung San:Men, Women, and Work in a Foraging Society*. Cambridge University Press, Cambridge.

Maloiy, M., Heglund, N., Prager, L., Cavagna, A., and Taylor, C., 1986, Energetic Cost of Carrying Loads: Have African Women Discovered an Economic Way? *Nature* 319:668–669.

McArdle, W., Katch, F., and Katch, V., 1991, *Exercise Physiology: Energy, Nutrition, and Human Performance*. Lea and Febiger, Philadelphia.

McHenry, H., and Coffing, K., 2000, *Australopithecus* To *Homo*: Transformations in Body and Mind. *Annual Review of Anthropology.* 29:125–146.

McMahon, T., 1984, *Muscles, Reflexes, and Locomotion*. Princeton University Press, Princeton, New Jersey.

Murdock, G., and Provost, C., 1972, Factors in the Division of Labor by Sex: A Cross-Cultural Analysis. *Ethnology* 12:203–25.

Napier, J., 1964, The Evolution of Bipedal Walking in the Hominids. *Archives de Biologie (Liege)* 75:673–708.

Napier, J., 1967, The Antiquity of Human Walking. *Scientific American* 216:56–66.

Potts, R., 1984, Home Bases and Early Hominids. *Scientific American* 72:338–347.

Potts, R., 1988, *Early Hominid Activities at Olduvai*. Aldine de Gruyter, New York.

Potts, R., 1991, Why the Oldowan? Plio-Pleistocene Toolmaking and the Transport of Resources. *Journal of Anthropological Research*. 47:153–176.

Potts, R., 1994, Variables Versus Models of Early Pleistocene Hominid Land Use. *Journal of Human Evolution* 27:7–24.

Rodman, P., and McHenry, H., 1980, Bioenergetics and the Origin of Hominid Bipedalism. *American Journal of Physical Anthropology* 52:103–106.

Rogers, M., Feibel, C., and Harris, J., 1994, Changing Patterns of Land Use by Plio-Pleistocene Hominids in the Lake Turkana Basin. *Journal of Human Evolution* 27:139–158.

Ruff, C., 1993, Climatic Adaptation and Hominid Evolution: The Thermoregulatory Imperative. *Evolutionary Anthropology* 2:53–60.

Ruff, C., and Walker, A., 1993, Body Size and Body Shape. In *The Nariokotome Homo erectus Skeleton*, edited by Walker, A., and Leakey, R., pp. 234–265. Harvard University Press, Cambridge.

Richmond, B., Aiello, L., and Wood, B., 2002, Early Hominin Limb Proportions. *Journal of Human Evolution* 43:529–548.

Saunders, J., Inman, V., and Eberhart, H., 1953, The Major Determinants in Normal and Pathological Gait. *Journal of Bone and Joint Surgery* 35A:543–558.

Schick, K., 1987, Modeling the Formation of Early Stone Age Artifact Concentrations. *Journal of Human Evolution* 16:789–808.

Semaw, S., Schick, K., Toth, N., Simpson, S., Quade, J., Rogers, M., Renne, P., and Stout, D., and Dominguez-Rodrigo, M., 2002, Recent Discoveries from Gona, Afar, Ethiopia. Paper presented at the annual meeting of the Paleoanthropology Society, Denver.

Senut, B., Pickford, M., Gommery, D., Mein, P., Cheboi, K., and Coppens, Y., 2001, First Hominid from the Miocene (Lukeino Formation, Kenya). *C. R. Acad. Sci. Paris, Sciences de la Terre et des planètes/Earth and Planetary Sciences* 332:137–144.

Stiner, M., 1991, Food Procurement and Transport by Human and Non-Human Predators. *Journal of Archaeological Science* 18:455–482.

Stiner, M., 1994, *Honor Among Thieves: A Zooarchaeological Study of Neandertal Ecology*. Princeton University Press, Princeton.

Stiner, M., and Kuhn, S., 1992, Subsistence, Technology, and Adaptive Variation in Middle Paleolithic Italy. *American Anthropologist* 92:306–339.

Soffer, O., Adavasio, J., and Hyland, D., 2000, The "Venus" Figurines. *Current Anthropology* 41:511–537.

Tonkinson, R., 1991, *The Mardu Aborigines: Living the Dream in Australia's Desert.* Wadsworth/Thomson Publishing, New York.

White, T., Suwa, G., and Asfaw, B., 1994, *Australopithecus ramidus* a New Species of Early Hominid from Aramis, Ethiopia. *Nature* 371:306–312.

White, T., Suwa, G., and Asfaw, B., 1995, *Australopithecus ramidus* a new species of early hominid from Aramis, Ethiopia. *Nature* (corrigendum) 375:306–312.

Chapter 2

Knuckle-Walking and the Origin of Human Bipedalism

DAVID R. BEGUN

Abstract. Functional analysis of fossil great apes and humans indicate that no known fossil taxon was a habitual knuckle-walker. However, the phylogenetic relations among hominoids suggests that the last common ancestor of the African apes and humans was in fact a knuckle-walker. Anatomical, fossil and molecular evidence of relations among the Hominoidea strongly suggest than *Pan* and *Homo* share a common ancestor not shared by any other living taxon. If this is correct, then knuckle-walking must have evolved once in the common ancestor of *Pan*, *Gorilla* and *Homo* (in which it was lost), or twice, independently in each African ape lineage. In addition to being less parsimonious, most multiple origins hypotheses for knuckle-walking also fail to account for characters shared by African apes and humans that are plausibly functionally related to increased terrestriality. These include vertebral, limb and intra-limb proportions, and limb long bone, carpal, tarsal and metapodial, and phalangeal morphology. In theory, knuckle-walking and obligate bipedalism could have evolved from an unknown type of terrestrial quadrupedalism, possibly associated with high frequencies of facultative bipedality. This would also account for characters and positional behavior shared by African apes and humans, the differences between *Pan* and *Gorilla*, and the apparent retention of primitive features of the trunk in humans. However, it implies a substantial increase in homoplasy as well as a hypothetical ancestral morphotype unkown in any group or out group. Functional anatomy and phylogeny together continue to suggest that humans are most likely to have evolved from a knuckle-walker. However, a functionally plausible, though less parsimonious facultative bipedality hypothesis is also possible. In this regard, hints of the unique functional morphology of *Ardipithecus ramidus* suggest that this taxon would serve to test these two alternatives.

INTRODUCTION

More than any other aspect of the morphology of fossil humans (hominins, or living humans and all fossil taxa more closely related to us than to any other species), the postcranial characters that are associated with bipedalism represent the defining features of our lineage. Masticatory robusticity in the earliest humans is duplicated in numerous lineages of Miocene hominoids, not to mention other primates and non-primates (Skelton and McHenry, 1992; Lieberman et al., 1996; Strait et al., 1997; Begun, 2001). Early human brains are also no larger relative to body size than are those of fossil or living apes and some other living mammals (Hartwig-Scherer, 1993; Kappelman, 1996; Marino, 1998; Sterling and Povinelli,

From Biped to Strider: The Emergence of Modern Human Walking, Running, and Resource Transport, edited by D. Jeffrey Meldrum and Charles E. Hilton. Kluwer Academic/Plenum Publishers, New York, 2004.

1999). Even so, that bipedalism defines our evolutionary lineage may still seem counterintuitive. Bipedalism is one of the oldest and most common forms of positional behavior among terrestrial tetrapods. The earliest documented biped is the fleet-footed bolosaurid parareptile *Eudibamus cursoris* from the early Permian (290 Mya) of Germany (Berman et al., 2000). In the 240 million year history of the Ornithosuchia (birds and dinosaurs), thousands of taxa in this group were and are bipeds (Benton, 1990; Middleton and Gatesy, 2000). But obligate bipedalism in humans is unique among primates, and the anatomical correlates of human bipedalism as an integrated whole are unique among tetrapods. As often noted, the evolution of bipedalism in humans is indeed central to the emergence of the human lineage. Understanding the process that contributed to the origin of human bipedalism will reveal much of the processes that account for the very origin of humans.

In order to be able to account for the origin of bipedalism it is necessary to know from what form of positional behavior human bipedalism evolved. For example, evolving from an orthograde, climbing arboreal ancestor implies selection favoring more terrestrial behaviors, for whatever reason. Evolving human bipedalism from an already terrestrial quadruped, such as a knuckle-walker, implies a different set of selection pressures. I hope to show here that the hypothesis that human bipedalism evolved from a semi-orthograde, more terrestrial form of quadrupedalism, knuckle-walking, is most consistent with the fossil and modern anatomical evidence, both from the point of view of evolutionary relations and functional anatomy. A lengthy period of terrestrial quadrupedalism preceding the emergence of human bipedalism has specific implications for developing scenarios to help explain human origins.

Function, Phylogeny, and Knuckle-Walking

Most researchers today consider one of three patterns as most likely to correctly describe relations among African apes and humans. Humans and chimps are either most closely related to one another, with gorilla as the sister clade, or chimpanzees and gorillas form a clade with humans as the sister clade, or the trichotomy is unresolved (Fig. 1). From a molecular point of view the majority of analyses favor a chimpanzee-human clade, with the

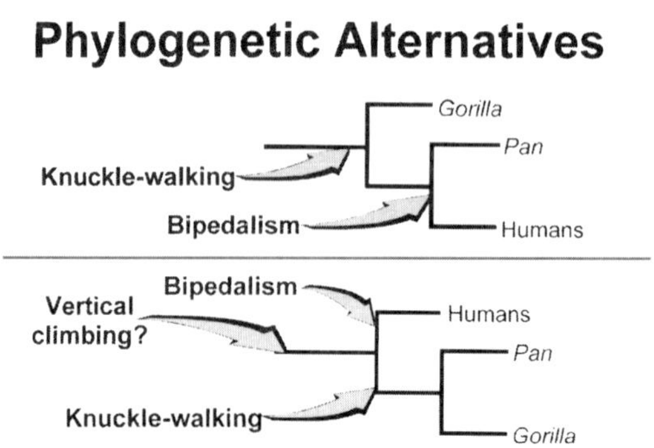

Figure 1. Two hypotheses of African ape-human relations, and their most parsimonious implications for the positional behavior of various ancestors. On top, a *Pan*-human clade suggests that each mode of positional behavior arose only once, knuckle-walking in the common ancestor of the African apes and humans and bipedalism in humans. The same holds under the hypothesis of a trichotomous split. On the bottom, knuckle-walking and bipedalism arise in the ancestors of the African apes and humans respectively, and either bipedalism, knuckle-walking, or a third plesiomorphic form of positional behavior, vertical climbing, characterized the common ancestor of all three. See text for discussion.

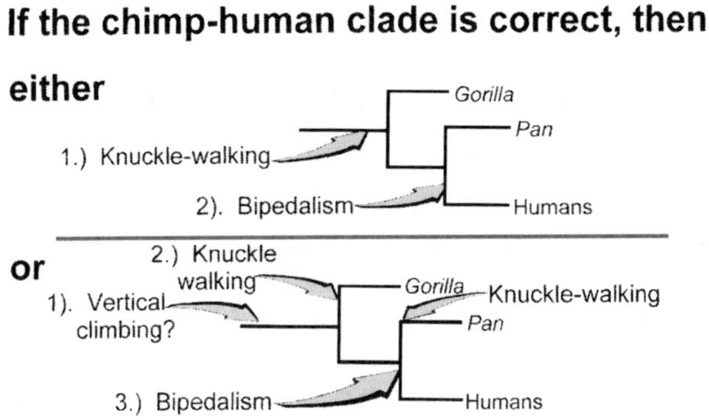

Figure 2. Assuming the majority view is correct, that *Pan* and humans form a clade that excludes *Gorilla*, the most parsimonious distribution of ancestral modes of positional behavior is depicted on top, as in figure 1. A noticeably less parsimonious explanation is required under the hypothesis that humans did not evolve from a knuckle-walker (bottom). See text for discussion.

strongest opposing arguments favoring a trichotomy (see reviews in Rogers, 1993; Ruvolo, 1994; Begun, 1999; Satta et al., 2000). Relatively few analyses today continue to suggest that there is strong evidence from molecular systematics for a chimpanzee–gorilla clade. From the point of view of morphology, fewer researchers are willing to make a commitment, but the most comprehensive analyses to date also point to a chimpanzee–human clade (Groves, 1986; Begun, 1992a, 1994; Begun et al., 1997; Shoshani et al., 1996).

Both the trichotomy and the chimpanzee–human clade hypotheses suggest as a *preliminary* hypothesis that knuckle-walking characterized the last common ancestor of African apes and humans, because if this were not the case then knuckle-walking would have to have evolved twice, either independently in the trichotomous chimpanzee and gorilla clades or independently in the dichotomous chimpanzee–human and chimpanzee–gorilla clades (Fig. 2). Dainton and Macho (1999) suggest that knuckle-walking did in fact probably evolve twice given differences in the ontogeny and kinematics of chimpanzee and gorilla hands. I will get back to that issue later. It is not legitimate to conclude that humans must have evolved from a knuckle-walker simply because of a certain pattern of evolutionary relationship and the logic of parsimony. However, it is a better initial explanation because it is simpler and easier to test. The fact that an explanation accounting for an observed pattern is simpler does not translate into the assumption that the processes that formed the pattern are uncomplicated. A simpler hypothesis requiring a smaller number of steps is methodologically more robust because it is easier to test and falsify than are complex hypotheses with built-in contingencies, such as homoplasy in evolutionary hypotheses. As an example, big bang theory is a simpler explanation for the origin and expansion of the observable Universe than is a multiple origins of individual galaxies hypothesis, but big bang theorizing is anything but simple in terms of process. On the opposite end of the spectrum, the hand of God is a *deus ex machina* that resolves the origin of everything. As an explanation it is simple, but at least at the proximate level probably incorrect and certainly uninformative from a scientific perspective with regard to a basic understanding of the Universe.

All phylogenetic analyses no matter how they are carried out reveal that the evolution of African apes and humans has been very complex, and has involved numerous homoplasies. But why start off with the hypothesis that 3 major shifts in postcranial anatomy and positional behavior evolved independently and in a very short time from a totally unknown hypothetical ancestral morphology and behavior *in the absence of compelling evidence to do so*? (Fig. 2). I would like to explore the simplest hypothesis first, knuckle-walking, to see if it makes sense from a functional perspective, before rejecting it in favor of a more complex, though not necessarily less likely hypothesis of multiple origins and hypothetical ancestral forms of positional behavior.

It has been argued on the basis of anatomical specialization and the absence in fossil humans of anatomical features associated with knuckle-walking in modern African apes that bipedalism did not evolve from knuckle-walking. With regard to the first objection, that of excessive specialization, it is difficult to accept *a priori* that an evolutionary process cannot occur because of a perception of specialization. Think of birds evolving from specialized theropod dinosaurs, or penguins or ostriches evolving from flying birds, in every case superimposing a specialized new morphology on an already existing highly specialized one (Middleton and Gatesy, 2000; Benton, 1990). Think also of climbing hyracoids that appear to have evolved from specialized cursorial perissodactyls by developing a unique, midcarpal mobility structure to compensate for a fused radius and ulna and other changes in the wrist (Fischer, 1986). Finally, among primates consider the example of the cat-sized indriids (vertical clingers and leapers) and their extinct relatives, the sometimes gorilla-sized palaeopropithecines (sloth lemurs) (Godfrey et al., 1997). These transformations often involve changes in limb proportions and morphology that recur many times, in other words, rampant homoplasy.

Similarly, the absence of characters in australopithecines that occur exclusively in knuckle-walkers means only that australopithecines were not knuckle-walkers, which is obvious from their anatomy, but not that they could not have evolved from a knuckle-walker (Begun, 1993). The absence of characters of the forelimb in suspensory hominoids that are found exclusively in pronograde anthropoids is not an indication that hominoids could not have evolved from an above branch arboreal quadruped, which is in fact almost certainly the case (Rose, 1988). It is difficult enough to reconstruct the behavior of fossil taxa from their anatomy, even when reasonably well known, as in australopithecines. It is almost impossible to infer the form of positional behavior of an unknown ancestor of a fossil taxon directly its anatomy.

A comprehensive functional and developmental analysis of human bipedalism and African ape knuckle-walking may ultimately reveal that it is unlikely that bipedalism evolved from knuckle-walking and more likely that African ape knuckle-walking evolved independently. However, a convincing case has yet to be made. Dainton and Macho (1999) contribute to this debate, but in contrast to their suggestion they do not show that it is more parsimonious to suggest that knuckle-walking evolved twice rather than once. These authors interpret growth differences in gorilla and chimpanzee wrist bones to be indicative of fundamental heterochronic shifts that suggest independent pathways to the development of knuckle-walking in each. However, these authors also show that there are marked similarities in the morphology and growth of African ape wrist bones. What accounts for these? And what accounts for the morphological similarities of the hands of African apes and humans that appear to be related to biomechanical properties of critical importance to knuckle-walkers (see below). Although it is true that the demonstration of developmental differences may help to distinguish between homology and homoplasy, there is no definitive developmental test to accomplish this goal. Dramatic differences are not only known to characterize the development of obviously homologous structures, like the vertebrate eye, they are to be expected (Hall, 1994). Developmental change is a principal engine of morphological evolution. It is hard to imagine significant amounts of evolution in a lineage without developmental differences. We know, for example, that australopithecines and modern humans grew very differently, in terms of rate, duration, and pattern within the skeleton (Bromage, 1987; Smith, 1991; Dean et al., 1993; Smith et al., 1995; Tardieu, 1997). Does this mean that bipedalism is not homologous in australopithecines and modern humans? Gibbons, orangutans, all African apes, and humans all exhibit differences in aspects of trunk and forelimb development, yet this is not taken to indicate that anatomical similarities among hominoids in these regions appeared independently. Instead, they are widely attributed to a suspensory heritage for all living hominoids. Developmental differences may help to explain how observed homoplasies happened, by pointing to different developmental pathways, but these are not sufficient to demonstrate homoplasy in the absence of additional evidence, such as independent evidence of phylogeny.

It is hard to see how the non-knuckle-walking hypothesis can be falsified because the hypothetical ancestral mode of positional behavior is not known. Any indication that human bipedalism could have evolved from knuckle-walking can always be countered with an *ad hoc* ancestral morphotype that accommodates proposed functional or developmental similarities among African apes and humans (for example, fusion of the os centrale to the scaphoid may be related to a hypothetical vertical climbing dominated mode of progression and just retained in knuckle-walkers and bipeds). Gebo (1992, 1996) has presented a convincing case for a lengthy terrestrial phase preceding the origin of bipedalism based on the terrestrial adaptations of the hindlimb of African apes and humans.

Meldrum (1993) and Schmitt and Larson (1995) dispute details of this analysis, but not to the extent that they directly affect the arguments presented here. Gebo (1992, 1993) argues that specific attributes of the African ape and human foot reflect plantigrady, or full heel-sole contact with the ground, in African apes and humans, while Meldrum (1993) argues that this condition is found in varying degrees in many primates and other mammals. Schmitt and Larson (1995) also argue that many of the kinematic features of African ape feet in plantigrady are also found in orangutans.

The focus of this chapter is forelimb morphology. Meldrum, Schmitt and Larson are correct in noting the complexity of the problem and the fact that the detailed morphology, and behavioral and evolutionary significance of these characters need to be more completely assessed. However, the issue of whether or not the characters Gebo cites are synapomorphies of the African ape and human clade or symplesiomorphies shared by a wider variety of mammals does not directly affect the premise of this contribution. Whether proximate or ultimate, African ape foot morphology clearly reflects a degree of terrestriality more strongly developed than in orangutans, the other great ape of body mass in the same range.

For the remainder of this chapter I will explore the anatomical traits of the hominoid forelimb that suggest that this terrestrial adaptation was essentially the same as is observable today in living African apes—knuckle-walking—the positional behavior from which bipedalism probably evolved.

FUNCTIONAL ANATOMY OF KNUCKLE-WALKING CHARACTERS IN AFRICAN APES AND HUMANS

African apes and humans share numerous characters of the forelimb that are plausibly related functionally to knuckle-walking, retained in humans because they are compatible with bipedalism and the suite of functions required of a human upperlimb (Table 1). In some cases these characters are described as non-functional (e.g., Collard and Aiello, 2000),

Table 1. Contrasting forelimb morphology in hominoids and its functional significance. Exceptions are noted in the text.

Characteristic	Asian apes	African apes	Fossil hominines[1]	Knuckle-walking function
Proportions				
Brachial index	large	smaller	smaller	reduce moments at the elbow
Metacarpal ln	long	shorter	shorter	reduce moments at the wrist
Proximal phalanx ln	long	shorter	shorter	reduce moments at the wrist
Intermediate phalanx ln	shorter	longer	longer	response to load
Shoulder and arm				
Humeral head	less torsion	more torsion	?	re-positioning of the elbow and hand
Zona conoidea	shallow	deep	deep	stable throughout flexion-extension
Lateral trochlear ridge	reduced	strong	strong	stable throughout flexion-extension
Wrist and hand				
Scaphoid notch	shallow	pronounced	pronounced	stability in extension
Radial dorsal margin	less projecting	more projecting	more projecting	close packed in extension
Scapho-lunate surface	angled	co-planar	co-planar	compression
Ulnar head	smaller	larger	larger	response to load
Meniscus	absent	present	present	load/shock absorbing
Scaphoid radial facet 1	smaller/rounded	larger/flat	larger/flat	response to load
Scaphoid radial facet 2	distal/proximal	dorsal/medial	dorsal/medial	re-positioned joint surfaces
Scaphoid capitate facet	smaller	proximal expansion	proximal expansion	restrict midcarpal abduction
Scaphoid beak	absent	present	present	stable in extension
Scaphoid size	smaller	larger	larger	response to load
Scaphoid trapezoid facet	disto-lateral	distal orientation	distal orientation	compression load
Os centrale	unfused	fused	fused	response to shear
Triquetrum	large	small	small	radial load dominance
pisiform	more distal	palmar/proximal	palmar/proximal	flexing the extended wrist
Captiate head	long, narrow	short. broad	short. broad	close packing in extension
Capitate waisting	absent	present	present	close packing in extension
Capitate mcl surface	flat	keeled	keeled	carpo-metacarpal shear
Capitate dorsal surface	smaller	expanded	expanded	limits midcarpal extension
Hamate triquetral facet	planar	helical	helical	close packing in extension
Hamate hamulus	less projecting	more projecting	more projecting	powerful carpal flexion
Hamate mcl surface	flat	keeled	keeled	carpo-metacarpal shear
Trapezoid size	smaller	larger	larger	stabilized Mcl 2 in extension
Trapezoid centrale facet	medial	proximal	proximal	compression load
Trapezoid mcl	surface flat	keeled	keeled	carpo-metacarpal shear
Mcl 2 carpal facets	broadly concave	notched	notched	carpo-metacarpal shear
Metacarpal curvature	greater	reduced	reduced	compression
Metacarpal head	smooth	flatter dorsally	smooth	close packed in hyperextension
Phalangeal curvature	greater	reduced	reduced	compression

Notes: 1) *Ardipithecus ramidus* and *Australopithecus anamensis*. Abbreviations: ln=length; Mcl=metacarpal.

but a better description would be re-functional; they are exaptations for the manipulative functions of the human forelimb derived from a knuckle-walking precursor. None of these characters are definitive indicators of positional behavior because they all occur in human bipeds and knuckle-walkers, and some are even primitive for the clade. However, most of these features have an apparent functional relationship to each other in that they tend to have 3 related effects: they reduce the moments between the elbow and the wrist, and between the wrist and the intermediate phalanges; they stabilize the wrist both transversely and in moderate hyperextension by close-packing in extended positions and by reducing the potential for joint translation through bone fusion, the development of more complex joint surfaces, and the presence of more strongly developed wrist flexor power arms, and; they reduce stress from vertically directed compressive loads through enlarged and repositioned articular surfaces and more robust bones.

In having these functional effects, these characters can be considered necessary but not sufficient features of either knuckle-walking or bipedalism. Each form of positional behavior incorporates characters of the forelimb not found in the other, though for knuckle-walkers, characters unique to them have been frustratingly difficult to identify. Most, it turns out, are variably expressed, mainly following gradients of age and body mass (Tuttle, 1969a; Inouye, 1992). This suggests, in contrast to prevailing opinion, that knuckle-walking is in fact fairly generalized, and is not associated with a suite of unusual characters known only within that group.

One probable unique character of knuckle-walkers is an increase in the radius of curvature of the dorsal half of the metacarpal head relative to the ventral half, which does not vary with body mass (Zylstra, 1998, 1999), but this is small in comparison to the anatomical complex that is recruited by knuckle-walkers. It almost seems as if, with a very minor modification of the basic African ape and human ground plan, knuckle-walking is possible, if not inevitable. Thus, the implication of the distribution of these characters in African apes and humans is that the common ancestor of these taxa loaded their limbs in such a manner that a premium was placed on reducing translation and limiting extension among a number of wrist bones and lowering moments to decrease angular momentum at the radiocarpal, metacarpophalangeal, and proximal interphalangeal joints. Knuckle-walking is one such form of positional behavior.

Shoulder and Arm

African apes and humans share characters of the proximal and distal humerus, and forelimb proportions not found in the two extant outgroups of that clade, *Pongo* and *Hylobates*. Humans share with African apes a greater degree of torsion of the head relative the transverse plane of the distal end (Larson, 1988, 1996), and a more strongly developed zona conoidea and lateral trochlear keel. Also in comparison to other apes, humans and African apes share a reduced brachial index. Larson suggests that increased humeral head torsion probably occurred independently in African apes and humans for different functional reasons (Larson, 1988, 1996). Nothing in the morphology of the human shoulder joint requires this more complicated explanation, and Larson (1988, 1996) does indeed suggest that on functional grounds it is possible that humeral head torsion on African apes and humans is related to knuckle-walking. Her strongest argument against this explanation is her series of estimates of three proximal humeri of Pliocene hominids, which are intermediate in humeral head torsion between *Pongo* and African apes in her analysis. This suggests that the higher level of humeral head torsion in modern humans converges on that of the African apes, having evolved independently from the intermediate morphology of Pliocene human ancestors (Larson, 1996). However, several facts suggest that the estimates of humeral head torsion in these Pliocene specimens cannot be considered definitive. Two of the three specimens (AL-288-1r and STS 7) are distorted, the former with a mildly crushed head and the later with a severely crushed proximal shaft. The range of estimates for the most intact specimen (OMO 119-73-2718) overlaps with the predicted angle of humeral head torsion for *Pan* and *Homo*. In addition, the ranges for the fossils are calculated using the mean percent prediction error (MPE). Use of the 95% CI of the estimate would likely result in more overlap between fossil and modern hominoid samples. Finally, the fossil specimens are toward the high end of the range of variation in size of the comparative sample, and, with only a few exceptions, variability (standard deviation) increases with size in her analysis, suggesting again that MPE may underestimate the range of likely values for the fossil specimens. In my view we will have to continue to wait to find more complete and less distorted humeri of Pliocene hominids to know whether or not they had humeral head torsion angles lower than those of modern African apes and humans. In the mean time, it is more parsimonious to suggest that this morphology evolved once in the knuckle-walking common ancestor of African apes and humans.

The strongly developed double keel of the African ape and human (australopithecine) elbow (a strong lateral trochlear keel coupled with a deep zona conoidea between the trochlea and the capitulum) functions to maintain a highly stable hinge joint throughout the range of flexion-extension (Rose, 1988). While *Pongo* usually lacks as strongly developed a keel-zona complex, hylobatids, *Oreopithecus, Sivapithecus* and *Dryopithecus* share this feature with African apes and fossil humans. Thus, while this character cannot be considered to have evolved first in a knuckle-walker, its presence in African apes and humans and absence in *Pongo* suggests differing patterns of positional behavior in the ancestors of *Pongo* and the African apes and humans. This contrasts with the view that the quadrumanous climbing *Pongo* represents a good model for positional behavior in the ancestor of human bipeds (Stern, 1975; Fleagle et al., 1981). Reduction in the length of the forearm relative to the upper arm (brachial index) in African apes and humans compared to *Pongo* and *Hylobates* is functionally consistent with an adaptation to reducing bending moments in the forearm given the extended wrist and digitigrade posture of a knuckle-walker, both of which tend to increase forearm functional length.

Wrist and Hand

The numerous shared characters of the African ape and human hand are all plausibly related to the functional requirements of knuckle-walking. Several characters of the distal end of the radius and the proximal carpal row shared by African apes and *Australopithecus anamensis* have recently been described, including a distally projecting dorsal ridge, a relatively large, dorsally oriented notch for the scaphoid, and flat and co-planar joint surfaces of the scaphoid and lunate for the radius (Richmond and Strait, 2000). These and a number of other features of the African ape and human wrist that appear to be related to resisting stresses in a slightly hyperextended wrist, such as are generated during the stance phase of knuckle-walking, are described in more detail elsewhere (Richmond and Strait, 2000; Richmond et al., 2001).

The analysis of Richmond and Strait (2000) has not gone unchallenged. To avoid a very lengthy digression I will only touch upon the key points of contention here. Several authors have noted that a number of the features isolated by Richmond and Strait are variable or difficult to discern in African apes and humans (Dainton, 2001) and may be impossible to verify in fossil specimens due to damage (Lovejoy, et al., 2001). There have also been concerns expressed about the multivariate statistical analysis Richmond and Strait (2000) used to support their analysis, and about the degree to which African ape wrists are actually "locked" or fully extended in typical knuckle-walking postures (Lovejoy, et al., 2001; Jenkins and Fleagle, 1975). Dainton's points regarding ontogeny have been addressed above, and in more detail in Richmond et al. (2001). Additional and more comprehensive statistical analysis using different methods of standardizing data (in response to specific criticisms), and expanded samples, including original fossil specimens, have also yielded results similar to those described by Richmond and Strait (2000) (Richmond et al., 2001). Lovejoy et al. (2001) present a developmental explanation for the distinctive morphology of the chimpanzee distal radius, but it is not clear how this relates to the similarities among hominines (African apes and humans). Richmond and Strait (2001a,b) and Richmond et al. (2001) respond in detail to all published concerns of their interpretations, and these sources should be consulted for additional information. In the context of this chapter, the characters presented by Richmond and Strait (2000) represent a small percentage of the total number of characters suggested here to imply a knuckle-walking stage of human evolution. While I agree with the interpretation of Richmond and Strait, the characters they analyze are not critical to the overall argument presented here.

Other features of the radioulnar and radiocarpal joints shared by African apes and humans include the presence of a large ulnar head and a well-formed meniscus between the ulna and proximal carpus. Both of these features are related to the use of the forelimb to transmit compressive stress from the ulna to the ulnar carpal row (triquetrum) while preventing the ulnar styloid process from articulating directly with the triquetrum and pisiform, as it does in most primates other than great apes and humans (Sarmiento, 1988; Beard, 1986). This arrangement allows for effective weight bearing while retaining a wide range of pronation and supination (Sarmiento, 1988).

Among the bones of the proximal carpal row African apes and humans share very frequent and ontogenetically early fusion of the os centrale to the scaphoid bone, comparatively large scaphoid radial and capitate facets, a small triquetrum, and a palmar and proximal pisiform (Marzke, 1971; Sarmiento, 1988; personal observations and see below) (Fig. 3–4). The triquetrum is small in hominoids generally compared to other anthropoids, because hominoids place an emphasis on radial-side loading of the wrist, and the triquetrum articulates with the ulna in most other primates (Lewis, 1969; Beard, 1986). But the triquetrum is much smaller in African apes and humans than in orangs, in which it is a robust, rectangular bone with large, flat joint surfaces for the hamate laterally, the lunate proximally and the pisiform palmarly. The robust and long triquetrum of *Pongo* serves in part to displace the facet for the pisiform distally and ulnarly, which apparently contributes to an increase in the leverage of the flexor carpi ulnaris when the wrist is flexed and adducted (Sarmiento, 1988). The large articulation of the triquetrum and the lunate in *Pongo* is also shifted such that the joint surface faces or is normal to the long axis of the metacarpals, which may be a response to loading in along this axis. The relatively large triquetrum of *Pongo* is wedged between the large hamate and lunate, and may also be a response to loading along the 4th and 5th metacarpals during suspensory positional behavior. African apes in contrast tend to load their last metacarpals less than the 2nd to 4th metacarpals, at least during knuckle-walking.

The triquetrum reduction in African apes and humans is also related to a repositioning of the pisiform proximally and palmarly, the latter possibly related to leverage for flexor

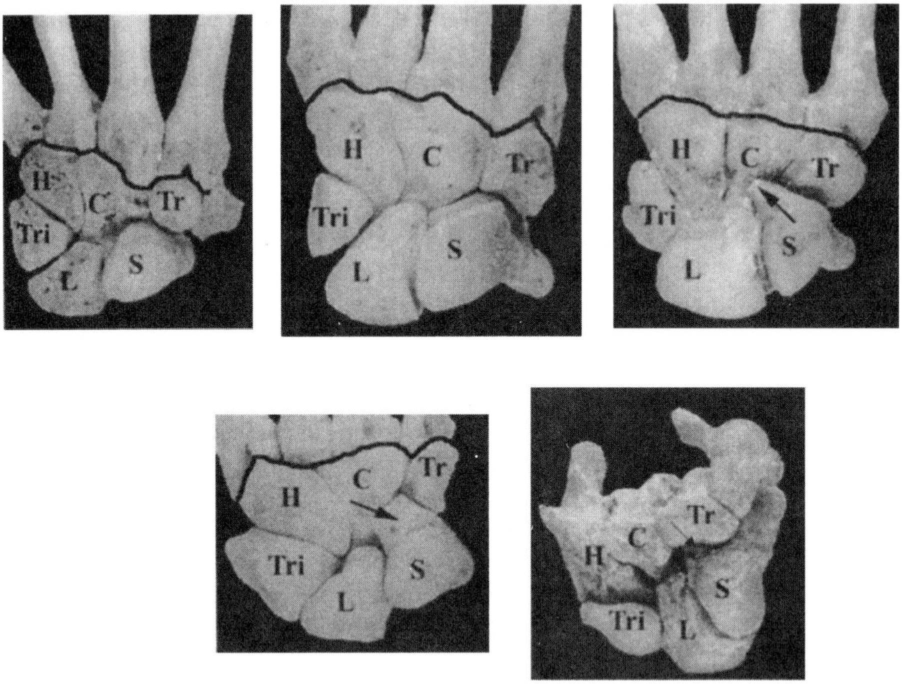

Figure 3. Dorsal views of some catarrhine left wrists in maximum extension at the midcarpal joint, all reproduced to the same maximum carpal breadth. Top row from the left, *Homo*, *Pan*, *Pongo*. Bottom row from the left, *Papio*, *Hylobates*. Note the dorsal beak on the *Pongo* scaphoid representing the fused os centrale (arrow). Os centrale fusion in *Pan* and *Homo* does not produce such a beak, instead reinforcing the joints between the scaphoid, and the capitate and trapezoid distally. Note also the irregular joint surfaces of the carpo-metacarpal joints in *Homo* and *Pan* and their more continuous contours in the other taxa. Finally, note the more dorsal orientation of the metacarpal surfaces of the distal carpal bones in *Hylobates*, reflecting a substantial range of extension at the midcarpal joint. Arrows point to the os centrale. S=scaphoid, L=lunate, Tri=triquetrum, H=hamate, C=capitate, Tr=trapezoid. See text for discussion.

carpi ulnaris in flexing the extended wrist (Sarmiento, 1988). The palmar orientation of the pisiform also contributes to the formation of a deeper carpal tunnel in African apes and humans compared to orangs, which may well be related to the restricted range of extension (stronger flexors) in African apes compared to Asian apes (Tuttle, 1967, 1969b; Sarmiento, 1988). Limiting the range of extension at the wrist in African apes is critical to maintaining a stable extended wrist posture during knuckle-walking. The small triquetrum of African apes and humans is medially displaced with lunate and hamate joint surfaces oriented oblique to the long axes of the forearm and metacarpals (Fig. 3).

The lunate is as large or larger than the scaphoid in most primates, including orangs, which have a very large lunate compared to other hominoids. In African apes and humans the scaphoid is the larger bone, not only due to fusion of the os centrale but also due to the expanded joint surface for the radius (Fig. 3–5). The expanded radial facet of the scaphoid in African apes and humans is oriented more dorsally and medially, and it is comparatively flat with a beak dorso-distally. This flat, beaked joint surface fits the medially oriented scaphoid joint surface of radius, the dorsal edge of which has a corresponding notch to receive the beak (the scaphoid notch). This morphology of the scaphoid and radius ensures that maximum joint congruence is achieved in mild hyperextension at the radio-carpal

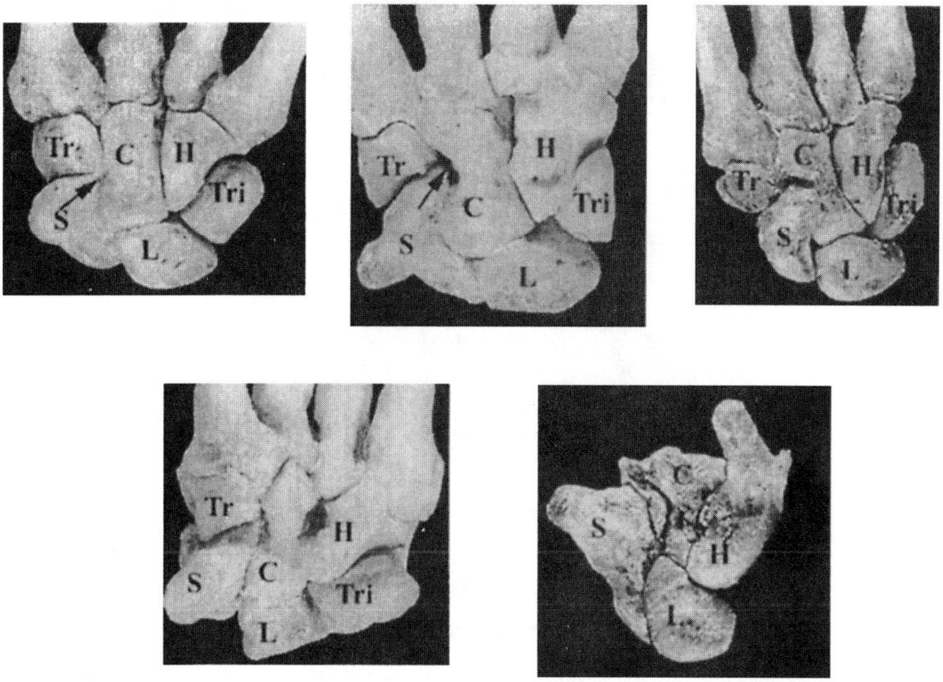

Figure 4. Palmar views of some catarrhine left wrists all reproduced to the same maximum carpal breadth. Arranged as in figure 3. Note the short, blocky capitates of *Homo* and *Pan* (strongly waisted in *Pan*), and the broad contribution of the hamate to the midcarpal ball and socket joint in *Pongo* and *Hylobates* (Pisiforms not included, nor the triquetrum in *Hylobates*). Arrows point to the fused centrale portion of the scaphoid in the embrasure of the capitate in *Pan* and *Homo*. Abbreviation as in Fig. 3.

joint, such as is necessary for wrist stabilization in a knuckle-walker (Fig. 5). The more rounded and distally displaced radial facet of the scaphoid in orangs permits more extension before maximum congruence at the radio-carpal joint is achieved. Hylobatids have a radial scaphoid facet that is similar to that of African apes but is oriented much more proximally, or less dorsally, such that joint congruence and stability is maximized in extension but not hyperextension.

The capitate facet of the scaphoid (centrale portion) dominates the midcarpal joint in African apes and humans, whereas the centrale, lunate and triquetrum contribute roughly equally to the socket of the midcarpal joint in other anthropoids. The difference is due mostly to proximal expansion of the scaphoid's capitate facet in African apes and humans, resulting in a steep-sided midcarpal socket with the radial wall facing completely medially that serves to restrict midcarpal abduction, another desirable characteristic of a knuckle-walker's wrist. The centrale portion of the scaphoid in African apes and humans also wedges itself between the tightly bound trapezoid and capitate in extension, offering stability to the wrist in this position (Fig. 3).

Of all the wrist bone similarities between African apes and humans, the fusion of the os centrale to the scaphoid has received the most commentary (Weinert, 1932; Schultz, 1936; Marzke, 1971; Jenkins and Fleagle, 1974; Lewis, 1974, 1985; Sarmiento, 1988; Begun, 1992a, 1994; Tuttle, 1992; Gebo, 1996; Schwartz and Yamada, 1998; Richmond and Strait, 2000), among many others. The os centrale or the centrale portion of the scaphoid in

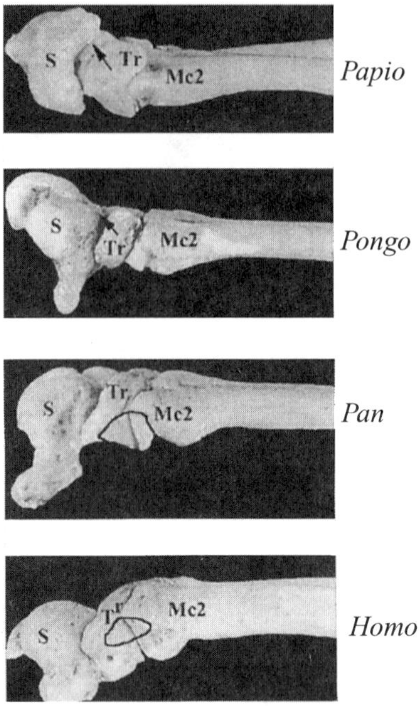

Figure 5. Lateral views of some catarrhine left wrists with the trapezia removed. From top to bottom, *Papio, Pongo, Pan*, and *Homo*. Note the more dorsally oriented radial joint surfaces and more hyperextended midcarpal postures in *Papio* and *Pongo*, and the small, flat trapezoid of *Pongo*. Note also the prominent, dorsally oriented keel on the Mc2 for the trapezoid, the continuous, concave facets for the trapezium on the Mc2 and trapezoid (marked), and the robust metacarpal shafts, all in *Pan* and *Homo*. Arrow points to the os centrale. Mc2=metacarpal 2. Other abbreviations as in Fig. 3.

anthropoids articulates distally principally with the capitate disto-medially and the trapezoid disto-laterally (Fig. 3). In most *Pongo, Hylobates* and cercopithecids the scaphoid is excluded from articulation with the capitate, and therefore does not contribute to the midcarpal joint (Fig. 3). In these taxa the os centrale articulates with the capitate medially and supports the trapezoid disto-laterally. The trapezoid, which supports the second metacarpal, is firmly attached to the trapezium, which supports the first metacarpal, so the os centrale is essentially functionally wedged between the first metacarpal laterally and the second and third metacarpal medially, thus offering stability in loading patterns coming from both these directions, i.e., grips and other loads between the thumb and the ring-middle fingers. Strong ligaments and a rough joint surface are sufficient to respond to the shear stress between the first and second-third metacarpals experienced by the scaphoid-centrale joint in most primates, which have not fused their scaphoids and centrales. However, in African apes and humans it appears as if shear across the scaphoid-centrale joint due to loading of the second metacarpal-trapezoid joint in compression has selected for fusion of these bones. With the fusion of the centrale to the scaphoid there is a solid bony wedge directly connecting the radius to the second and third metacarpals via the incisure of the capitate, into which the centrale portion of the scaphoid fits, and the second metacarpal, via the trapezoid. Knuckle-walkers load their hands and wrists in compression and experience shear stress between the bones of their wrists as their intermediate phalanges strike the ground, first with their fourth digits, and then rolling onto their third and second digits. As weight is transferred from fourth to third to second digit in most knuckle-walkers there is shear stress across the joints of the carpus that principally connect each metacarpal to the wrist (hamate to fourth, capitate to third, and trapezoid to second metacarpal). The trapezoid facet of the scaphoid in African apes and humans (the centrale portion) is oriented normal to the long axis of the 2nd metacarpal, suggesting a reorientation in response to compression along the axis of this bone (Fig. 3). The trapezoid, to which the 2nd metacarpal primarily attaches, is also highly modified in African apes and humans compared to Asian apes (see below). Shear stress between the distal row carpals and between these and the metacarpals in knuckle-walkers has resulted in other changes in the joints between these bones that are discussed below. With regard to the proximal carpals, fusion of the os centrale to the scaphoid in knuckle-walkers is probably related to shear across the joint in an extended wrist loaded in compression with rolling from the fourth to the second ray. Retention of a fused os centrale and scaphoid in humans is most likely related to the benefits of an effective mechanism to respond to shear that maximizes power grip between the first and second-third digits.

Schultz (1936) and others have noted that hylobatids and *Pongo* occasionally fuse their os centrales to their scaphoids but only relatively late in ontogeny, certainly after they have been loaded in their species typical patterns for a number of years (Fig. 3). In contrast, African apes and humans nearly always fuse these bones before they are loaded, in most cases in utero or perinatally. Fusion of these bones in African apes and humans is thus an aspect of their developmental programming, and is not dependent on habit, and instances of non-fusion are in fact rare. While Schultz (1936) is correct in noting the significance of the difference in timing of fusion of the centrale and scaphoid in African apes vs. humans, this difference (humans tend to fuse the bones earlier) is probably less important than the fact that other hominoids do not fuse these bones until they load them, and then only rarely. The fact that Asian hominoids occasionally fuse these bones under different circumstances is not an indication that early fusion of these bones in African apes and humans is not homologous. And phylogenetically, it is more parsimoniously interpreted as homologous than homoplasious.

Some indriids have also fused the os centrale to the scaphoid, as noted by many authors (e.g., Schwartz and Yamada, 1998; Hamrick et al., 2000) (Fig. 6). These authors note that because the fusion of the os centrale is not universal in indriids, and even varies within one indriid (two of nine *Paleopropithecus* described by Hamrick et al. (2000) did not fuse these bones) this developmental event must have little phylogenetic or functional significance. However, to say that fusion of these bones occurs randomly in primates (Schwartz and Yamada, 1998; Hamrick et al., 2000) is an overstatement. If that were the case one would expect it to have been documented in more than just a hand-full of indriids and hominids. *Paleopropithecus* is unusual in having a relatively high frequency of each condition (fused and unfused), although the small sample size makes the significance of this finding unclear. All other primates that commonly fuse the centrale rarely have this element unfused (*Avahi, Indri, Lepilemur, Babakotia, Pan, Gorilla, Homo*). While Schwartz and Yamada (1998) dismiss the difference in timing of fusion between African and Asian apes, and imply that there is no real difference in frequency, the reality is that fusion is much more rare in Asian apes and confined to older individuals (Schultz, 1936 and personal observations). In most indriids os centrale fusion to the scaphoid is associated with a radial orientation of the facet for the trapezoid, which may be a response to compression along the long axis of the first metacarpal, which is large in these vertical climbers and leapers (Fig. 6). It is not known, to my knowledge, at what point in ontogeny this fusion occurs, but the fact that it is not reported to vary in the indriids that have this condition suggests that it is probably a feature that appears early, and it may well be a similar response to that seen in African apes and

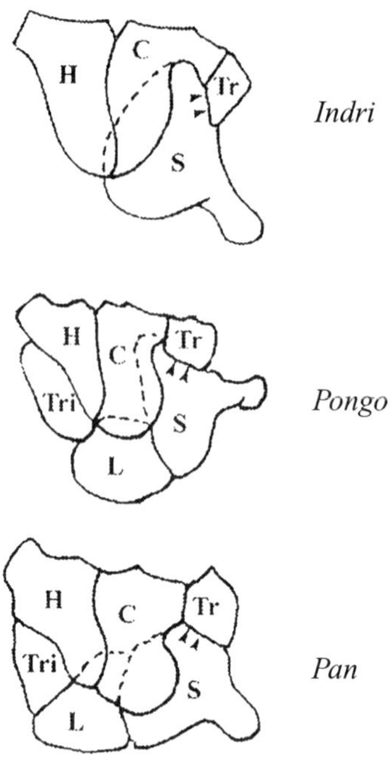

Figure 6. Schematic drawings of dorsal views of the left wrists of an indriid (*Indri*), *Pongo* with a fused os centrale, and *Pan*, all to the same carpal breadth. Dotted lines represent dorsal superpositioning of bones. Arrows point to the joint surface between the scaphoid and the trapezoid. Abbreviatiations as in Fig. 3. See text for discussion.

humans, in which the fused os centrale stabilizes the trapezoid along the long axis of the second metacarpal. In indriids, the fused os centrale re-orients the trapezoid facet radially, probably in response to transverse shear, such as would be generated by a powerful, strongly divergent pollex opposing the other metacarpals in a powerful grasp. *Paleopropithecus* may sometimes lack this feature because the articular surface for the trapezoid is small (Hamrick et al., 2000), and may have been less often loaded to peak levels of stress than other joints that have enlarged for this reason. This in turn may relate to the presence of a number of wrist and hand specializations associated with the development of hook-like hands in these primates, and the possible reduced relative importance of pollical grips.

So, another reason to fuse the os centrale to the scaphoid in primates may be to stabilize the radial side carpals in order to maintain a very powerful grip generated by a large and highly divergent pollex, as seen in indriids. And in fact, fusion of the os centrale to the scaphoid serves this function in humans (Marzke, 1971). However, the absence of other indriid forelimb characters in African apes and humans and the absence of a large pollex in African apes suggests that the indriid model does not effectively account for the suite of features seen in African apes, just as the absence of African ape and human characters in indriids suggests that a knuckle-walking ancestry is not a good explanation for in indriid forelimb morphology.

It is well established that taxa that usually do not fuse the os centrale to the scaphoid occasionally do so (*Pongo* and *Hylobates*), and those that usually fuse the two bones occasionally fail to do so (African apes and humans) (Schultz, 1936; Bergman et al., 1988). It is legitimate to conclude that African apes and humans normally fuse the os centrale and the scaphoid before or shortly after birth, and before significant amounts of load has been applied across the wrist, whereas in other apes these bone never fuse before they are loaded.

There are reasonable explanations for fusion of the os centrale in primates that generate high levels of shear stress between the scaphoid and the centrale, whether proximo-distally (African apes and humans) or transversely (most indriids) (see above), but the more interesting question may be why is it variable in *Paleopropithecus* and why does it not occur in indriids that otherwise have similar hands to those that do fuse the os centrale (*Propithecus*). Many vertebrates fuse or lose carpals, in association with a reduction in digital number (ungulates) or as an effect of reduced body mass (Alberch and Alberch, 1980). We also do not know why the os centrale fuses sometimes in *Pongo* and *Hylobates* with age, but most times not (but see below). These questions and others that relate to the ontogeny of the carpus in indriids require further analysis. However, the premise that fusion of adjacent carpal bones related to elevated shear is to be expected in mammals is illustrated below in the example of a non-primate that has converged on the African ape pattern anatomically, functionally and behaviorally (see below).

The distal carpals have also undergone considerable changes related to knuckle-walking in the common ancestor of African apes and humans. The capitate is the most obviously modified bone, with a broad, short head, strongly developed waisting between the head and distal portion of the bone, and a complex joint surface for the third metacarpal (Fig. 4). Dorsally the non-articular or subcutaneous surface of the capitate in African apes and humans is expanded, such that the lunate cannot ride up dorsally on the capitate to the same extent as is possible in Asian apes or even in digitigrade monkeys like baboons, which also must place a premium on limiting extension at the midcarpal joint (Fig. 3). The head of the capitate, which forms most of the ball of the ball-and-socket midcarpal joint in African apes and humans, is broad transversely and, in African apes and fossil humans, separated from the distal portion by a deep notch, contributing to a waisted morphology (Bush et al., 1982; McHenry, 1983; Ward et al., 1999) (Fig. 3). The centrale portion of the scaphoid fits into the embrasure formed by the capitate notch medially and the trapezoid laterally, and is close-packed in dorsiflexion (extension) (Lewis, 1985). This configuration contributes to limiting extension at the mid-carpal joint in African apes and humans. As with the hamate and the trapezoid, the capitate's metacarpal articular surface is more complex than in Asian apes and other catarrhines (see below). McHenry (1983) stresses the chimpanzee-like morphology of australopithecines capitates, and this is supported by newer material described in Ward et al. (1999). McHenry (1983) interprets australopithecine chimpanzee-like capitate characters to be retained primitive characters and at least one of these features, a waisted capitate neck, is functionally consistent with knuckle-walking.

The hamate is firmly connected to the capitate by interosseus ligaments and both essentially function as a unit in primates. In most anthropoids the proximal portion of the medial articular surface, which articulates with the triquetrum, is broad and rounded, following the contours of the head of the capitate and contributing to the medial portion of the ball of the midcarpal ball and socket joint (Fig. 4). In African apes and humans a comparatively small portion of the hamate's medial articular surface follows the contours of the capitate head, the remaining more distal and medial portion of the articular surface being strongly helical, twisting proximo-distally from a more medial to a more palmar orientation. The most distal end of the facet for the triquetrum on the hamate faces proximally, nearly normal to the long axis of the 5th metacarpal and parallel to the facet for this bone on the hamate (Fig. 4). In flexion, the small African ape and human triquetrum rides along this helical facet during conjunct pronation that accompanies flexion at the midcarpal joint (supination accompanies extension at the same joint (Lewis, 1985)). While Jenkins and Fleagle (1975) noted that they failed to identify conjunction movements during flexion and extension in chimpanzee knuckle-walking under cineradiographic observation, Lewis finds these movements present in manipulations of wet specimens, and I have made similar observations. As Lewis (1985) notes, and as Jenkins and Fleagle (1975) also suggest, technical difficulties and the fact that the chimp subjects were immature may have contributed to their failure to observe conjunction midcarpal movements. While Lewis (1985) concludes that these movements are related to close packing a midcarpal joint loaded in tension, this is no more likely than the interpretation that it is a response to close packing in compression, and it may in fact be analogous to the "screwing home" mechanism of the human knee joint in extension that also increases joint stability in compression. Other aspects of the African apes and human wrist are more understandable in terms of a response to compression. As Jenkins and Fleagle (1975) have noted, why would one find more elaborate adaptations to suspensory positional behaviors in African apes than in Asian apes that are in fact more suspensory? It makes more sense to view these changes as adaptation to quadrupedalism.

The most prominent feature of the hamate in hominoids is the hook, or hamulus, which forms part of the medial wall of the flexor carpal tunnel, along with the pisiform. In African apes and fossil humans the hamulus is very prominent and projected distally and proximally, which may serve via its attachment to the pisiform to increase the leverage of the flexor carpi ulnaris in flexion of the wrist (Bush et al., 1982; Sarmiento, 1988; Ward et al., 1999). The large size of the hamulus is certainly related to the size of the carpal tunnel and reflects the large size of the digital flexor tendons and associated intrinsic hand muscles.

Finally, the distal carpal bones have modified distal articular surfaces for the bases of the metacarpals. The trapezoid is a relatively large bone in African apes and humans and, in addition to having a proximally oriented proximal surface for the centrale portion of the scaphoid, has a strongly keeled distal articular surface that fits tightly into a deep notch on the base of the second metacarpal (Fig. 3). This notch-keel morphology is further reinforced by the facet on the trapezium that articulates medially with the enlarged lateral tubercle of the second metacarpal. *Pongo* has a flat distal trapezoid joint surface that is continuous with a relatively large, concave facet for the second metacarpal on the trapezium, which together match the broad, transversely convex joint surface of the orang second metacarpal (Fig. 3 and 5). While hylobatids and other anthropoids have a more restrictive trapezoidal-second metacarpal joint than orangs, none have the complex keel-notch pattern seen in African apes and humans.

This theme is continued medially. The large facet for the third metacarpal on the capitate is keeled along the dorsal half of the joint surface in African apes and fossil humans while it is broadly rounded in *Pongo* and other anthropoids (Fig. 3). McHenry (1983) describes a distal cupping morphology of this surface, which is probably related to the keeling described here, and he attributes this morphology to a mechanism to stabilize the wrist in knuckle-walking postures (see below). He also notes that "cupping" does not occur in the australopithecines he examined (*A. afarensis* and *A. africanus*) though it does appear to have been present in the earlier *A. anamensis* (Ward et al., 1999). While there are other important differences in the capitate distal articular surfaces between fossil humans and African apes (Bush et al., 1982; Marzke, 1983; McHenry, 1983; Marzke et al., 1992; Ward, et al., 1999), they all share this keeled morphology, which is lost in later humans. The distal articular surface of the hamate is also keeled in African apes and fossil humans, separating the facets for the 4th and 5th metacarpal bases, which are thus somewhat divergent (Bush et al., 1982; Marzke, 1983; Ward et al., 1999). *Pongo* and other anthropoids have more smoothly concave joint surfaces with less or no distinction between the metacarpal joint surfaces (Fig. 3).

A keeled articular surface for the metacarpal bases in African apes and fossil humans, and a corresponding series of notches on the metacarpals contributes to a transversely "jagged" carpo-metacarpal joint that is probably related to resistance to twisting at these joints, such as would be generated during the stance phase of knuckle-walking as body mass transfers across the carpo-metacarpal joints (Marzke, 1983; McHenry, 1983).

In African apes and humans, unlike Asian apes but more like other catarrhines, the metacarpal and proximal phalangeal shafts are shorter and less curved (Susman, 1979; Begun, 1988, 1993; Inouye, 1992. Shorter bones reduce bending moments at the carpo-metacarpal and metacarpo-phalangeal joints, of functional significance to quadrupeds loading their hands in compression, whether in knuckle-walking, digitigrade or palmigrade hand postures. Increased curvature in more arboreal forms is a typical response to loading regimes dominated by tension, such as experienced by suspensory forms (Preuschoft, 1973). However, unlike Asian apes and primate quadrupeds, African apes and humans have relatively long intermediate phalanges compared to the lengths of their proximal phalanges. Most primate quadrupeds have short intermediate phalanges, but Asian apes have elongated their proximal phalanges. Between specimens of *Pongo* and *Pan* with intermediate phalanges of the same length the proximal phalanges of the *Pongo* specimen are much longer than those of *Pan*. The fact that African apes and humans retain long intermediate phalanges while reducing the length of their proximal phalanges compared to their more arboreal relatives may be related to the fact that the intermediate phalangeal shafts support body mass along their lengths in knuckle-walking. Washburn (1971) noted that the absence of hair on the intermediate segments of African apes and human hand digits was an indication of knuckle-walking ancestry for humans, an interpretation that Tuttle (1969a) has criticized. However, the comparatively long intermediate phalangeal segment of African ape and human digits may make a more compelling case for knuckle-walking, along with the suite of characters described here.

DISCUSSION

A number of the semi-digitigrade Old World monkeys, such as *Papio* and *Erythrocebus* share some of the characters found in African apes and humans, such as reduced length and increased the robusticity of their phalanges, but they have not undergone the same changes in the wrist and they have not reduced the lengths of other limb segments, probably to maintain the potential to produce rapid bursts of acceleration. Semi-digitigrade hand posture in these primates increases functional forelimb length, but also loads the phalanges very strongly in compression, torsion and bending, leading to increases in robusticity. These monkeys are for the most part much smaller than African apes, and probably do not experience the same stresses in the wrist that occur in African apes. Their hand postures are also very different from those of African apes, with their metacarpals less vertically oriented, and their phalanges in hyperextension at the metacarpophalangeal joint, and extended at the interphalangeal joints, in contrast to the vertical metacarpals, less extended metacarpophalangeal joint, and flexed interphalangeal joints of African apes.

Amazingly, there is another example of what happens when a hand that retains most of the primitive elements of mammalian forelimbs becomes modified into a hook-like appendage with features for terrestrial quadrupedalism. It is found in the morphology of

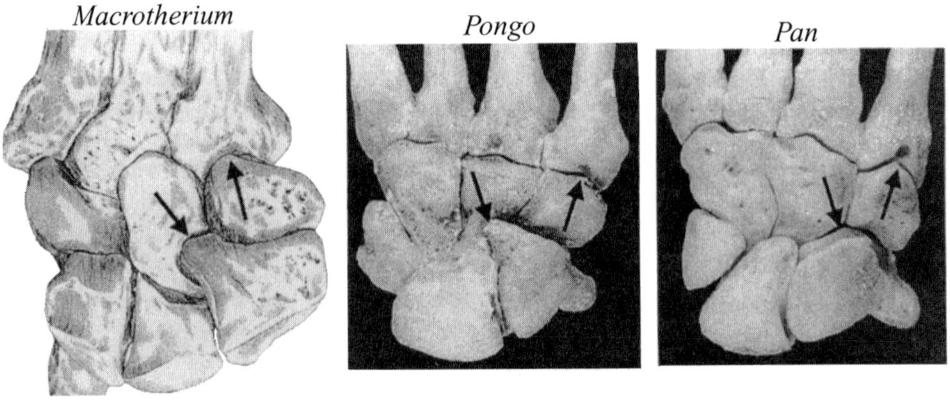

Figure 7. Dorsal views of the left wrists of *Macrotherium*, *Pongo* and *Pan* showing a number of convergent features. Downward pointing arrows indicate the fused os centale and upward pointing arrows the trapezoid keel. See text for discussion. View of *Macrotherium* modified from Zapfe (1979).

chalicotheres, an unusual family of extinct perissodactyls. Zapfe (1979) in his exceptional monograph on *Chalicotherium grande* (now attributed by many to *Macrotherium grande*) illustrates the complete carpus, articulated manus and articulated skeleton, and the parallels to the skeletal morphology of African apes and humans are striking (Figure 7).

There are of course many dramatic differences between the hands of chalicotheres and hominoids related to differences in evolutionary history and functional anatomy. Like other "hook-handed" ungulates (e.g., South American ground and tree sloths, order Xenartha), *Macrotherium* has long metacarpals, very short proximal phalanges, and greatly enlarged and elongated terminal phalanges. Xenartha are superficially similar to African apes and distinct from *Marcotherium* in having elongated intermediate phalanges, but in the case of the South American taxa, unlike hominoids, the intermediate phalanges are as long or longer than the metacarpals. In both ungulate groups it is the terminal phalanges that bear the bulk of body mass, in contrast to African apes in which it is the proximal and especially the intermediate phalanges. Other critical differences from African apes include the fact that *Macrotherium* had only three hand digits (rays 2 to 4), as in *Bradypus* (the three toed sloth; *Choloepus*, the two-toed sloth has also lost the 4th ray) and a further reduction or fusion of carpal bones (*Macrotherium* lacks a trapezium and in *Choloepus* it is fused to the first metacarpal (Mendel, 1985)). However *Macrotherium*, which appears to have loaded its forelimb in a flexed digitigrade posture similar to African apes (Zapfe, 1979), has a number of surprising similarities to African apes not found in the highly suspensory sloths.

The wrist of *Macrotherium* retains the same bones with roughly similar morphology as found in African apes and humans (except the trapezium, not surprising given that chalicotheres had no thumbs). Among the more detailed similarities to African apes and humans are the comparatively large scaphoid, the distally projected and robust hamate hamulus, the distal orientation of the facets of the proximal carpals for the distal carpals and the irregular morphology of the facets of the distal carpals for the metacarpals (Fig. 7). In dorsal view a deep facet on the capitate articulates dorso-laterally with the centrale portion of the scaphoid, a mechanism very similar to that seen in African apes and humans to restrict excessive extension at the midcarpal joint. The "jagged" morphology of the carpo-metacarpal joint is also obvious in dorsal view, including such detailed similarities to African apes and humans as the strongly keeled trapezoid-deeply notched second metacarpal and the strongly divergent facets for the third metacarpal on the capitate. The hamate receives the base of the third and fourth metacarpal in this perissodactyl that lacks the fifth digit, but is otherwise functionally similar to the African ape and human hamate with divergent facets for the fourth and fifth metacarpals. The capitate is not waisted mediolaterally as in African apes and fossil humans. African apes and fossil humans have transversely very broad capitate heads, probably to maintain joint congruence in wide ranges of pronation and supination. *Macrotherium* had much less mobility in the wrist, as is typical of ungulates, and lacked a transversely broad capitate, but achieved the same stabilizing effect with proximodistal waisting (the capitate of *Macrotherium* has deep notches proximally and distally providing a structural link between the lunate and third metacarpal (Zapfe, 1979). Perhaps the most striking convergence between African ape/human wrists and those of chalicotheres is the fusion of the os centrale to the scaphoid, represented by a prominent dorsal process (Fig. 7). Interestingly, when *Pongo* fuses the os centrale to the scaphoid a very similar morphology results, with a prominent, pointed process positioned dorsally over the capitate (Fig. 7). Other basic similarities to hominoids in the skeleton of *Macrotherium* include semi-erect vertebral column, reduced number of lumbar vertebrae, broad ilium, long forelimbs, short hindlimbs (intermembral index = 141), reduced brachial index (83), increased range of elbow extension, supination set at the wrist, and short, stout metacarpals (compared to forelimb long bones, but not compared to the phalanges) (Zapfe, 1979).

CONCLUSIONS

Three lines of evidence suggest that humans evolved from knuckle-walkers. It is suggested by the evolutionary relations among the taxa, by the biomechanical implications of forelimb characters shared uniquely among African apes and humans, and by evidence that a very similar set of features is known to have developed independently in another mammal that loaded its hands in a broadly similar manner. In addition, the presence in all African apes and humans of characters of the feet related to terrestriality (Gebo, 1992) also argues strongly for a shared ancestry of terrestrial positional behavior, most likely knuckle-walking. This is not a new conclusion, and has been suggested often in the classic literature (reviewed in Tuttle [1969a, 1974, 1975]) and more recently backed up by detailed analysis (Marzke, 1971; Corruccini, 1978; Gebo, 1992, 1996; Begun, 1994; Richmond and Strait, 2000, 2001; Corruccini and McHenry, 2001; Richmond et al., 2001) (Fig. 8).

Knuckle-walkers retain obvious anatomical specializations related to suspensory positional behavior in both their forelimbs and hindlimbs, and all knuckle-walkers are in fact adept climbers as well. Vertical climbing and suspensory positional behavior, while preferred by many as the ancestral form of positional behavior from which human bipedalism evolved (e.g., Tuttle, 1969a; Stern, 1975; Fleagle et al., 1981), is not inconsistent with a knuckle-walking ancestry for bipedalism. Early fossil humans, like African apes, retain anatomical features related to climbing, and were probably more adept in the trees than are modern humans (Stern and Susman, 1983). But they were also clearly adapted to terrestrial bipedalism. The common ancestor of African apes and humans may have evolved knuckle-walking as an efficient means of locomoting on the ground while retaining well developed arboreal skills. Elsewhere I have suggested that the common ancestor of African apes and humans evolved from a Eurasian ancestor closely related to *Dryopithecus* and *Ouranopithecus* (Begun, 1994, 2001). This ancestor appears to have entered Africa from Eurasia between 8 and 10 Ma during a time of global cooling resulting in drier and more seasonal climates

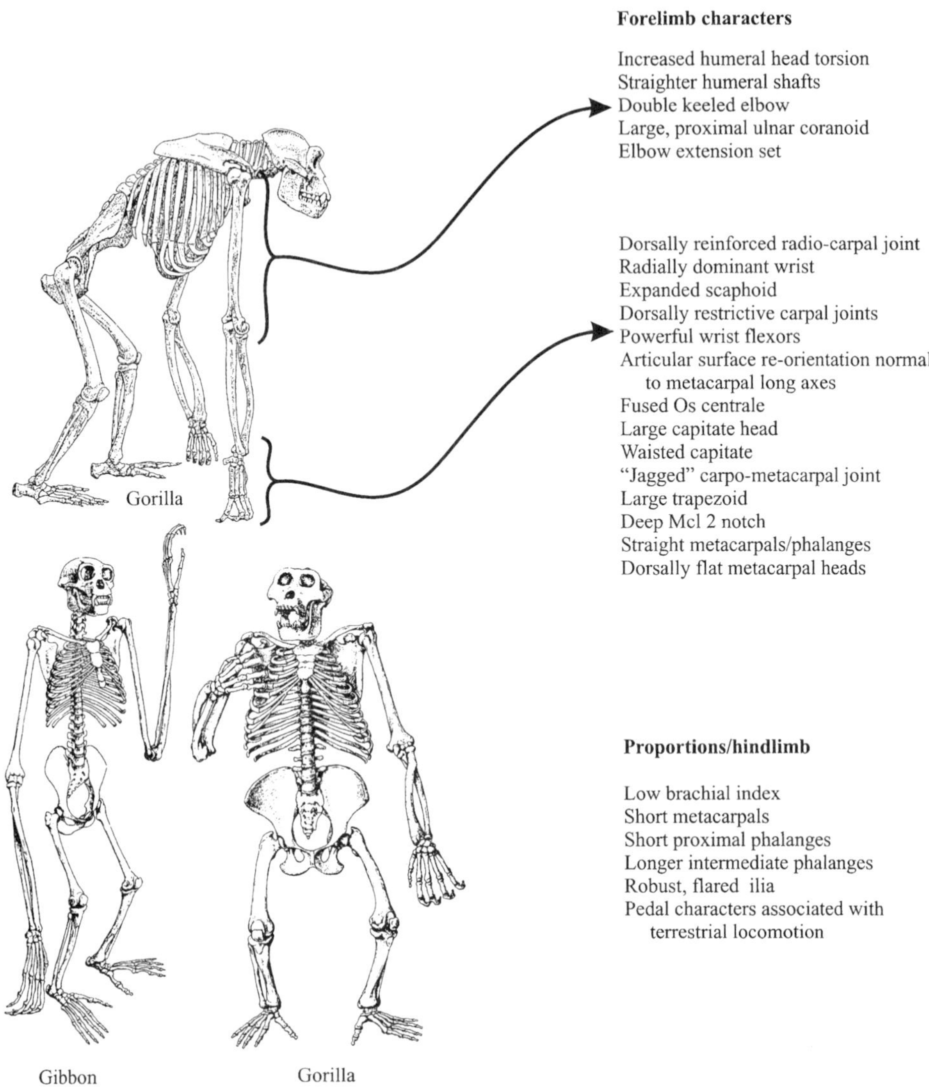

Figure 8. Summary of characters functionally related to knuckle-walking, most of which are also present in hominine bipeds. Above: lateral view of a gorilla in typical knuckle-walking pose (modified from Napier, 1974). Below: Gibbon and gorilla in bipedal postures. Modified from Schultz (1969). Note that short, robust trunk and low brachial index of the gorilla compared with the gibbon.

in Eurasia (Cerling et al., 1997; Begun, 2001). We know from the anatomy of *Dryopithecus* and that of the living outgroup to the African ape and human clade (*Pongo*) that this ancestor was probably arboreal and suspensory (Begun, 1993, 1994). It is possible that the climate changes that forced hominoids from most of Eurasia, which are associated with the development of modern grasslands (Cerling et al., 1997), may have produced selection pressures favoring individuals that could efficiently bridge relatively open-country gaps within forests. Modern *Pan* may represent the best model for the first knuckle-walker, having ubiquitous ecological preferences, in terms of habitat and diet, and in having very large ranges that they cover primarily on the ground. *Gorilla* is also ecologically diverse compared to *Pongo*, but less so than *Pan*. Gorillas *may* have become isolated early on in more closed forested conditions, and its terrestrial ancestry may have allowed *Gorilla* to become extremely large by alleviating a size constraint imposed by arboreality. Although *Pongo* is large and highly arboreal, it is highly specialized, probably descendent from a primarily or exclusively arboreal ancestor, and considerably smaller than *Gorilla*. Early human bipeds are probably the result of the trend toward terrestriality that began with the evolution of knuckle-walking. In the same way that *Pongo* has become a specialized, essentially committed arboreal hominid, humans have become a specialized, essentially committed terrestrial hominid. The differing phylogenetic trajectories of each were determined primarily by accident of evolutionary history.

This view suggests that there was some selection for increases in frequency, change of mode, or both, of terrestrial locomotion in the immediate ancestors of humans, although the precise pattern remains unknown. Human bipedalism is essentially emergent from knuckle-walking, and represents an extreme along a terrestrial continuum. Tuttle (1969a) calls this the "troglodytian" stage of human evolution, of which he is critical though not completely dismissive (see above). Tuttle (1969a, 1974, 1975) provides a comprehensive intellectual history of the "troglodytians", and makes clear the fact that they did not, in the end, provide exhaustive anatomical evidence of a link between African ape and human positional behavior. I have tried to show here that there are numerous anatomical characters of the wrists of African apes and humans that are plausibly related to knuckle-walking, and have come to be exapted for other functions in humans. It is not clear whether this transformation was related to selection for changes in forelimb function (tool use, carrying, etc.), hindlimb use (bipedal locomotion, long distance travel, open country habitats) or aspects of both, and nearly every possible alternative has been explored in the vast literature on bipedal origins. The interpretation supported here is that human bipedalism is an expectable outcome of hominid (great ape and human) terrestriality. Lovejoy et al. (1999) suggest that the dramatic morphological changes in the lower limb separating fossil and living humans from other hominids, which are so extensive that they tend to be considered to have developed over an extended period of time, may in fact be explicable by a relatively small number of developmental events. Evolving a bipedal ape from an already terrestrial one may thus be feasible genetically as well as ecologically and functionally. Given this pattern, I would speculate that trunk and lower limb changes were probably the initial targets of selection for further increases in terrestriality, culminating in a commitment in humans to terrestrial bipedalism, and not selection for alternative functions for the forelimb. Selection for efficient locomotion on two limbs may well have also been associated with a reduction in mass of the forelimb, as in flightless birds, theropod dinosaurs and many marsupials. Pleiotropic effects on the hand from genetic changes resulting from selection for human bipedal foot morphology may also explain some of the changes in early human hands without requiring that they were specific targets of selection (Webb and Fabiny, 2001). Selection relaxed for arboreal adaptations of the early human forelimb before renewed selection produced transformations of the human forelimb related to manipulative capabilities, possibly actin on a forelimb of reduced mass given the kinematics of bipedalism. This is reflected in the fossil and modern comparative anatomical record. Traces of arboreality remain to this day, in humans, but are especially well developed in fossil humans, while direct or indirect evidence of selection for increased manipulative skills (tools, big brains, etc.) postdates the origin of bipedalism in humans by at least 2 Mya.

Continued research on the paleoecology of the latest Miocene (e.g., Leakey et al., 1996; Cerling, et al., 1997; Pickford et al., 2001) will help to test this hypothesis, although the strongest evidence will have to come from the recovery of fossil hominids from Africa between about 5 to 7 Ma in age. The best-known hominid from this time period, *Orrorin tugenensis*, does not show any unambiguous indications of bipedalism, despite the claims of its describers (Senut et al., 2001; Richmond et al., 2001). The morphology of the *Orrorin* femur and its dimensions are consistent with both *Pan* and *Australopithecus* in most respects, including neck length, orientation and femoral head size. The so-called "intertrochanteric groove" cited by Senut et al. (2001) possibly refers to the groove for the tendon of the obturator externus muscle on the back of the femur. This groove is not the clear indication of bipedalism suggested by Senut et al. (2001). Many higher primates express this trait variably, including *Pan* and many monkeys. It is present on the femora of *Griphopithecus* (*Equatorius*) from Maboko, *Paidopithex* from Eppelsheim, *Turkanapithecus* from Kalodirr and *Epipliopithecus* from Děvínská Nová Ves, none of which were bipeds (Le Gros Clark and Leakey, 1951; Zapfe, 1960; Leakey and Leakey, 1989; Begun, 1992b). Whatever the positional behavior of the earliest hominines was, they appear to be associated with locally forested environments (Pickford, et al., 2001; WoldeGabriel, et al., 2001), which is consistent with a chimp-like ancestor (Richmond et al., 2001).

At present the fossils needed to test the knuckle-walking ancestry of human bipedalism hypothesis remain elusive.

ACKNOWLEDGEMENTS. I would like to thank Jeff Meldrum and Charles Hilton for organizing the symposium and this volume, and for asking me to participate in both. I am grateful to them and to Russell Tuttle and an anonymous reviewer for helpful suggestions that improved this manuscript. Thanks also to Carol Ward, Brian Richmond and Dave Strait for stimulating and ongoing discussions on wrist morphology and the evolution of positional behavior, and to Jens Franzen for information on *Macrotherium*. This work was supported by grants from NSERC and Wenner-Gren.

REFERENCES

Alberch, P. and Alberch, J., 1980, Heterochronic mechanisms of morphological diversification and evolutionary change in the neotropical salamander, *Bolitoglossa occidentalis* (Amphibia, Plethodontidae). *Journal of Morphology* 167:249–264.

Beard, K. C., Teaford, M. F., and Walker, A., 1986, New wrist bones of *Proconsul africanus* and *Proconsul nyanzae* from Rusinga Island, Kenya. *Folia Primatologica* 47:97–118.

Begun, D. R., 1988, Catarrhine phalanges from the Late Miocene (Vallesian) of Rudabánya. *Journal of Human Evolution* 17:413–438.

Begun, D. R., 1992a, Miocene fossil hominids and the chimp-human clade. *Science* 257:1929–1933.

Begun, D. R., 1992b, Phyletic diversity and locomotion in primitive European hominids. *American Journal of Physical Anthropology* 87:311–340.

Begun, D. R., 1993, New catarrhine phalanges fron Rudabánya (Northeastern Hungary) and the problem of parallelism and convergence in hominoid postcranial morphology. *Journal of Human Evolution* 24:373–402.

Begun, D. R., 1994, Relations among the great apes and humans: New interpretations based on the fossil great ape *Dryopithecus*. *Yearbook of Physical Anthropoogy* 37:11–63.

Begun, D. R., 1999, Hominid family values: morphological and molecular data on relations among the great apes and humans. In *The Mentalities of Gorillas and Orangutans. Comparative perspectives*, edited by S. T. Parker, R. W. Mitchell, and H. L. Miles, pp. 3–42. Cambridge University Press, Cambridge.

Begun, D. R., 2001, African and Eurasian Miocene hominoids and the origins of the Hominidae. In *Hominoid Evolution and Environmental Change in the Neogene of Europe*, edited by P. Andrews, G. Koufos, and L. d. Bonis, pp. 231–253. Cambridge University Press, Cambridge.

Begun, D. R., Ward, C. V., and Rose, M. D., 1997, Events in hominoid evolution. In *Function, Phylogeny and Fossils: Miocene Hominoid Origins and Adaptations.*, edited by D. R. Begun, C. V. Ward, and M. D. Rose, pp. 389–415. Plenum Press, New York.

Benton, M. J., 1990, Origin and interrelationships of Dinosaurs. in *The Dinosauria*, edited by D. B. Weishampel, P. Dodson, and H. Osmólska, pp. 11–30. University of California Press, Berkeley.

Bergman, R. A., Thompson, S. A., Afifi, A. K., and Saadeh, F. A., editors, 1988, *Compendium of Human Anatomic Variation*. Urban & Schwarzenberg, Baltimore.

Berman, D. S., Reisz, R. R., Scott, D., Henrici, A. C., Sumida, S. S., and Martens, T., 2000, Early Permian bipedal reptile. *Science* 290:969–972.

Bromage, T. G., 1987, The biological and chronological maturation of early hominids. *Journal of Human Evolution* 16:257–272.

Bush, M. E., Lovejoy, C. O., Johanson, D. C., and Coppens, Y., 1982, Hominid carpal, metacarpal, and phalangeal bones recovered from the Hadar formation: 1974–1977 collections. *American Journal of Physical Anthropology* 57:651–677.

Cerling, T., Harris, J. R., MacFadden, B. J., Leakey, M. G., Quade, J., Eisenmann, V., and Ehleringer, J. R., 1997, Global vegetation change through the Miocene/Pliocene boundary. *Nature* 389:153–158.

Collard, M., and Aiello, L., 2000, From forelimbs to two legs. *Nature* 404:339–340.

Corruccini, R. S., 1978, Comparative osteometrics of the hominoid wrist, with special reference to knuckle-walking. *Journal of Human Evolution* 7:307–321.

Corruccini, R. S., and McHenry, H. M., 2001, Knuckle-walking hominid ancestors. *Journal of Human Evolution* 40:507–511.

Dainton, M., 2001, Did our ancestors knuckle-walk? *Nature* 410:324–325.

Dainton, M., and Macho, G. A., 1999, Did knuckle walking evolve twice? *Journal of Human Evolution, Vol.* 36:171–194.

Dean, M. C., Benyon, A. D., Thackery, J. F., and Macho, G. A., 1993, Histological reconstruction of dental development and age at death of a juvenile *Paranthropus robustus* SK 63, from Swartskrans, South Africa. *American Journal of Physical Anthropology* 91:401–419.

Fischer, M., 1986, Die Stellung der Schliefer (Hyracoidea) im phylogenetischen System der Eutheria. Zugleich ein Beitrag zur Anpassungsgeschichte der Procaviidae. *Cour. Forsch.-Inst. Senckenberg* 84:1–132.

Fleagle, J. G., Stern, J. T., Jungers, W. L., Susman, R. L., Vangor, A. K., and Wells, J. P., 1981, Climbing: A biomechanical link with brachiation and with bipedalism. In *Vertebrate Locomotion*, edited by M. H. Day, pp. 359–375. 48. Symposium of the Zoological Society of London, London.

Gebo, D. L., 1992, Plantigrady and foot adaptation in African apes: Implications for hominid origins. *American Journal of Physical Anthropology* 89:29–58.

Gebo, D. L., 1993, Reply to Dr. Meldrum. *American Journal of Physical Anthropology* 91:382–385.

Gebo, D. L., 1996, Climbing, brachiation and terrestrial quadrupedalism: Historical precursors of hominid bipedalism. *American Journal of Physical Anthropology* 101:55–92.

Godfrey, L., Jungers, W. L., Wunderlich, R. E., and Richmond, B. G., 1997, Reappraisal of the postcranium of *Hadropithecus* (Primates, Indroidea). *American Journal of Physical Anthropology* 103:529–556.

Groves, C. P., 1986, Systematics of the great apes. In *Comparative Primate Biology*, vol. 1, edited by D. W. Swindler and J. Erwin, pp. 187–217. Alan R. Liss, New York.

Hall, B. K., 1994, *Homology: the hierarchical basis of comparative biology*. Academic Press, San Diego.

Hamrick, M. W., Simons, E. L., and Jungers, W. L., 2000, New wrist bones of the Malagasy giant subfossil lemurs. *Journal of Human Evolution* 38:635–650.

Hartwig-Scherer, S., 1993, Body weight prediction in early fossil hominids: Towards a taxon "independent" approach. *American Journal of Physical Anthropology* 92:17–36.

Inouye, S. E., 1992, Ontogeny and allometry of African ape manual rays. *Journal of Human Evolution* 26:459–485.

Jenkins, F. A., and Fleagle, J. G., 1975, Knuckle-Walking and the Functional Anatomy of the Wrists in Living Apes. In *Primate Functional Morphology and Evolution*, edited by R. H. Tuttle, pp. 213–227. Mouton, The Hague.

Kappelman, J., 1996, The evolution of body mass and relative brain size in fossil hominids. *Journal of Human Evolution* 30:243–276.

Larson, S. G., 1988, Subscapularis function in gibbons and chimpanzees: Implications for interpretation of humeral head torsion in hominoids. *American Journal of Physical Anthropology* 76:449–462.

Larson, S. G., 1996, Estimating humeral head torsion on incomplete fossil anthropoid humeri. *Journal of Human Evolution* 31:239–257.

Le Gros Clark, W. E., and Leakey, L. S. B., 1951, The Miocene Hominoidea of East Africa. *British Museum of Natural History: Fossil Mammals of Africa* 1:1–117.

Leakey, M. G., Feibel, C. S., Bernor, R. L., Harris, J. M., Cerling, T. E., Stewart, K. M., Storrs, G. W., Walker, A., Werdelin, L., and Winkler, A. J., 1996, Lothagam: a record of faunal change in the Late Miocene of East Africa. *Journal of Vertebrate Paleontology* 16:556–570.

Leakey, R. E. F., and Leakey, M. G., 1989, A New Miocene small-bodied ape from kenya. *American Journal of Physical Anthropology* 16:369–387.
Lewis, O. J., 1969, The hominoid wrist joint. *American Journal of Physical Anthropology* 30:251–268.
Lewis, O. J., 1974, The wrist articulations of the Anthropoidea. In *Primate Locomotion*, edited by F. A. Jenkins, pp. 143–169. Academic Press, New York.
Lewis, O. J., 1985, Derived morphology of the wrist articulations and theories of hominoid evolution. Part I. The lorisine joints. *Journal of Anatomy* 140:447–460.
Lieberman, D. E., Wood, B. A., and Pilbeam, D. R., 1996, Homoplasy and early *Homo*: an analysis of the evolutionary relationships of *H. habilis senso stricto* and *H. rudolfensis*. *Journal of Human Evolution* 30:97–120.
Lovejoy, C. O., Heiple, K. G. and Meindl, R. S., 2001, Did our ancestors knuckle-walk? *Nature* 410:325–326.
Marino, L., 1998, A comparison of encephalization between odontocete cetaceans and anthropoid primates. *Brain, Behavior and Evolution* 51:230–238.
Marzke, M. W., 1971, Origin of the human hand. *American Journal of Physical Anthropology* 34:61–84.
Marzke, M. W., Wullstein, K. L., and Viegas, S. F., 1992, Evolution of the power ("squeeze") grip and its morphological correlates in hominids. *American Journal of Physical Anthropology* 89:283–298.
McHenry, H. M., 1983, The capitate of *Australopithecus afarensis* and *A. africanus*. *American Journal of Physical Anthropology* 62:187–198.
Meldrum, D. J., 1993, On plantigrady and quadrupedalism. *American Journal of Physical Anthropology* 91:379–381.
Mendel, F. C., 1985, Adaptations for suspensory behavior in the limbs of two-toed sloths. In *The Evolution and Ecology of Armadillos, Sloths, and Vermilinguas.*, edited by G. G. Montgomery, pp. 151–162. Smithsonian Institution Press, Washington, D.C.
Middleton, K. M., and Gatesy, S. M., 2000, Theropod forelimb design and evolution. *Zoological Journal of the Linnean Society* 128:149–187.
Napier, J., 1974, The antiquity of human walking. In *Vertebrate Structures and Functions: Readings from Scientific American*, pp. 48–58. W. H. Freeman and Company, San Francisco.
Pickford, M., and Senut, B., 2001, Contexte géologique et faunique des restes d'hominidés du Miocène supérieur de Lukeino, Kenya. *Comptes Rendus de l' Académie des Sciences. Paris—Series IIA—Earth and Planetary Science* 332:145–152.
Preuschoft, H., 1973, Functional anatomy of the upper extremity. *The Chimpanzee* 6:34–120.
Richmond, B. G., Begun, D. R., and Strait, D. S., 2001, Origin of human bipedalism: The knuckle-walking hypothesis revisited. *Yearbook of Physical Anthropology* 44: 70–105.
Richmond, B. G., and Strait, D. S., 2000, Evidence that humans evolved from a knuckle-walking ancestor. *Nature* 404:382–385.
Richmond, B. G. and Strait, D. S., 2001, Knuckle-walking hominid ancestors: A reply to Corruccini & McHenry. *Journal of Human Evolution* 40:513–520.
Rogers, J., 1993, The phylogenetic relationships among *Homo*, *Pan*, and *Gorilla*: A population genetics perspective. *Journal of Human Evolution* 25:201–215.
Rose, M. D., 1988, Another look at the anthropoid elbow. *Journal of Human Evolution* 17:193–224.
Ruvolo, M., 1994, Molecular evolutionary processes and conflicting gene trees: The hominoid case. *American Journal of Physical Anthropology* 94:89–113.
Sarmiento, E. E., 1988, Anatomy of the hominoid wrist joint: Its evolutionary functional implications. *International Journal of Primatology* 9:281–345.
Satta, Y., Klein, J., and Takahata, N., 2000, DNA archives and our nearest relative: The trichotomy revisited. *Molecular Phylogenetics and Evolution* 14:259–275.
Schmitt, D. and Larson, S. G., 1995, Heel contact as a function of substrate type and speed in primates. *American Journal of Physical Anthropology* 96:39–50.
Schultz, A. H., 1936, Characters common to higher primates and characters specific to man. *Quarterly Review of Biology* 11:425–455.
Schultz, A. H., 1969, *The Life of Primates*. New York: Universe Books.
Schwartz, J. H., and Yamada, T. K., 1998, Carpal anatomy and primate relationships. *Anthropological Science* 106 (suppl.):47–65.
Senut, B., Pickford, M., Gommery, D., Mein, P., Cheboi, K., and Coppens, Y., 2001, Premier hominidé du Miocène (formation de Lukeino, Kenya). *Comptes Rendus de l' Académie des Sciences. Paris—Series IIA—Earth and Planetary Science* 332:137–144.
Shoshani, J., Groves, C. P., Simons, E. L., and Gunnel, G. F., 1996, Primate phylogeny: Morphological vs. molecular results. *Molecular Phylogenetics and Evolution* 5:101–153.

Skelton, R. R., and McHenry, H. M., 1992, Evolutionary relationships among early hominids. *Journal of Human Evolution* 23:309–350.

Smith, B. H., 1991, Dental development and the evolution of life history in the Hominidae. *American Journal of Physical Anthropology* 86:157–174.

Smith, R. J., Gannon, P. J., and, Smith, B. H., 1995, Ontogeny of australopithecines and early *Homo*: Evidence from cranial capacity and dental eruption. *Journal of Human Evolution* 29:155–168.

Sterling, E. J., and Povinelli, D. J., 1999, Tool use, aye-ayes, and sensorimotor intelligence. *Folia Primatologica* 70:8–16.

Stern, J. T., 1975, Before bipedality. *Yearbook of Physical Anthropology* 19:59–68.

Stern, J. T., and Susman, R. L., 1983, The locomotor anatomy of *Australopithecus afarensis*. *American Journal of Physical Anthropology* 60:279–317.

Strait, D. S., Grine, F. E., and Moniz, M. A., 1997, A reappraisal of early hominid phylogeny. *Journal of Human Evolution* 32:17–82.

Susman, R. L., 1979, Comparative and functional morphology of hominoid fingers. *American Journal of Physical Anthropology* 50:215–236.

Tardieu, C., 1997, Femur ontogeny in humans and great apes: Heterochronic implications for hominid evolution. *Comptes Rendus de l'Académie des Sciences. Paris—Series IIA – Earth and Planetary Science* 325:899–904.

Tuttle, R. H., 1967, Knuckle-walking and the evolution of hominoid hands. *American Journal of Physical Anthropology* 26:171–206.

Tuttle, R. H., 1969a, Knuckle-walking and the problem of human origins. *Science* 166:953–961.

Tuttle, R. H., 1969b, Quantitative and functional studies on the hands of the Anthropoidea. I. The Hominoidea. *Journal of Morphology* 128:309–364.

Tuttle, R. H., 1974, Darwin's apes, dental apes, and the descent of man: Normal science in evolutionary anthropology. *Current Anthropology* 15:389–398.

Tuttle, R. H., 1975, Parallelism, brachiation, and hominoid phylogeny. In *Phylogeny of the Primates*, edited by W. P. Luckett and F. S. Szalay, pp. 447–480. Plenum Press, New York.

Tuttle, R. H., 1992, Hands from newt to Napier. In *Topics in Primatology, Vol. 3: Evolutionary Biology, Reproductive Endocrinology and Virology*, edited by S. Matano, R. H. Tuttle, H. Ishida, and M. Goodman, pp. 3–20. University of Tokyo Press, Tokyo.

Ward, C. V., Leakey, M. G., Brown, B., Brown, F., Harris, J., and, Walker, A., 1999, South Turkwel: A new Pliocene hominid site in Kenya. *Journal of Human Evolution* 36:69–95.

Ward, C. V., Leakey, M. G. and Walker, A., 2001, Morphology of *Australopithecus anamensis* from Kanapoi and Allia Bay, Kenya. *Journal of Human Evolution* 41:255–368.

Washburn, S. L., 1971, The study of human evolution. In *Background for Man: Readings in Physical Anthropology*, edited by P. Dolhinow and V. Sarich, pp. 82–117. Little, Brown & Co., Boston.

Webb, D., and Fabiny, S., 2001, Human hands are adapted for bipedalism, not tool use. *American Journal of Physical Anthropology* Suppliment 32:161.

Weinert, H., 1932, *Ursprung der Menschheit. Uber den engeren Anschkuss des Menschengeschlechts an die Menschenaffen.* Enke, Stuttgart.

WoldeGabriel, G., Haile-Selassie, Y., Renne, P. R., Hart, W. K., Ambrose, S. H., Asfaw, B., Heiken, G., White, T., 2001, Geology and palaeontology of the Late Miocene middle Awash valley, Afar rift, Ethiopia *Nature* 412:175–178.

Zapfe, H., 1960, Die primatenfunde aus der Miozänen spaltenfullung von Neudorf an der March (Dévinská Nová Ves), Tschechoslovakei. Mit anhang: Der primtenfund aus dem Miozän von Klein Hadersdorf in Niederosterreich. *Schweizerische Paleontologische Abhandlungen* 78:1–293.

Zapfe, H., 1979, *Chalicotherium grande aus der miozänen Spaltenfüllung von Neudorf an der March (Devinská Nová Ves), Tschechoslowakei*. Vol. 2. *Neue Denkschriften des Naturhistorischen Museums Wien*. Ferdinand Bergerr & Söhne, Wien.

Zylstra, M., 1998, Metacarpophalangeal joint morphology in non-human primates. *American Journal of Physical Anthropology* Suppl. 26:240.

Zylstra, M., 1999, *Functional morphology of the hominoid forelimb: Implications for knuckle-walking and the origin of hominid bipedalism*. Ph.D. dissertation, University of Toronto. University Microfilms, Ann Arbor.

Chapter 3

A New Hypothesis on the Origin of Hominoid Locomotion

YVETTE DELOISON

Abstract. Australopithecine foot bones from Hadar (Ethiopia) as well as ten fossil footprints from Laetoli (Tanzania) exhibit arboreal anatomical characteristics. About 3 Mya there were two genera in South Africa: a more human-like genus with bipedal locomotion and one *Australopithecus* genus with a more ape-like morphology of mixed features, i.e. bipedal for the talus and the distal epiphysis of the tibia, and arboreal for the other foot bones. The human hand is a primitive unspecialized organ whose axis lies on the third ray, suggesting that the hand was never used for specialized locomotion. A common ancestor of humans and apes and australopithecines is proposed. This ancestor was neither arboreal nor quadrudepal, but bipedal. Primitive bipedalism was a characteristic feature of "Protohominoidea."

INTRODUCTION

In earlier work on the fossil foot bones from Hadar, Ethiopia, I demonstrated that *Australopithecus afarensis* almost certainly engaged in arboreal locomotion and was also capable of bipedal locomotion as indicated by certain features of the pelvis (Abitbol, 1995) and the dorsal surface of the talus (Deloison, 1993, 1997). My analysis of the Laetoli footprints in Tanzania led to the same conclusion (Deloison, 1992). In this paper, my analysis is extended to foot bones found recently at Sterkfontein, South Africa, leading to an overall synthesis on the locomotion of Hominoidea, particularly, australopithecines, apes and *Homo*.

MATERIAL AND RESULTS

Let us first consider the 3.3 Mya StW 573 foot bones Ron Clarke found in 1995 in Member 2, at Sterkfontein. In a dorsal view of the StW 573 left talus (Fig. 1 and 12), the edges of the trochlea are not parallel as in *Homo* but diverge toward the neck, and the articular surface encroaches onto the neck of the talus, in a manner similar to *Pan*. The strong

From Biped to Strider: The Emergence of Modern Human Walking, Running, and Resource Transport, edited by D. Jeffrey Meldrum and Charles E. Hilton. Kluwer Academic/Plenum Publishers, New York, 2004.

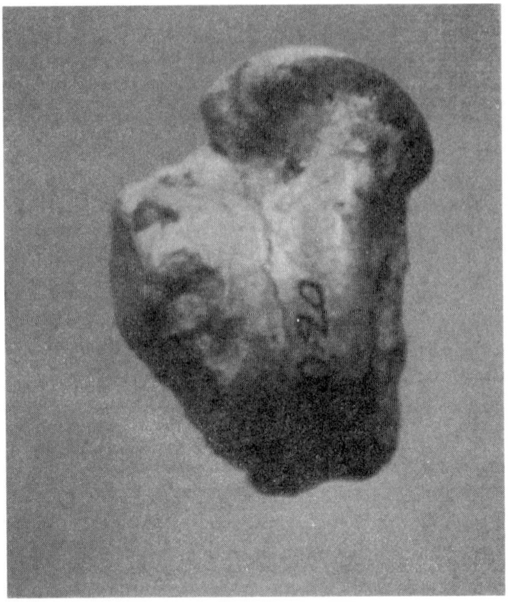

Figure 1. StW 573 talus plantar view.

curvature of the tibial facet of this hominid indicates a greater range of ankle movement relative to *Homo*. On the plantar aspect of this talus (Fig. 2), the facies articularis calcanea posterior tali is narrower. The angle between the greater axis of this facet and the axis of the talus body is markedly different from the small angle in *Homo*. On the proximal view of this talus (Fig. 3), the edges of the trochlea have almost the same height. Moreover,

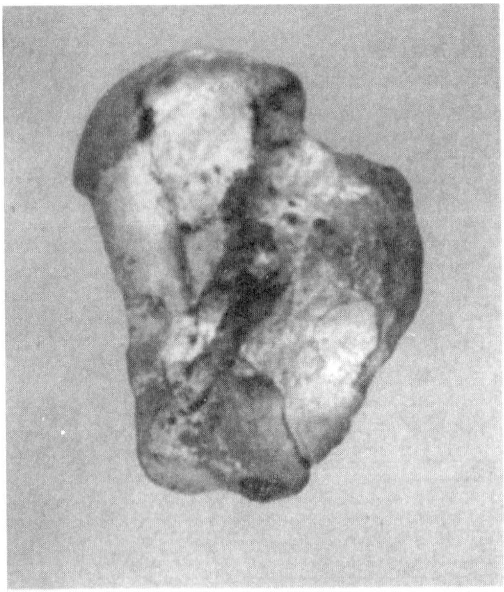

Figure 2. StW 573 talus dorsal view.

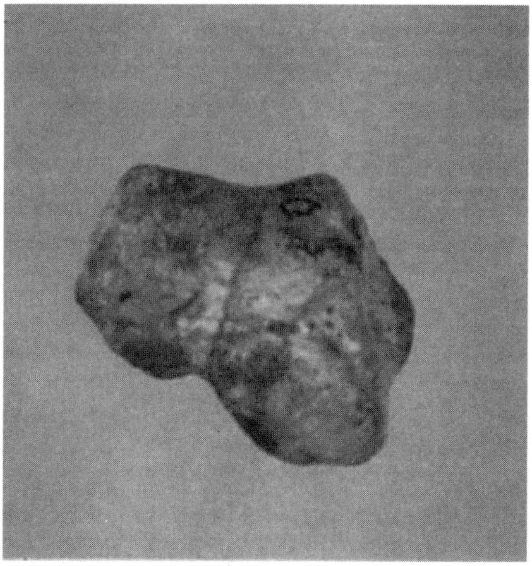

Figure 3. StW 573 talus distal view.

the angle of the flexor hallucis longus groove is 50°, which is more vertical than in *Pan* (20° in *Pan* and 70° in *Homo*). This view shows that the bone articulates with the tibia and the fibula in a vertical orientation similar to that seen in *Homo* and *Pongo*. On the distal view of this talus (Fig. 4) the angle of the axis of the head is 30°, which is smaller than in *Homo* (45°) and more like that seen in *Pan* (25°). The width of the talus head is greater dorsally than ventrally and thus nearer to that of *Pan* than to *Homo*.

The thinness of the dorsal part of the left navicular of StW 573 (Fig. 5, 6 and 13), is an ape-like character, different from OH 8 and *Homo*. This bone is strongly curved. Although the medial side of the joint is lacking, the proximal view (Fig. 5) shows this joint surface

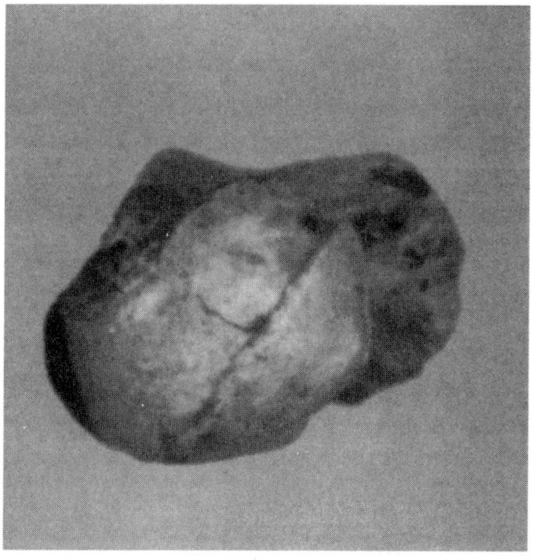

Figure 4. StW 573 talus proximal view.

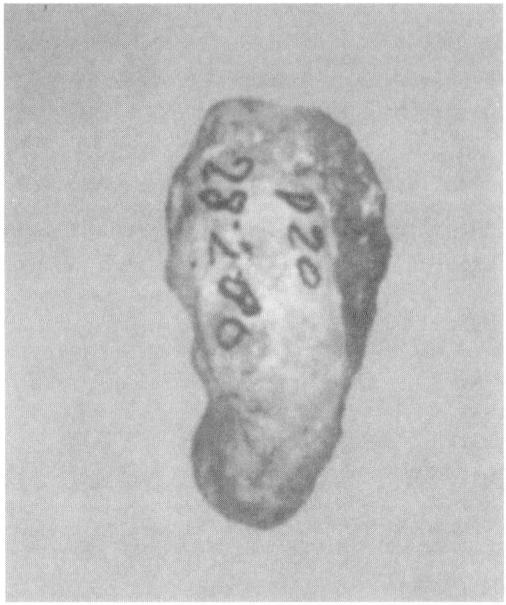

Figure 5. StW 573 navicular proximal view.

has a long dorso-plantar direction. A clear angle exists between the facets for the lateral and medial cuneiforms. This bone exhibits different arboreal features but it is not the same as the *Pan* bone. It has its own specific, or generic, morphology.

The distal facet of the StW 573 left medial cuneiform (Fig. 7 and 14) exhibits convexity. In a lateral view, this bone is less square-shaped than that seen in *Homo* and closely resembles

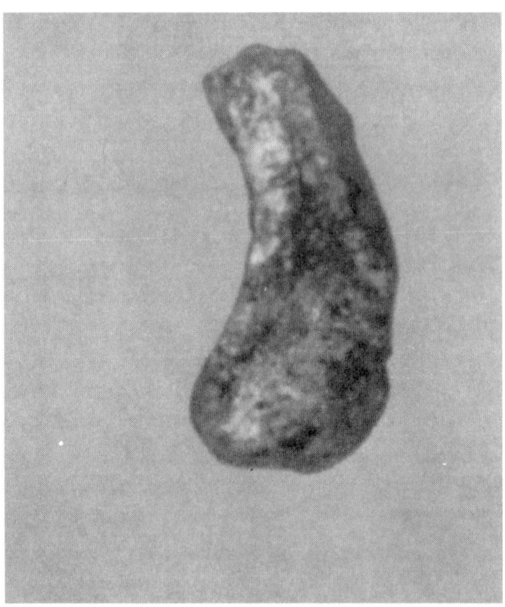

Figure 6. StW 573 navicular medial view.

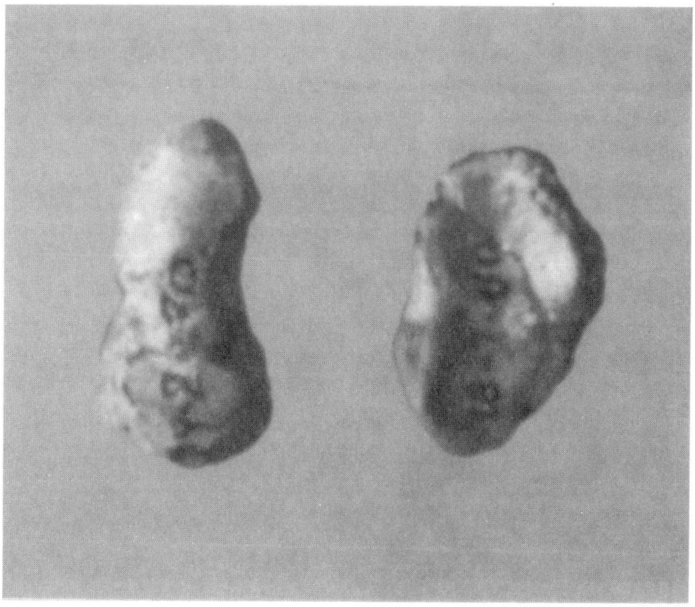

Figure 7. StW 573 medial cuneiforms distal view at left and first metatarsal proximal view.

the configuration seen in *Pan*. It has an L-shaped facet for the intermediate cuneiform as in other hominids, while *Pan* has two separate facets. The distal facet extends onto the medial surface as in *Pan*. This overlap is greater in StW 573 than in OH 8 and *Homo* and is an indication of greater mobility in the hallux. Abduction of the first metatarsal on the medial cuneiform would make a spiral movement similar to that seen in *Pan*. The StW 573 foot exhibits greater medial divergence of the first metatarsal on the medial cuneiform relative to that seen in OH8 and in *Homo*. On the hallux, the markedly convex distal medial cuneiform facet connotes a highly mobile joint at the base of the great toe, as in *Pan*, but unlike the flattish facet and limited mobility of this joint in humans. The plantar view is like A.L. 333-28. The medial facet is reniform and convex in a medio-lateral direction and faces the medial side. The plantar part of this facet has a small projection restricting the opening of the first commissure as in *Pan* and in A.L. 333-28. Clarke and Tobias (1995) reported similar observations.

The left first metatarsal of StW 573 (Fig. 15) preserves only the proximal portion. The thickness of the diaphysis is intermediate between *Pan* and *Homo*. In the plantar view, there is a strongly marked impression of the peroneus longus muscle, which probably played an important part in adducting the hallux (Deloison, 1993, p. 147). The proximal aspect is formed by the articular facet for the medial cuneiform. Its vertical axis forms a 50° angle with the lateral aspect of the bone, a *Pan*-like feature; in *Homo* this angle is 0°. This articular surface exhibits a marked medio-lateral concavity, in correspondence to the convex surface of the medial cuneiform. This morphology indicates a rather mobile first cuneo-metatarsal joint. The proximal facet is indented on both the medial and lateral margins (Clarke and Tobias, 1995). The form of this surface implies a special type of trochoid articulation, as observed in *Pan* but never found in modern humans. This first metatarsal has a morphology characteristic of an abductable first metatarsal, like that of the apes suggesting some arboreality.

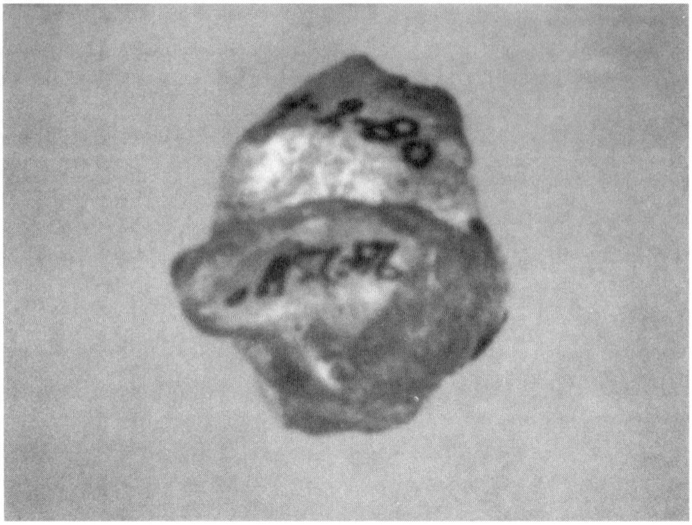

Figure 8. StW 573 calcaneus distal view.

The lateral and medial aspects of the left second cuneiform of StW 573 exhibit specific features, but the cuneo-metatarsal joints call to mind more arboreal than bipedal locomotion.

The left lateral cuneiform of StW 573 also has its own particular features: a plantar tuberosity more lateral than in *Pan*—a small facet for the second cuneiform—the articular facet for the navicular bone is smaller than in *Pan* and *Homo* and the distal facet is different. But the morphology is more ape-like with plantar zone, lateral, medial, proximal and distal aspects that are narrow in the low portion and cut out.

The calcaneus of StW 573 (Fig. 8) is fragmentary but does preserve the articulation for the cuboid. This facet has the same conical shape as in *Pan*, thereby indicating a more *Pan*-like degree of mobility in the mid-tarsus region, especially between the calcaneus and the cuboid.

Like *Pongo* (personal observation), StW 573 could have had a vertically aligned leg, but its entire foot was specialized for arboreal locomotion as much as in apes especially that of *Pan*.

In Member 4 of Sterkfontein, dating between 2.6–2.8 MYA, there are some foot bones showing ape-like characters but others that are more similar to modern humans.

The right calcaneus of StW 352 (Fig. 9 and 16) is missing the proximal part of its greater tuberosity, as seen in the dorsal view. It is a very robust bone with a longitudinal axis more similar to that of modern humans than apes. The sustentaculum tali is very large. In the lateral view (Fig. 10), this bone has the same downward facing direction of the distal articular surface, as do humans. It seems likely, that this calcaneus had an elevated anterior part and the StW 352 foot probably possessed a human-like plantar arch. Other bones seem to indicate bipedalism: the third left metatarsal StW 496, the left third metatarsal StW 477, and the third right metatarsal StW 435.

On the other hand, the profile of the lateral aspect of the first metatarsal StW 562 (Fig. 17) is more concave than that of *Homo*. The distal epiphysis of this bone is very rounded, like that of *Pan*, and it is low and wide as in chimpanzees. Its proximal articular

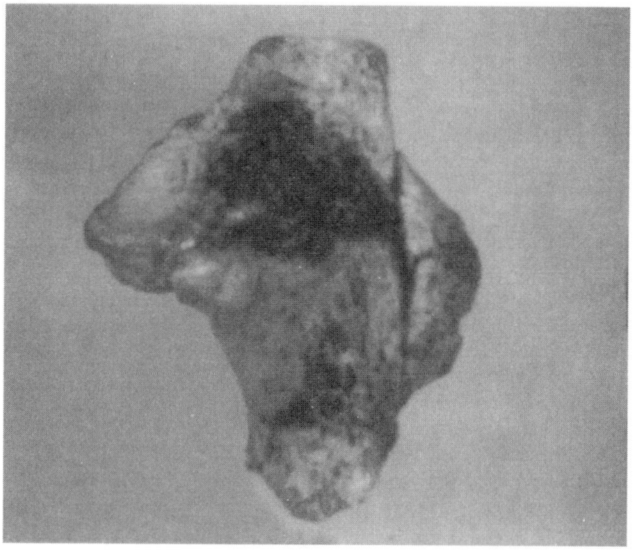

Figure 9. StW 352 right calcaneus dorsal view.

surface (Fig. 11) is very concave and the axis has a strong curvature like that of *Pan*. This first metatarsal presents features suggesting a certain degree of abductability and of arboreal climbing. The same observations can be made for the right first metatarsal StW 595. Thus, the left second metatarsal StW 377, and the left fourth metatarsal StW 485 have mixed features; these left metatarsals may belong to the same foot. The same may be said for the right third StW 388, the fourth StW 596 and the fifth StW 114/115 metatarsals which could be from the same foot.

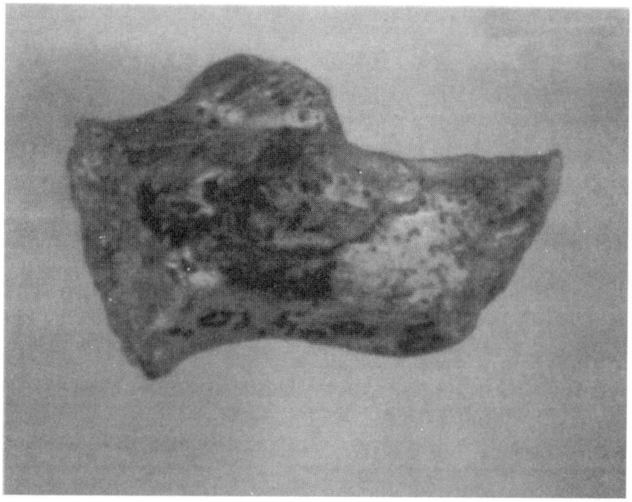

Figure 10. StW 352 right calcaneus lateral view.

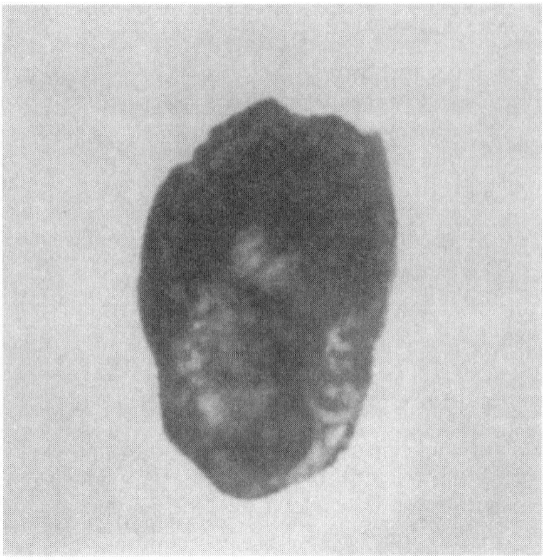

Figure 11. StW 562 first metatarsal proximal view.

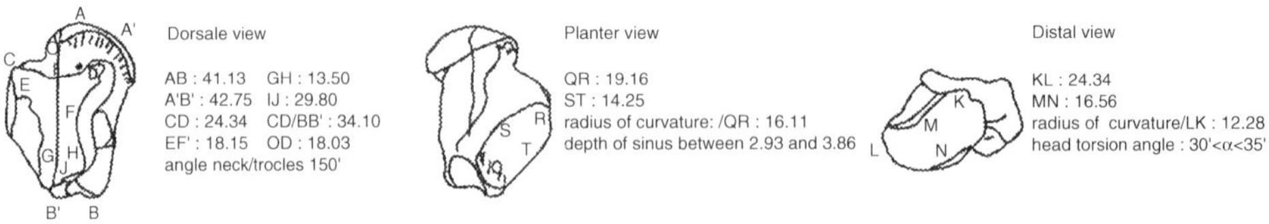

Figure 12. StW 573 left talus.

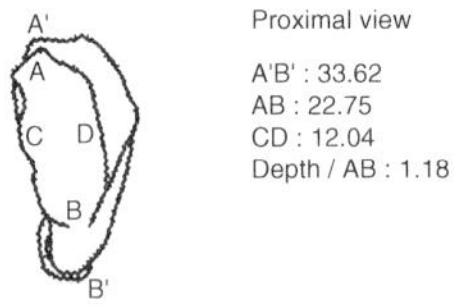

Figure 13. StW 573 left navicular.

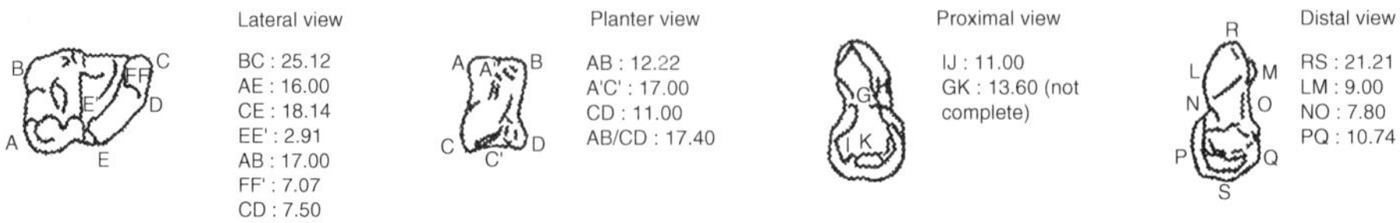

Figure 14. StW 573 left medial cuneiform.

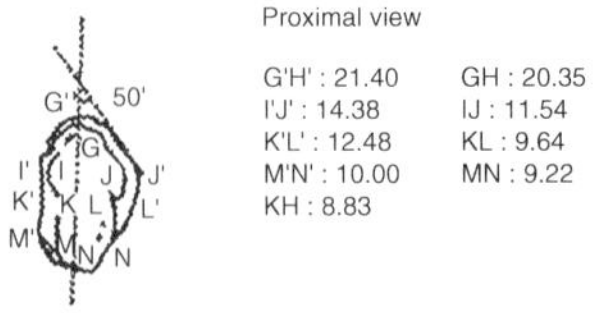

Figure 15. StW 573 left first metatarsal.

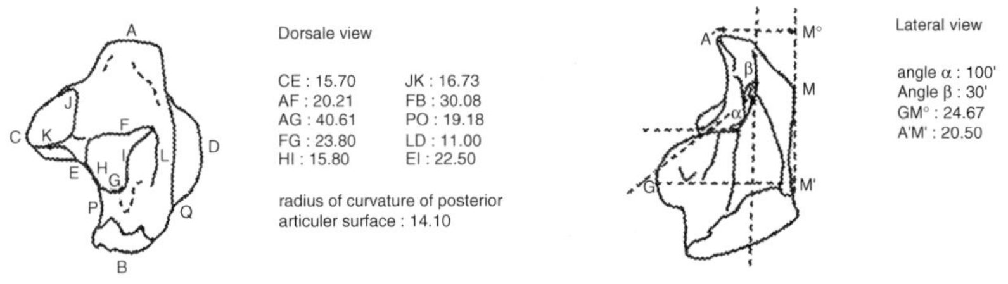

Figure 16. StW 352 right calcaneus.

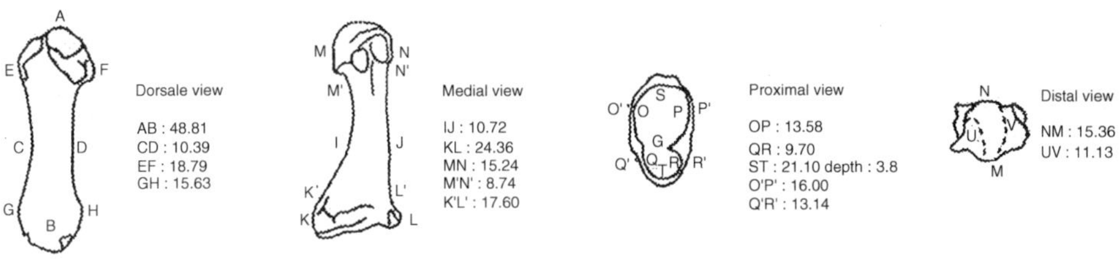

Figure 17. StW 562 right first metatarsal.

All the foot bones from Sterkfontein exhibit either bipedal or mixed, arboreal and bipedal, features. The bones with mixed features belong to the Australopithecines.

To sum up, the Hadar (Ethiopia) *Australopithecus afarensis* has the characteristics of an arboreal foot. Two genera are found in South Africa at the same period about 3 Mya. One genus was more human-like with bipedal locomotion as seen in StW 352. The other genus was an *Australopithecus*, probably *africanus*, whose metatarsal examples exhibit mixed features: a bipedal tibio-talus joint and arboreal talus and foot bones with a more ape-like morphology.

DISCUSSION

Australopithecines, apes and *Homo*, have the same bone morphology in their feet because they are primates with the same ancestor. Studying all the foot bones of the different australopithecines available and especially those of the almost complete Stw 573, enabled us to better understand the evolution of hominid locomotion. Currently, it is accepted that human bipedalism is the expression of an evolutionary accomplishment from primitive tetrapods to upright humans. According to this theory, even the larger size of the brain is a consequence of bipedalism which freed the hand and induced greater brain capacity.

Leading authorities in anthropology, such as Howells (1953) and Washburn (1978), developed a hypothesis that humans evolved from an arboreal primate, suggesting that upright posture originated in trees and not on the ground. This hypothesis therefore implies that the human foot is derived from the foot of an arboreal primate. Huxley (1868, quoted by Grassé, 1980) was one of the first to develop this concept.

Like Stern (2000), I "assert that the relatively long toes of *A. afarensis* were compatible with use of the foot for some kind of prehension in trees, and would also have increased the length of the foot in a way that would have affected the kinematics of bipedal swing phase."

In earlier work I reconstructed the length of the *A. afarensis* foot using the foot bones from Hadar (Deloison, 1993). The following proportions were found:

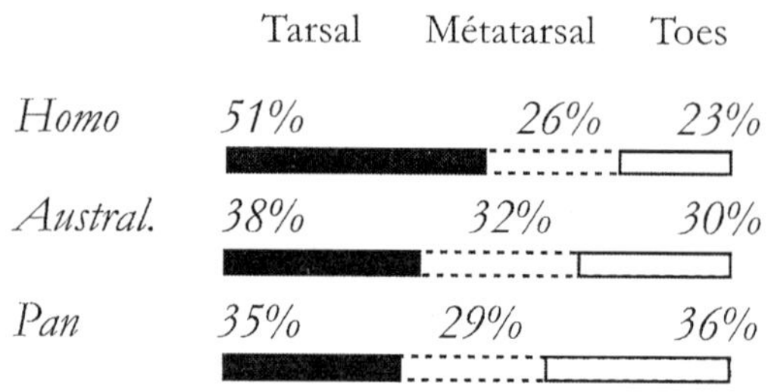

Since we do not have any complete distal phalanges, the length of the *A. afarensis* toes could have been much longer and more than 30%. It is noteworthy that *A. afarensis* lies closer to *Pan* than *Homo*.

It seems quite reasonable to consider 1) that apes which use arboreal and occasionally bipedal locomotion have an anatomy (e.g., long and curved bones in the hand and in the foot, etc.) specialized for these activities, and 2) that humans who use obligate bipedal locomotion and occasional climbing activities have an anatomy (long legs, etc.) specialized for erect posture. So why do some authors consider australopithecines, which had apelike features, to be "perfect bipeds"? This is not logical. Australopithecines had specific anatomy with many ape-like characteristics such as their intermembral index and their curved fingers and toes. Their common locomotion was arboreal with knuckle-walking on the ground; when they used bipedal locomotion, it was a very different kind of bipedalism than human bipedalism.

In their paper on the origin of human bipedalism, Richmond et al. (2001), decide in favor of a knuckle-walking ancestor. Both Stern (2000) and I agree with Duncan et al. (1994) who state that "every species is composed of characteristics that reflect both its ancestry as well as its unique evolutionary pathway; understanding the overall functional pattern of the organism requires an equal consideration of all its anatomical features, regardless of whether they are apomorphies, plesiomorphies, or homoplasies . . . ".

The human hand has maintained a structure close to the morphology of the primitive autopode (Grassé, 1967). Among other primitive features, the axis of the human hand lies on the third ray and unlike an ape hand it is not specialized for particular functions. I have studied ape hands and observed in gorillas a bony margin on the dorsal aspect of the head of the metacarpals, also found inconsistently in chimpanzee hands. This bony ridge limits extension of the fingers in dorsal flexion when the forelimb is extended as during knuckle-walking. The curvature of the hand phalanges is a common feature shared with the foot. Apes exhibit opposition of the first toe to the lateral toes, due to the abduction capacity of the hallux, but not opposition of the thumb to the lateral fingers. Prehension is not a characteristic feature of the human foot.

In humans, the carpus and metacarpus are both different from those of apes. In my opinion, characteristic features linked to knuckle-walking, as proposed by Corruccini (1978) and Inouye (1994a, b), cannot be found in the human hand. The human hand has retained its undifferentiated anatomical state and is neither specialized for suspension or locomotion. In contrast, the human foot is highly specialized for our bipedalism. One could view the human hand as isotropic (note the isotropic space of the carpal bone during chondrification) and the foot as anisotropic, i.e. its properties depend on direction. The human foot does not have any of the arboreal characteristics observed among apes such as regression and abduction of the first toes as well as the long and convex lateral toes.

In addition to the trochoid shape of the first cuneometatarsal joint, the Stw 573 foot, attributed to *Australopithecus*, possesses long and strongly curved phalanges, a typical adaptation in arboreal limbs consistent with the other limb bones of the Sterkfontein skeleton. Thus, confirming its capacity for arboreal locomotion. Like any other specialized organ, this foot must have developed from a primary condition with a wide range of potential, i.e. from a non-specialized organ. The foot of the human ancestor followed an evolutionary pathway distinct from that of the australopithecines. Recently, Clarke (1999) noted that the distal ends of the StW 573 radius and the ulna are more similar to those of apes. Together with the similarity of the distal humeri and proximal ulna with those of *Pongo* indicates an upper limb used for arboreal locomotion. The shape which gave rise to these respectively different evolutionary lineages was undifferentiated.

In a radiographic study of six-, nine- and twelve-month old chimpanzee feet in comparison to that of a ten-month old human foot, I noted that a six-month-old chimpanzee foot is completely ossified except for the growth plate allowing for lengthening (Deloison, 1993). Ossification comes later in the human foot. At six months, the chimpanzee foot has a rounded rather wide talus with a thick calcaneus extending forward to the forefoot. At nine months, the simian talus and calcaneus are flatter and longer. The calcaneus then turns upward posteriorly, unlike the human calcaneus which exhibits an upward growth in its anterior portion at this same age. An analysis of the embryology of the human foot (Deloison, n.d.) indicates that all the tarsal and metatarsal bones are well in place in the 43-day-old human fetus. The lateral tubercle of the calcaneus, a specific feature of the human adult, can be identified at the end of fetal life. It is also noteworthy that infant apes resemble humans more than adult apes and the characteristic features of apes develop during growth.

Summarizing, the human hand has a primitive morphology that does not exhibit evidence indicative of a past role in specialized locomotor behavior. Conversely, the human foot is a very specialized organ that arose from a non-specialized ancestral morphology. Likewise, it was a non-specialized ancestral organ that gave rise to the very (though differently) specialized ape foot. Indeed, as clearly demonstrated by the law of irreversibility of evolution (Dollo Law, W. Howells, 1953), an organ cannot lose its specialization (or adaptation) and return to a primitive state; it can cease its evolution at different stages, but never revert. For example, a horse will not recover fingers lost in the course of the evolutionary process, nor will *Pongo* recover its atrophied thumbs. In the same manner, the human foot cannot proceed from a prehensile foot like the ape foot nor from an australopithecine-like foot. The foot of the ancestor of the primates probably had a morphology close to that of the non-specialized primitive autopode.

I suggest a common ancestor for hominoids: protohominoid. Protohominoid had a small slender body with thin bones (not found now because they fossilized poorly and because we have not looked for them). It walked erect and was bipedal and plantigrade. Its anatomy showed little differentiation: a round head, almost equal limbs, pentadactylism and battledoor-shaped hands and feet. It must have lived near a river or lake or swamp. It was able to climb in the trees but favored bipedalism over climbing. I suggest protohominoid gave rise to three lineages.

One lineage adopted arboreal life, perhaps because it was safer. These subjects acquired bodily specializations: long forelimbs and short hindlimbs (high humerofemoral index), hands for suspension, feet for prehension. They used their long forelimbs and short hindlimbs when moving on the ground with a new kind of locomotion where the second phalanges of the hands were applied to the ground: knuckle-walking. This lineage gave rise to the apes.

A second lineage coming from the bipedal protohominoid also climbed in trees, perhaps for the same reason as the first lineage. These beings acquired climbing specializations but departed earlier and did not have enough time to complete their climber characteristics. This lineage gave rise to australopithecines whose lifestyle was probably little different from the ancestors of apes, explaining their specific ape-like anatomy.

Finally, a third lineage retained the ancestral bipedalism and acquired a very specialized morphology, especially for feet but also for limb proportions, with long hindlimbs and short forelimbs, etc. This lineage, which survived despite the lack of natural means of defense (e.g., horns, claws, etc.) because of the quality of its brain, led to more human-like forms.

It was not bipedalism that freed hands and consequently man. It was the sitting position, which leaves the hands free for manual activity. But many mammals sit down and use their hands, yet have not become human. Neither free hands nor bipedalism made man.

The ancestrality of bipedalism can be illustrated by the following schema:

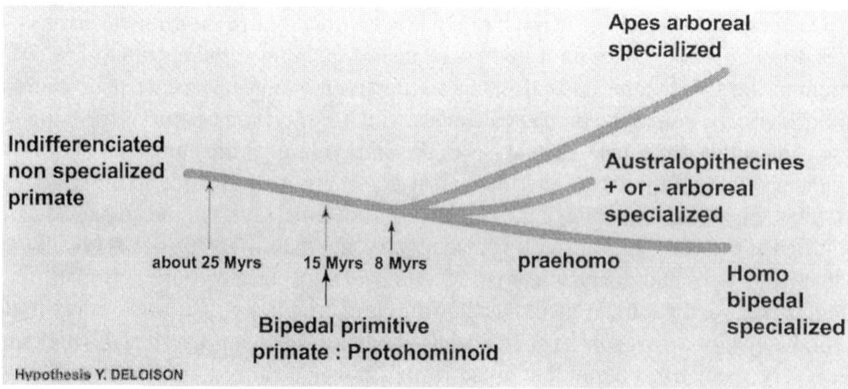

Further to interesting features of the foot bones from South Africa, and discussions fostered by different studies on australopithecines, these data provide a new basis for thought for all concerned about the fascinating search for the origins of mankind.

ACKNOWLEDGEMENTS. I am grateful to Pr Tobias and Dr Ron J. Clarke for allowing me to examine the fossil foot bones and especially R.J. Clarke to study the StW 573 specimen. Thanks to Pr Y. Coppens, Director of the Paleoanthropology and Prehistory Department of the Collège de France and to CNRS UPR 2147 which supported this work. We would also like to thank Monique Tersis for preparing the manuscript and Danièle Fouchier for the illustrations.

REFERENCES

Abitbol, M., 1995, Lateral view of *Australopithecus afarensis*: primitive aspects of bipedal positional behaviour in the earliest hominids, *Journal of Human Evolution* 28:211–229.

Clarke, R. J., and Tobias, P., 1995, Sterkfontein Member 2 foot bones of the oldest South African hominid, *Science*, 269:521–524.

Clarke, R. J., 1998, First ever discovery of well preserved skull and associated skeleton of Australopithecus, *South African Journal of Science* 94:460–463.

Clarke, R. J., 1999, Discovery of complete arm and hand of the 3.3 million-year-old Australopithecus skeleton from Sterkfontein, *South African Journal of Science* 95:477–480.

Corruccini, R. S., 1978, Comparative osteometricsof the hominoid wrist joint, with special reference to knuckle-walking, *Journal of Human Evolution* 7:307–321.

Deloison, Y., 1992a, Articulation cunéo-métatarsienne de l'hallux, considérée comme un des éléments déterminants de la forme de locomotion à partir de son anatomie osseuse. Comparaison entre l'Australopithèque, l'Homme et le Chimpanzé. *C. R. Acad. Sci. Paris* 314, sér. II:379–1385.

Deloison, Y., 1992b, Empreintes de pas à Laetoli (Tanzanie). Leur apport à une meilleure connaissance de la locomotion des Hominidés fossiles. *C.R. Acad. Sci. Paris*, 1992, t. 315, sér. II:103–109.

Deloison, Y., 1993, Relations multivariées portant sur les os de pieds de dix genres de Primates actuels et deux fossiles, OH 8 et AL 333-115 sont comparés à l'Homme et au Chimpanzé. *Cahiers d'Anthropologie et de Biométrie humaine*, t. XI, n° 1–2:55–76.

Deloison, Y., 1997a, The foot bones from Hadar, Ethiopia and the Laetoli, Tanzania, footprints. Locomotion of *A. afarensis*. *American Journal of Physical Anthropology*. Supplement 24:101.

Deloison, Y., 1997b, Description d'un *talus* fossile de Primate et sa comparaison avec des astragales de Chimpanzés, d'Hommes et d'Hominidés fossiles: *Australopithecus* et *Homo habilis*. Note présentée par Yves Coppens, *C.R. Acad. Sci. Paris*, t. 324, série II a, p. 685 à 692.

Deloison, Y., 1999, L'Homme ne descend pas d'un Primate arboricole ! une évidence méconnue. *Biom. Hum. et Anthropol.* 17:47–150.

Deloison, Y., 1999, New hypothesis on hominoid bipedalism. *American Journal of Physical Anthropology* Supplément 30:137.

Duncan, A. S., Kappelman, J., Shapiro, L., 1994, Metatarsophalangeal joint function and positional behavior in *Australopithecus afarensis, American Journal of Physical Anthropology* 93:67–81.

Grassé, P. P., 1967, *Traité de Zoologie*, volume sur le squelette; Masson ed.

Grassé, P. P., 1980, *L'Homme en accusation*. Albin Michel ed.

Howells, W., 1953, *Préhistoire et histoire naturelle de l'Homme*, Payot.

Inouye, S., 1994a, Ontogeny of knuckle-walking hand posture in African apes, *Journal of Human Evolution* 26:459–485.

Inouye, S., 1994b, *The Ontogeny of the knuckle-walking behavior and associated morphology in the African apes*, Phd. Dissertation, Northwestern University, University Microfilms, Ann Arbor.

Richmond, B. G., Begun, D. R., and Strait, D. S., 2001, Origin of human bipedalism: the knuckle-walking hypothesis revisited, *Yearbook of Physical Anthropology*, 44:70–105.

Stern, J., 2000, Climbing to the top: a personal memoir of *Australopithecus afarensis*, *Evolutionary Anthropology*, 113–133.

Washburn, S., 1978, *Préhistoire et histoire naturelle de l'Homme*, Payot.

Chapter 4

Functional Interpretation of the Laetoli Footprints

PETER SCHMID

Abstract. Examination of the Laetoli hominid footprints has already lead to a diversity of opinion on different topics. Recent reconstruction of the locomotor apparatus of australopithecines led to the hypothesis that their locomotor repertoire included a significant proportion of arboreal activity. While the structure of the foot of the earliest hominids may be partially gleaned from the study of its preserved skeletal parts from sites in Eastern and Southern Africa, such fossils lack information about the weight transfer during progression.

The most important external force that acts on the body during progression is the force exerted on the ground. This force is due to the reaction to gravity and to the momentary acceleration of the body. The relative distribution of pressure between the lateral and the medial portions of the foot is a crucial matter, not only for the functioning of the foot, but also for interpreting the fossil footprints fixed in the ash layers at Laetoli. In bipedal movement the fore-limb is no longer used to support against gravity or to accelerate the body. The forward swing of the arm contributes angular momentum to neutralize partially that of the advancing leg of the opposite side. The arms do not behave as pendulums in typical walking, but are subject to muscle action. The swinging of the arms is consequently not a purely incidental accompaniment of forward movement, but is an integral part of the dynamics of progression.

Consequently, we analysed the weight transfer of an individual in walking normally and also in walking with different constraints on arm and trunk motion during bipedal locomotion. The differences in the course of the gravitational force in the footprint can give some insight for the particular footprint morphology in the Laetoli tracks.

INTRODUCTION

With a few exceptions, quadrupedal locomotion marks the body construction of all non-human primates. In general, climbing and crossing gaps in the canopy by leaping dominate the locomotor spectrum of these forms. Most of the large-bodied primates (apes and atelines) suspend their bodies below the branches to remain in a stable equilibrium. Therefore, they increased the mobility of their shoulder girdle and are engaged in brachiating. The anatomy of modern humans differs from all non-human primates in an adaptation to upright posture and bipedal locomotion. Humans diverge in several complexes from the basic quadrupedal pattern. A lordotic curvature of the lumbar spine, a shortening of the distance between

---

From Biped to Strider: The Emergence of Modern Human Walking, Running, and Resource Transport, edited by D. Jeffrey Meldrum and Charles E. Hilton. Kluwer Academic/Plenum Publishers, New York, 2004.

the iliosacral and the iliofemoral joints as well as the valgus position of the knee joints accompany a significant change in the construction of the foot.

The human foot is adapted for walking on the ground. Elftman and Manter considered the fact that the human foot bears a closer resemblance to the ape foot as another evidence of hominid arboreal ancestry (Elftman and Manter, 1935a, b). They also suggested that the essential features of man's foot were acquired at an early stage of his terrestrial existence, rather than after long apprenticeship on the ground. The chimpanzee foot—still primarily adapted to arboreal life—has not undergone any noticeable evolutionary changes toward the human type of terrestrial foot. Although the gorilla resembles man more closely than does the chimpanzee in the relative shortness of the lateral digits, it shows no indication of the more fundamental changes, which are essential for the development of thc human condition. The latter condition is characterized by the stabilisation of the transverse tarsal joint in a position of plantarflexion. This immobilization allies the human foot more closely to the ape foot as used in arboreal rather than in terrestrial progression (Elftman and Manter, 1935b). This crucial factor in the evolution of the human foot accounts for the longitudinal arch and partially for the alignment of the lateral metatarsals with the long axis of the calcaneus. The appearance of a new process—the antero-lateral process of the calcaneus—prevents an extensive eversion in the lower ankle joint. The trochlea of the talus is rotated with respect to the body of the bone, and the medial border of the trochlea is raised. Furthermore, this is correlated with the changed position of the tibia in upright locomotion, and with the knees held near the midplane of the body during movement. A torsion of the metatarsal heads—allowing flexion in the metatarsophalangeal joints perpendicular to the ground—eliminates the anterior transverse arch as a functional factor and allows the distribution of pressure over the ball of the foot. The permanent adduction of the hallux forms a strong anterior support for the longitudinal arch. In addition the shortening of the four lateral metatarsals and their phalangeal series, the increase in relative length of the cuboid, navicular, first and second cuneiforms as well as the increase in stoutness and length of the calcaneus are further important changes in proportion.

These important differences can already be recognized in the lower limbs of the Hadar hominids (Latimer and Lovejoy, 1989, 1990a, b; Latimer and Lovejoy, 1990b; Latimer et al., 1987). Latimer concluded that these modifications have a negative effect upon climbing ability and remove the stabilizing selection for arboreal competence in the lower limb of *Australopithecus afarensis* (Latimer, 1991). Based on the same material Deloison suggested a foot with prehensile abilities, some laxity in the foot joints, a varus stance, and an abducted hallux (Deloison, 1991). Several authors came to the conclusion that the leg and foot of *A. afarensis* retain *Pan*-like features and connote an intermediate degree of adaptation (Berillon, this volume; Christie, 1977; Day and Wickens, 1980; Meldrum, this volume; Senut and Tardieu, 1985; Stern Jr. and Susman, 1983). New material from Sterkfontein (South Africa) is also considered to show evidence that the earliest australopithecines, while bipedal, included arboreal climbing in their locomotor spectrum (Clarke and Tobias, 1995).

ADDITIONAL EVIDENCE

Additional evidence to that of the study of fossil foot bones was referred to by several authors cited above. This evidence was based on fossilized footprints from East Africa. In 1976, in an area of northern Tanzania called Laetoli, A. Hill and D. Western, members of a team led by archaeologist Mary D. Leakey noticed animal tracks in a layer of exposed tuff. Then in 1978, P. Abell located a hominid heel impression and subsequent excavations of the

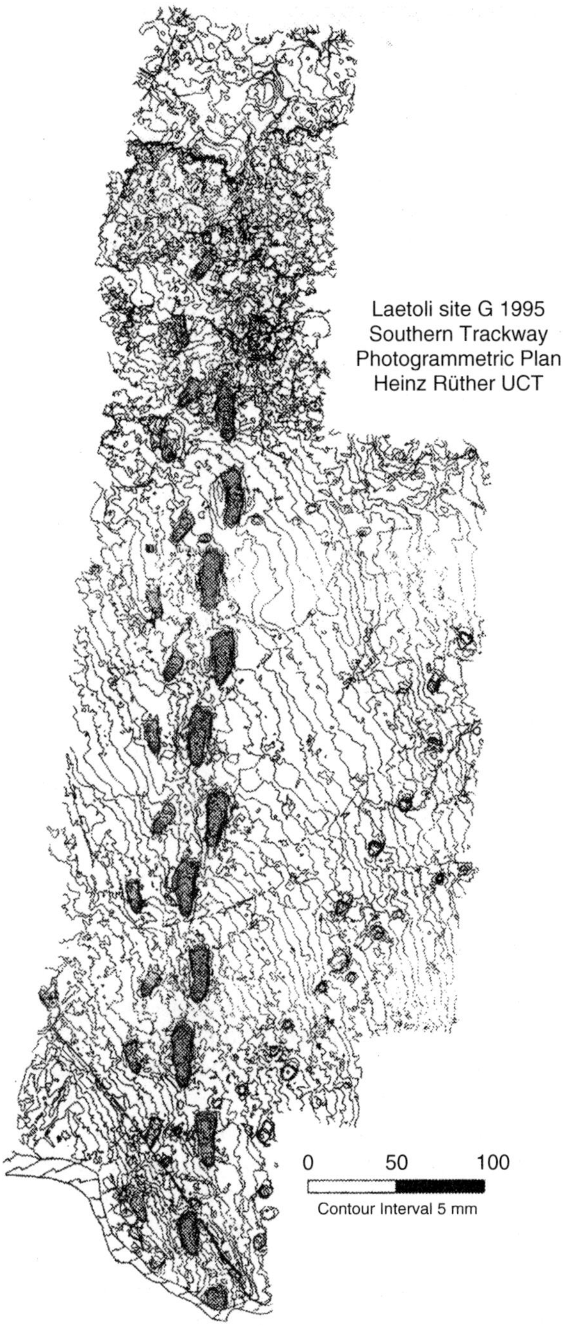

Figure 1. A photogrammetric plan of the Laetoli site G showing the southern trackway. The plan was done by Heinz Rüther (University of Capetown).

footprint tuffs at site G in 1978 and 1979, revealed two parallel trails of hominid footprints extending some 27 meters (Fig. 1). The volcanic sediments were dated radiometrically to be between 3.4 million and 3.8 million years old (Leakey, 1978; Leakey and Harris, 1987; Leakey and Hay, 1979). After the discovery, it was immediately apparent that the prints were made by hominids (Leakey, 1978, 1979, 1987; Leakey and Hay, 1979). At another locality, site A, footprints were always very dubiously hominid, or bipedal. The last interpretation invokes the hypothetical existence of an animal (ursid) (Tuttle, 1984). At Laetoli G, it has been considered that three bipedal individuals produced two closely juxtaposed trails (G-l and G-2/3). There are footprints of a small individual (G-l) and prints of two other individuals (G-2 and G-3). G-3 partially overprinted the track way of G-2, producing a single trail. However, R. J. Clarke who excavated the trail and Y. Deloison who studied some casts of the prints consider there was no G-3 individual (Clarke, 1999; Deloison, 1991).

The trails contained a total of 54 footprints that could be clearly identified as hominid tracks. Partially the tracks are heavily weathered and in this section the footprints were poorly preserved. But in the less weathered part the preservation was good, allowing clear recognition of soft-tissue anatomical features such as heel, a longitudinal arch and big toe impression.

The discovery proved that early hominids had an erect posture and were bipedal long before the advent of stone tool-making or the expansion in size of the brain. Because the track provided information about the soft tissue of the hominids' feet and the length of their strides—information that cannot be ascertained from fossil bones alone—the Laetoli footprints attracted a huge amount of attention from scientists as well as the general public. Fieldwork on the Laetoli tracks ended with the 1979 season, and the site was covered with local river sand to protect the site.

Over the following decade, trees grew to heights of over two meters. Scientists who occasionally visited the Laetoli area were concerned that the roots from these plants would eventually destroy the footprint tuff. A team of conservators and scientists of the Getty Conservation Institute (Los Angeles) completed a project to protect the footprints from erosion, plant growth and other causes of deterioration (Agnew and Demas, 1998; Agnew et al., 1993).

Few researchers had ever seen the exposed footprints and most of the scientific literature is based on casts and photographs. For this reason, the Tanzania's Antiquities Department invited a group of scientists to reexamine the trackway while the conservation work was going on. Bruce Latimer, curator of physical anthropology at the Cleveland Museum of Natural History, Craig S. Feibel, a geologist at Rutgers University, and the author, were nominated by specialists in the field of paleoanthropology to do research at Laetoli.

As for fossil skeletal remains, the Laetoli footprints, in the short time since their discovery, have already been employed as evidence to support contradictory interpretations of early hominids. One point in dispute concerned the number of hominids that made the two parallel trails. In one trail, the footprints were small and well defined, but in the other the prints were larger and less clear. Some scientists speculated that the trails were made by two hominids—possibly a female and a male—walking abreast or close to each other. Clarke claimed that because the footprints in the eastern trail had so many consistent features, they must have been made by a single large hominid. The larger footprints were more poorly defined than the smaller ones, Clarke argued, because the feet of the large hominid had slid more in the rain-slickened ash (Clarke, 1979; Clarke, 1985).

The footprints in the westernmost trail were much smaller and more clearly defined than the prints in the eastern trail. Actually we are convinced that the eastern trail was made by two individuals, with the second hominid stepping in the tracks of the first (B. M. Latimer and C. S. Feibel, pers. comm.). The fact that the stride length does not differ between the G-1 trail and the G-2/3 trail indicate an almost equal body height of the three individuals.

Important features of gait, such as relative stride length, relative speed, velocity, and cadence, depend on estimates of stature or speed or both for the Laetoli G creatures. The patterns thus far reported from the ashfall tuff of Laetoli are patterns consistent with what is called "strolling" (Charteris, 1981, 1982). Normal humans, strolling at around 0.4 relative speed, often exhibit the same features. They caution that their conclusions may not be valid if some environmental or motivational constraint was acting on the hominids to walk cautiously. Given the environmental circumstances of a barren plain covered with deep, soft ash described above, this caution may be particularly appropriate.

If it is assumed that Laetoli hominids had foot proportions like modern humans, the individuals represented were adult, the footprints are good indicators of foot length, and the reported measurements are accurate, then stature could be estimated. Based on studies of modern *Homo sapiens*, the individuals have been estimated to be between 112 and 149 cm tall (Tuttle, 1987). The size difference of these individuals may not be very big because the step-length of the two tracks are almost identical. When walking together, at a speed calculated in standard units, a tall and a short subject have very different step-lengths and cadences. The shorter individual has to increase one and/or the other, in order to "keep up". In addition, the assumptions mentioned before are not backed up by the prints in the tuff layers. Especially length measurements are only approximate and without surveying the originals in Laetoli it is in some cases inapropriate to interpret the photographs and casts. This was clearly demonstrated by the recognition of the extensive effects of bioturbation subsequent to initial reburial. The traces of some kind of burrowing invertebrates disrupted and modified the outline of the foot impression in several cases as well as some structures in the front part of the prints. The ash layers of the footprint tuff were subsequently modified by a variety of postdepositional processes for at least a few thousand years. The tuffs were weakly cemented by calcite resulting from reaction of carbonate with water. Dung and plant materials were probably replaced by calcite at this time (Feibel et al., 1996).

Estimates of body proportions of these early hominids are mainly based on the Hadar partial skeleton A.L. 288-1. It was proposed that *Australopithecus afarensis* had relatively short hind limbs and that this both "implies substantial kinematic differences in bipedal gait from the modern condition" and reduces relative and absolute stride length (Jungers, 1982). Jungers suggested that these findings would account for the relatively short stride "without having to resort to hypothetical 'strolling' behaviour". Jungers claimed that his results counter "evidence of a fully modern adaptation to bipedalism" in Pliocene hominids. A relatively short hip-to-heel length compared to foot length could produce significantly lower foot lengths per stride than humans. Alternatively, their ratios of foot length per stride may simply reflect the fact that the Laetoli G bipeds were walking slowly (Tuttle et al., 1992).

For Tuttle the abundant pedal specimens of *A. afarensis* from Hadar, Ethiopia, sport features, such as long down-curved toes, which are functionally incompatible with the humanoid footprints at Laetoli (Tuttle, 1985). Although in the footprints of the chimpanzee the lateral toes do not project far ahead of the hallux. This is due to the fact that the four lateral toes are curled under and toward the hallux. This position of the toes compensates to a certain extent for the great length of the phalanges, which makes walking with the toes extended more awkward (Elftman and Manter, 1935a). White and Suwa saw no evidence of toe curling impressions in the prints (White and Suwa, 1987) but Deloison has suggested the toes were curled underneath (Bourdial, 1999). It is clear that unflexed toes should leave shallower but more discrete impressions. Unfortunately, individual digit morphology is not retained in the prints. Moreover, the before mentioned postimprinting distortion as a result of the adjacent invertebrate burrowing and calcite exfoliation adds further uncertainity to these findings.

White and Suwa demonstrated, by showing that a small *A. afarensis* foot based on Hadar bones metrically and morphologically conforms to the Laetoli prints (White and Suwa, 1987). In their discernible gross morphology, the Laetoli footprints are apparently more like *H. sapiens* than they are like any modern pongid. This does not, of course, rule out the possibility that bipedal locomotion practised by the Pliocene creatures who left the prints differed in some ways from that characterising modern humans.

Stern and Susman studied the Hadar hominids and casts of some Laetoli prints (Stern and Susman, 1983). They concluded that the Tanzanian prints and the Ethiopian skeletal material conform in their "transitional morphology." Stern and Susman also attributed the Laetoli prints to *A. afarensis* but contended that this taxon exhibited a mode of bipedality distinct from that of modern humans. They point to the footprints as indicating that "the medial part of the ball of the foot was poorly, if at all, developed in these hominids and that the weight was not transferred to the medial part of the foot at the end of the stance phase as it is in humans" (Susman et al., 1984).

WEIGHT TRANSFER

On my first visit to the G site, I was puzzled by the fact, that although I wished to see differences of the Laetoli footprints in regard to human impressions I couldn't observe an indication of an abducted first ray comparable to ape conditions. A clear arch-like elevation in medial part of the planta was present, and the imprints of the digital region were proportionally short in relation to the metatarsal and tarsal region. The anatomy of the foot was comparable to the conditions of modern humans. One important character of bipedal support is the broad foot contact. Man has the possibility of using the ball of the foot for support, giving up the arboreal adaptation of his toes. This is especially important in the contraction of the calf muscles, used in walking for propelling the leg forward. Not the slightest similarity to a prehensile foot structure could be recognized. An abducted hallux is clearly missing. In addition, an analysis of soft tissue morphology is more than hypothetical because no consistent feature can be followed-up over more than one print. This is clearly the case for the supposed bulging of an abductor muscle for the hallux (Deloison, 1991).

In normal human walking, the heel first strikes the ground to break the locomotion. Initially, the pressure is very slight, the weight being still largely carried by the contralateral foot. As weight is increasingly transferred to the foot, the pressure under the heel increases. The forward moving body causes the heel to roll forward until the ball of the foot also comes into contact. As forward movement continues, the pressure transmitted by the heel decreases, that on the ball of the foot increases. The transfer of weight to the ball of the foot when the heel rises is made possible by the longitudinal arch of the foot—a human peculiarity. The arch is put under strain, and its supporting ligaments get some active assistance from intrinsic muscles of the foot. The longitudinal arch or vault of the foot presents considerable amount of variations. When the inner arch is low, the record shows a transmission of pressure to the substratum in this region. In advanced cases of flat foot there is almost an even distribution of pressure over the foot. Extreme variation is to be found in the distribution of pressure in the ball of the foot. The lateral portion of the longitudinal arch may be low and carry an important portion of the body pressure, without involvement of the medial arch. In this case the base of the fifth metatarsal bone may make a distinct impression on the record, and the area of the foot underlying the entire fifth metatarsal transmits notable pressure. The importance of the transverse arch or vault is considerable at the heads of the metatarsals as the weight of the body acts on this region of the foot. When the heel of the contralateral foot has made contact with the ground, the weight is normally transferred towards the great toe, resulting in decreased pressure on the part of the standing foot. When the ball of the foot leaves the ground, the toes alone exert pressure.

In the Laetoli footprints primary weight-bearing pressure was kept on the central and lateral half of the heel. The hominids apparently had a functionally stable longitudinal arch structure. The height of the medial side of the arch indicates that weight bearing pressures were not placed on that side of the arch. Weighted pressure was therefore shifted forward along the lateral half of the arch, along metatarsal III, IV, and V. From the deeply impressed lateral part of the ball, weight-bearing pressure was immediately shifted across and forward on to the pad of the hallux. On inspection of the original footprints during the 1995 field season, we could realize that some prints show deep impressions below the metatarsal V heads, as was already mentioned by Stern and Susman (Stern and Susman, 1983). Apparently, the transfer of weight along the foot, which can be inferred from the differential depressions along the foot prints was different to the normal human condition. This is not an uncommon condition in modern humans as White and Suwa noticed in experimental prints (White and Suwa, 1987). The major reason for the variant of weight transfer in modern humans is the placement of the foot on the ground. With toeing out, pressure is concentrated on the lateral portion of the heel at first and when transferred to the ball of the foot is concentrated on the inner portion, frequently resulting in a concentration of pressure under the head of the first metatarsal. When the second metatarsal is longer than the first, it participates in this increased pressure. On the contrary, when the individual toes in, the inner border of the heel bears the greater weight, and it is transferred to the outer border of the ball of the foot and a deep impression is presented on the head of the fifth metatarsal. In this case the lateral toes may be the last to leave the ground. A peculiarity in the fossil footprints is the fact that the foot is placed in a toeing out condition, but still the impression is deepest on the lateral side of the distal metatarsal region (Fig. 2). After photogrammetric analyses of Laetoli prints, Day and Wickens concluded that the hominids transmitted their body weight and the forces of propulsion to the ground in a manner very similar to that of modern man (Day, 1985; Day and Wickens, 1980). We also recognize this similarity to the human condition where the the hallux takes a major part in the last phase of the weight transfer, but we are puzzled by the emphasis on the impression at the lateral region of the distal metatarsal joints.

THE ROLE OF THE UPPER BODY IN BIPEDAL LOCOMOTION

The evidence shows that around 3.6 million years ago an early hominid species was capable of bipedal locomotion. The relative distribution of pressure between the lateral and the medial portions of the foot is a crucial matter, not only for the functioning of the foot, but also for interpreting the fossil footprints fixed in the ash layers at Laetoli and subsequently the mode of locomotion of the individuals that crossed the ashes at this place. It is reasonable to assume that the tracks were made by *Australopithecus* sp. There is a growing consensus that there are differences between *Australopithecus* sp. and *Homo* sp. in morphology and stature. The australopithecines have been interpreted by some authors as "fully bipedalised locomotor homunculi" (Gomberg and Latimer, 1984; Latimer, 1991; Latimer et al., 1987;

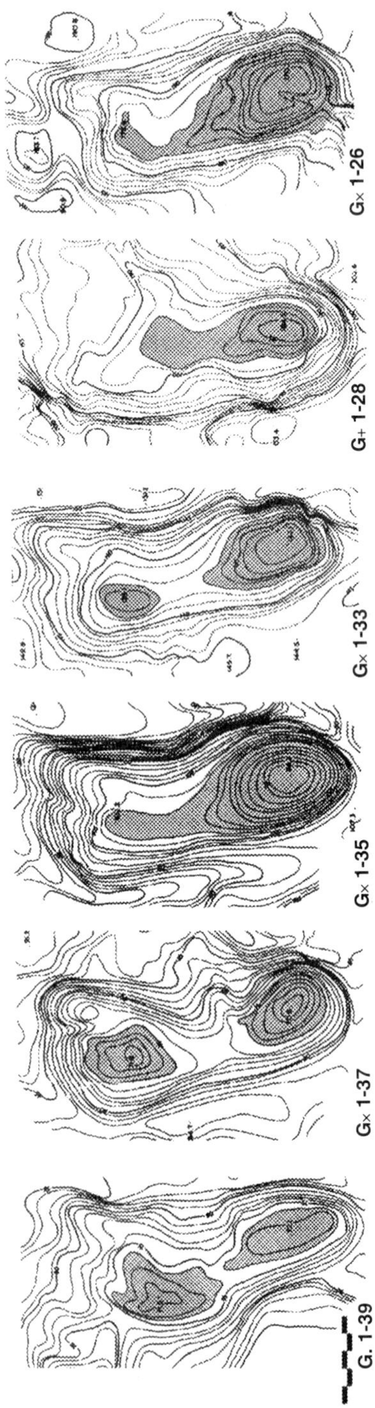

Figure 2. Several photogrammetric plans of the hominid prints in the G-1 trackway (after Jones. 1987). The deepest areas are shaded.

White and Suwa, 1987), but there are increasing indications of functional compromise morphology, such that they are more appropriately interpreted as biped-climbers. Its skeleton is a perplexing mosaic of an apelike upper body combined with hips, legs and feet showing a mixture of features, some ape-like, some frankly unique, and some similar to those of modern humans.

In bipedal movement the fore-limb is no longer used as a support against gravity or with acceleration of the body, but is primarily concerned with the rotation that the body undergoes (Elftman, 1939). The forward swing of the arm contributes angular momentum to neutralize partially that of the advancing leg of the opposite side. An analysis of the dynamics of the arm movements during walking shows that they decrease the rotation of the body as a whole about the vertical axis during the interval in which only one foot is on the ground and allow a more gradual change in the rotation of the body than would be possible without them. The arms also modify the rotation of the body about the direction of progression. The effect of the movements of the arms depends on the amplitude of their swing and on the positions through which the arms swing. The arms do not behave as pendulums in typical walking but are subject to muscle action. The swinging of the arms is consequently not a purely incidental accompaniment of forward movement but is an integral part of the dynamics of progression.

In 1983, the partial skeleton of *Australopithecus* from the Hadar Formation in Ethiopia (A.L. 288-1), was reconstructed and various anatomical features of the cranial and postcranial elements were reconsidered (Schmid, 1983). It was proposed that the rib cage was in fact still chimpanzee-like, indicating a funnel-shaped thorax, and that it was consistent with laterally oriented iliac blades. The broad ilia would have provided the origin for a strong m. latissimus dorsi, thus indicating a high proportion of climbing behaviour. The reconstruction of the entire skeleton of A.L. 288-1 also permitted the first accurate estimate of stature. The inferred body height of 105 cm—one of the smallest known for the genus *Australopithecus*—was subsequently confirmed by several different authors (Aiello, 1992; Jungers, 1988a, b; McHenry, 1991, 1992).

Following this reconstruction, our own research in functional morphology has focused mainly on the locomotor apparatus of early hominids. In particular the little-studied trunk region of fossil hominids was revealed to be an important source of inferences with respect to upright bipedal locomotion (Schmid, 1991). In apes, strong muscles help to bridge the relatively small gap between the thorax and the cranially elongated iliac blades, such that there is no waist in an ape's trunk. This arrangement increases the rigidity of the whole trunk, thus fulfilling the requirements for the use of emancipated upper extremities in suspension and quadrupedal locomotion. In *Australopithecus*, both the outline of the iliac crest and the broad, massive lower rib-cage indicate that the m. obliquus internus and externus occupied a more dorsal position, which would be inefficient for torsion in the abdominal region. As in apes, a waist would also have been lacking. Running in a modern human manner would have been impossible because the muscles of the abdominal wall would have made no important functional contribution in the absence of a waist.

The funnel-shaped thorax of *Australopithecus* further indicates a shoulder complex resembling that of apes, such that free mobility of a descended shoulder is improbable. Swinging of the arms, an important component in stabilizing the body in running humans, is unlikely. The form of the upper part of the thorax of the australopithecines in fact indicates certain differences in respiratory efficiency in comparison to modern humans. The construction of the shoulder girdle would have restricted elevation of the upper part of the rib-cage for respiratory function, as in apes. Heavy breathing, which is permitted through elevation of the upper part of the thorax in modern humans, would not have been possible. Breathing remained relatively constrained because it was mainly dependent on movements of the diaphragm.

Subsequently, analytical research into bipedal gaits in living primates has been conducted in order to develop the contribution of functional morphological studies to theories of humanization (Müller, 1994; Schmid and Piaget, 1994). Using kinesiological analysis, 3-dimensional studies were undertaken to explore inter-taxon differences in bipedal walking among living primates. Synchronous videotapes were taken of humans and gibbons (*Hylobates pileatus*) performing bipedal locomotion on a level surface. The films were analyzed frame by frame and a 3-dimensional stick diagram sequence, rotable in space, was generated (using a specially written computer program) in order to visualize body motion. This novel 3-dimensional approach permitted identification of similarities and differences in bipedal locomotion of humans and gibbons.

Although the lumbar spine is not shortened as in the great apes (Martelli and Schmid, 2000, in press), the trunk of the gibbon remains rigid. Angular measurements between the pelvic axis and the shoulder line revealed small differences in walking humans and gibbons. Whereas clear differences between running humans and fast walking gibbons were obvious. In the latter, torsion within the trunk is almost absent. This is associated with a slight difference in biomechanical strategies used by these different bipeds during walking and bipedal running represents an exclusively human locomotor pattern.

Human running involves not only the lower extremities. An additional factor is the mobility of the lowered shoulders, arm swinging, and a torque in the trunk made possible by the flexible waist and the muscles of the lateral abdominal wall. The morphology of the trunk therefore, is indicative for the ability to run and to use the upper extremities as important balancing structures. The absence of an effective torsion in the bulky trunk of australopithecines as well as the absence of a so-called deltoid complex in the hip (mm. gluteai) is indicative of the australopithecines having been incapable of bipedal running.

EXPERIMENTS

The most important external force, which acts on the body during progression, is the force exerted on it by the ground. This force is due to the reaction to gravity and to the momentary acceleration of the body. Its magnitude can consequently be ascertained if the weight of the body and the acceleration of its center of gravity are known.

For the discussion of the influence of different trunk and arm movements on the weight transfer in the foot several measurements were conducted during normal bipedal walking and during bipedal walking with amble movement of the arms. Not knowing the body weight of the Laetoli individuals we used five-to seven-year-old children with a comparable foot length to the size of the prints in the volcanic ashes. They had to walk over a inked rubber mat to document the sum of the foot pressure. By placing an inked rubber mat with pyramidal projections upon a paper, it is possible to study the distribution of pressure in the foot of an individual walking by noting the increase in area of each pyramid in contact with the ground (Fig. 3).

The movements of the point of the ground reaction force were printed on the podogram to trace the pathway of this force on the supporting area. For the interpretation of the path of

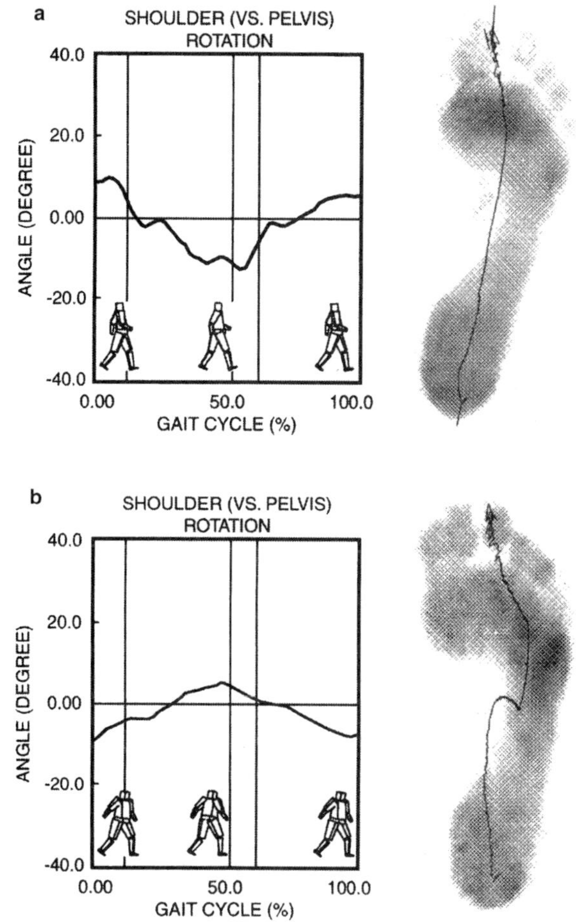

Figure 3. Different locomotor pattern analyzed during the experiments, (a) normal gait with torsion in the trunk, (b) amble gait where the shoulder followed ipsilaterally the movements of the pelvis. Usually, the test person overturned the shoulder movement as can be seen in the graph (b). The podogram of the right foot contains the line of the weight transfer.

the force point it must be noted that the foot moves with respect to the ground. The outline of the footprint represents the total area with which the foot comes into contact during the entire step, not the outline of the footprint at any particular moment. Because of the rolling of the heel when contact is established, the heel print is longer than the area of contact when the foot is fully planted. For similar reasons, the toe prints are elongated.

Two different exercises were performed: (a) normal bipedal gait and (b) bipedal gait with a fixed trunk where we tried to eliminate torsional movements within the trunk by inhibiting contralateral twist of the shoulder horizontal axis versus the pelvic horizontal axis. At the same time force plate analyses and dynamic podograms were made. In addition, a motion analysis with the Vicon System was performed. Twenty-one self-reflecting markers were fixed to mark the body segments (shoulder, elbow, wrist, sacrum, iliac crest, trochanter major, knee joint, talocrural joint, heel, head of metatarsals). The body segments were therefore defined by three marker points, which allowed the calculation of the joint angles. Seven synchronized infrared cameras filmed the three-dimensional movements of the body segments (50 frames/sec). The analyzing software calculated the following measurements:

Shoulder rotation versus pelvis
Knee flexion/extension
Center of body mass
Stride length [meter]
Floor contact [second]
Reaction force trajectory
Pelvic rotation versus laboratory
Rotation of the foot versus laboratory
Velocity [meter/second]
Cadence [stride/minute]
Reaction force (z-axis)

Of course, a lot of variation resulted in this "artificial amble gait" but some selection followed the experiment to include only movements where the shoulder axis and the pelvic axis moved in more or less the same direction. The experiments are considered as preliminary but gave an interesting insight in the consequences of different trunk movements (rotation versus torsion) in relation to the weight transfer in the foot. Although the heterogeneity of the individual gait pattern was obvious, this seems to be trivial because we already are able to recognize an individual person with only looking at the way she moves. As a consequence, a normal human gait pattern can only be defined superficially. For this reason, no statistical treatment was undertaken.

Against all odds, we observed a clear difference between two movements strategies (torsion, versus rotation) in the way the weight is transmitted on the ground. All test persons produced a greater pressure on the region of the lateral ball (head of the metatarsal V) when moving with rotational amble technique in the trunk.

CONCLUSIONS

The Laetoli footprints were made by a bipedal primate, the foot morphology of which is comparable to the human condition. This includes a relatively adducted hallux which lies in line with the other rays, short toe impressions, and a clear longitudinal arch construction. There is no information about individual toe morphology except the one of the great toe. Information about special abductor muscle configurations are almost hypothetical contrary to Deloison, (1991).

Although gait pattern is dominated by an extreme variability—hence characteristic individuality determines the typical movement of a person—the weight transfer in the foot exhibits charcteristic patterns. The latter are clearly influenced by the behaviour of trunk and arm movements.

The fossilized prints in the volcanic ash at Laetoli demonstrate clear signs of rotational movements of the upper body parts. The reason for this could have been: (1) an ape-like morphology of the trunk, (2) individuals carrying bigger objects, or (3) pathological conditions. We prefer the first hypothesis, although we cannot falsify the remaining possibilities. In consequence, it seems that the Laetoli hominids were habitual bipeds, but they probably moved in amble gait pattern and therefore would not have been capable of running, an important character of the genus *Homo*.

ACKNOWLEDGEMENTS We thank the United Republic of Tanzania for permission to do research at Laetoli, the Getty Conservation Trust and the Leakey Foundation for financial support, Dr. S. Waane, Director of Antiquities, Tanzania and A. J. F. Mgina, Conservator, Ngorongoro Conservation Authority, for their help. S. Odermatt and R. Müller conducted the experimental work at the Laboratory of biomechanics of the ETH (Director E. Stüssi). H. Rüther (University of Capetown, South Africa) generated the marvelous digital maps of the individual footprints. B. M. Latimer and C. S. Feibel added helpful discussions in the field. S. Martelli carefully corrected the manuscript. To all these persons we are extremely grateful. We much appreciate the helpful comments and criticism of R. J. Clarke and D. J. Meldrum as well as some unknown reviewers.

REFERENCES

Agnew, N., and Demas, M., 1998, Preserving the Laetoli Footprints. *Scientific American 279*:26–37.

Agnew, N., Demas, M., Lin, P., Marshall, F., McDonnel, K., Moon, T., Podany, J., and Tebble, T., 1993, *Laetoli Project: Conservation of the hominid trackway site at Laetoli, Tanzania.* The Getty Conservation Institute, Marina del Rey, CA.

Aiello, L. C., 1992, Allometry and the analysis of size and shape in human evolution. *Journal of Human Evolution 22*:127–147.

Bourdial, I., 1999, Adieu Lucy. *Science et Vie 980*:52–62.

Charteris, J., 1981, Functional reconstruction of gait from Pliocene hominid fottprints at Laetoli, northern Tanzania. *Nature 290*:496–498.

Charteris, J., 1982, Pliocene Hominid Gait: New Interpretations Based on Aivalable Footprint Data From Laetoli. *American Journal of Physical Anthropology 58*:133–144.

Christie, P. W., 1977, Form and Function of the Afar Ankle. *American Journal of Physical Anthropology 47*: 123.

Clarke, R. J., 1979, Early hominid footprints from Tanzania. *South African Journal of Science 75*:148–149.

Clarke, R. J., 1985, Comments on the Laetoli footprints: Taung Diamond, pp. 148–149.

Clarke, R. J., 1999, Discovery of complete arm and hand of the 3.3 million-year-old Australopithecus skeleton from Sterkfontein. *South African Journal of Science 95*:477–480.

Clarke, R. J., and Tobias, P. V., 1995, Sterkfontein Member 2 Foot Bones of the Oldest South African Hominid. *Science 269*:521–524.

Day, M. H., 1985, Hominid locomotion—from Taung to the Laetoli footprints. In Hominid Evolution: Past, Present, and Future. Taung Diamond Jubilee International Symposium. edited by P. V. Tobias, pp. 115–127. New York.

Day, M. H., and Wickens, E. H., 1980, Laetoli Pliocene hominid footprints and bipedalism. *Nature 286*:385–387.

Deloison, Y., 1991, Les australopithèques marchaient-ils comme nous? In Origine(s) de la Bipédie chez les Hominidés, Cahier de Paléoanthropologie, edited by Y. Coppens and B. Senut. Editions du CNRS, Paris.

Elftman, H., 1939, The Function of the Arms in Walking. *Human Biology 11*:529–535.

Elftman, H., and Manter, J., 1935a, Chimpanzee and human feet in bipedal walking. *American Journal of Physical Anthropology 20*:69–79.

Elftman, H., and Manter, J., 1935b, The evolution of the human foot, with especial reference to the joints. *Journal of Anatomy 70*:56–67.

Feibel, C. S., Agnew, N., Latimer, B., Demas, M., Marshall, F., Waane, S. A. C., and Schmid, P., 1996, The Laetoli Hominid Footprints: A Preliminary Report on the Conservation and Scientific Restudy. *Evolutionary Anthropology 4*:149–154.

Gomberg, D. N., and Latimer, B., 1984, Observations on the transverse tarsal joint of *A. afarensis* and some comments on the interpretation of behavior from morphology. *American Journal of Physical Anthropology 63*:164.

Jungers, W. L., 1982, Lucy's limbs: skeletal allometry and locomotion in *Australopithecus afarensis*. *Nature 297*:676–678.

Jungers, W. L., 1988a, Lucy's length: stature reconstruction in *Australopithecus afarensis* (A.L. 288-1) with implications for other small-bodied hominids. *American Journal of Physical Anthropology 76*:227–232.

Jungers, W. L., 1988b, New estimates of body size in australopithecines. In *Evolutionary History of the "Robust" Australopithecines*, edited by F. E. Grine, pp. 115–126. Aldine de Gruyter, New York.

Latimer, B., and Lovejoy, C. O., 1989, The Calcaneus of *Australopithecus afarensis* and Its Implications for the Evolution of Bipedality. *American Journal of Physical Anthropology 78*:369–386.

Latimer, B., and Lovejoy, C. O., 1990a, Hallucal Tarsometatarsal Joint in *Australopithecus afarensis. American Journal of Physical Anthropology 82*:125–133.

Latimer, B., and Lovejoy, C. O., 1990b, Metatarsophalangeal Joints of *Australopithecus afarensis. American Journal of Physical Anthropology 83*:13–23.

Latimer, B. M., 1991, Locomotor adaptations in *Australopithecus afarensis*: the issue of arboreality. In Origine(s) de la Bipédie chez les Hominidés, Cahier de Paléoanthropologie, edited by Y. Coppens and B. Senut. pp. 169–176. CNRS, Paris.

Latimer, B. M., Ohman, J., and Lovejoy, C. O., 1987, Talocrural joint in African hominoids: Implications for *Australopithecus afarensis. American Journal of Physical Anthropology 74*:155–176.

Leakey, M. D., 1978, Pliocene footprints at Laetoli, Northern Tanzania. *Antiquity 52*:133.

Leakey, M. D., 1979, Footprints in the ashes of time. *National Geographic 155*:446–457.

Leakey, M. D., 1987, Introduction. In Laetoli—A Pliocene Site in Northern Tanzania, edited by M. D. Leakey and J.M. Harris, pp. 1–22. Clarendon Press, Oxford.

Leakey, M. D., and Harris, J. M., 1987, *Laetoli: A Pliocene Site in Northern Tanzania*. Clarendon Press, Oxford.

Leakey, M. D., and Hay, R. L., 1979, Pliocene footprints in the Laetolil Beds at Laetoli, northern Tanzania. *Nature 278*:317–323.

Martelli, S. A., and Schmid, P., 2000, Functional morphology of the lumbar vertebral column in modern *Homo sapiens*, great apes and australopithecines. *American Journal of Physical Anthropology*, *Suppl. 30*:221.

Martelli, S. A., and Schmid, P., (in press), Functional Morphology of the Lumbar Spine in Hominoids. *Courier Forschungsinstitut Senckenberg*.

McHenry, H. M., 1991, Femoral Lengths and Stature in Plio-Pleistocene Hominids. *American Journal of Physical Anthropology 85*:149–158.

McHenry, H. M., 1992, Body Size and Proportions in Early Hominids. *American Journal of Physical Anthropology 87*:407–431.

Müller, G., 1994, 3-D-Bewegungsanalyse bei Gibbons. Diplomarbeit, Universität Zürich.

Schmid, P., 1983, Eine Rekonstruktion des Skelettes von A.L. 288-1 (Hadar) und deren Konsequenzen. *Folia Primatologica 40*:283–306.

Schmid, P., 1991, The trunk of the australopithecines. In *Origine(s) de la Bipédie chez les Hominidés, Cahier de Paléoanthropologie*, edited by Y. Coppens and B. Senut, pp. 225–234. Editions du CNRS, Paris.

Schmid, P., and Piaget, A., 1994, Three-dimensional kinematics of bipedal locomotion. *Zeitschrift für Morphologie und Anthropologie 80*:79–87.

Senut, B., and Tardieu, C., 1985, Functional aspects of Plio-Pleistocene hominid limb bones: implications for taxonomy and phylogeny. In E. Delson (ed.): *Ancestors: the Hard Evidence*, edited by E. Delson, pp. 193–201. Alan R. Liss, Inc., New York.

Stern, Jr., J. T., and Susman, R. L., 1983, Locomotor anatomy of *Australopithecus afarensis. American Journal of Physical Anthropology 60*:279–317.

Susman, R. L., Stern, Jr., J. T., and Jungers, W. L., 1984, Arboreality and bipedality in the Hadar hominids. *Folia Primatologica 43*:113–156.

Tuttle, R. H., 1984, Bear facts and Laetoli impressions. *American Journal of Physical Anthropology 63*:230.

Tuttle, R. H., 1985, Ape footprints and Laetoli impressions: A response to the SUNY claims. In PV Tobias (ed.): *Hominid Evolution: Past, Present, and Future. Taung Diamond Jubilee International Symposium*, edited by P. V. Tobias, pp. 129–133. Alan R. Liss, New York.

Tuttle, R. H., 1987, Kinesiological inferences and evolutionary implications from Laetoli bipedal trails G-1, G-2/3, and A. In *Laetoli: A Pliocene Site in Northern Tanzania*, edited by M. D. Leakey and J. M. Harris, pp. 503–523. Clarendon Press, Ozford.

Tuttle, R. H., Webb, D. M., Tuttle, N. I., and Baksh, M., 1992, Footprints and Gaits of Bipedal Apes, Bears, and Barefoot People: Perspective on Pliocene Tracks. In *Topics in Primatology: Evolutionary Biology, Reproductive Endocrinology, and Virology*, edited by S. Matano, R. H. Tuttle, H. Ishida and M. Goodman, pp. 221–242. University of Tokyo Press, Tokyo.

White, T. D., and Suwa, G., 1987, Hominid footprints at Laetoli: facts and interpretations. *American Journal of Physical Anthropology 72*:485–514.

Chapter 5

Fossilized Hawaiian Footprints Compared with Laetoli Hominid Footprints

D. JEFFREY MELDRUM

Abstract. Fossilized footprints preserved in volcanic ash tuffs in the Hawaii Volcanoes National Park afford an exceptional opportunity to study footprints of traditionally unshod modern human pedestrians, laid down under conditions relatively similar to those associated with footprints made by Pliocene hominids at the east African site of Laetoli, Tanzania. The Hawaiian footprint area encompasses about 13 km^2 of the Ka'u Desert. Two footprint-bearing ash layers date to 200 and 400 years before present (ybp) respectively. The track localities were surveyed with GPS coordinates and photographed, including stereophotography. In many instances footprint clarity and preservation are exceptional, revealing anatomical details of foot shape, toe conformation and arch development. Footprints were contrasted with stereophotos and casts of the Laetoli tracks and with footprints made by a captive chimpanzee in a sand trackbox. The modern human footprints were distinct in the marked development of the ball of the foot and toe pads, indicative of differential weight bearing or traction by these structures. There is consistent expression of a well-developed arch and no indication of midtarsal flexion. In contrast, the Laetoli hominid tracks do not exhibit these adaptations. Instead, they display a tapering proximal heel outline, relatively narrow distal metatarsus, indistinct ball and poorly differentiated impressions of the lateral toes. Preservation of the ash layer is patchy throughout the Ka'u footprint area; however, several examples of tracks permit limited documentation of spontaneous gait parameters. Contrasts with the Laetoli tracks emphasize the short step length of the fossil hominid gait and the highly variable angle of gait and step width in both ancient hominid and modern human trackways.

INTRODUCTION

Tremendous attention has been afforded the initial evolution of bipedalism in early human ancestors, and considerable debate has ensued about the specific pattern and process of its emergence (e.g. Coppens and Senut, 1991; Agnew and Demas, 1998). The earliest direct evidence of hominid bipedalism is found at Laetoli, where fossilized footprints record the bipedal procession of three hominids (Leakey and Hay, 1979; however, others have concluded that the eastern trackway was produced by a single individual, e.g. Clarke, 1999). This discovery and the ensuing analyses exposed a dearth of comparative data on both the footprints of hominoids and unshod human populations. Exceptions were studies

From Biped to Strider: The Emergence of Modern Human Walking, Running, and Resource Transport, edited by D. Jeffrey Meldrum and Charles E. Hilton. Kluwer Academic/Plenum Publishers, New York, 2004.

Figure 1. Stereopair of a deeply impressed, 200 year old footprint in the Ka'u desert, Hawaii (scale in cm).

by Hoffman (1905), James (1939), and Stewart (1970). As a result, interpretations regarding the degree to which, and the manner in which, these hominids were bipedal have varied considerably, from fully human-like to considerably ape-like (e.g. Susman et al., 1984; Tuttle, 1985; Lovejoy, 1988; Deloison, 1992; Berillon, this volume; Schmid, this volume). This has prompted renewed interest in the nature of gait and footprint dynamics in native populations by anthropologists (Tuttle, 1985; Tuttle et al., 1990; Hilton, 1997). It has also drawn attention to unappreciated features of the Laetoli tracks that indicate a distinct functional anatomy, while retaining some apish aspects (Deloison, 1992; Meldrum and Wunderlich, 1998; Meldrum, 1999, 2002).

One persistent question that has dogged the interpretation of the Laetoli tracks has been what effect did the nature of the ash substrate have on the dynamics of the hominids' gait, and even more fundamentally, on the very appearance and preservation of the footprints themselves. The presence of modern unshod pedestrian human footprints preserved in the Hawaii Volcanoes National Park affords an exceptional opportunity to examine and contrast modern unshod human footprints laid down under conditions of substrate relatively similar to those encountered by the Laetoli hominids (Fig. 1). The clarity of detail in the ashen Hawaiian footprints suggests that what is seen at Laetoli is not simply artifactual of the substrate but reflects real distinctions of morphology and gait dynamics.

GEOLOGICAL CONTEXT

The footprint area, on the Big Island of Hawaii, encompasses about 13 km^2 of the Ka'u Desert, southeast of the Kilauea caldera. At least nine layers of volcanic ash, the earliest being tens of thousands of years old, are exposed south of Kilauea's summit (Mastin et al.,

Figure 2. (a) Exposure of the two principal footprint-bearing tuffs; (b) older tuff (400 ybp) characterized by large desiccation polygons and (c) variable-sized accretionary lapilli; (d) younger tuff (200 ybp) with no desiccation polygons and (e) uniform lapilli.

1999). These range from less than one meter to tens of meters in thickness. The youngest is the Keanakako'i Ash, which covers an area of over 190 km^2. This ash resulted from a series of explosive eruptions that ended about 200 ybp. There are at least two significant footprint-bearing tuffs, separated by varying thickness of interbedded eolian sands (Fig. 2a). The younger corresponds to the 1790 eruption; carbon dating of the older indicates an age

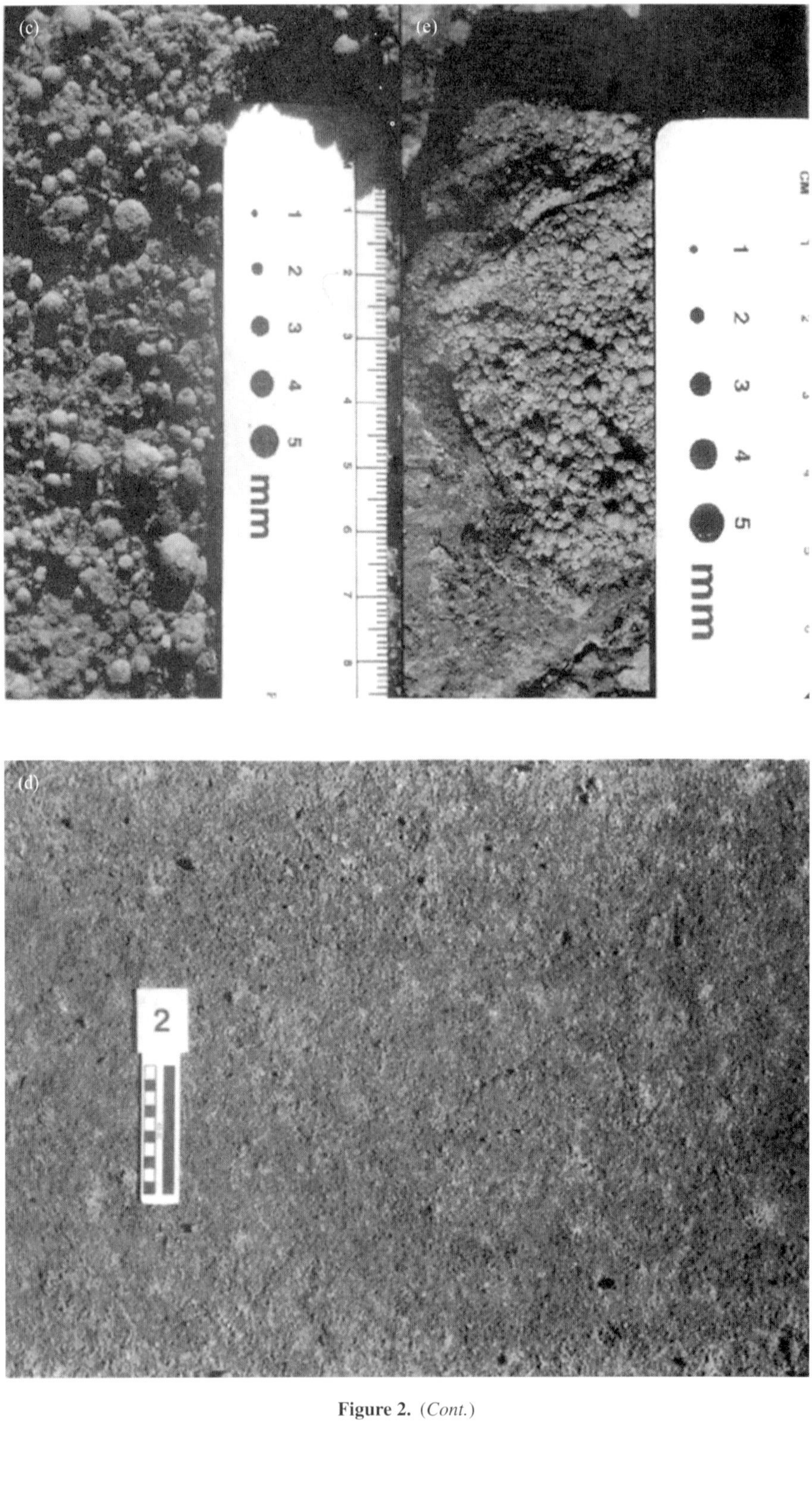

Figure 2. (*Cont.*)

of approximately 400 ybp (Swanson and Christiansen, 1973; Swanson, pers. comm.). The tuffs can be distinguished in the field on the basis of acceretionary lapilli and desiccation polygons. The older tuff has large distinct desiccation polygons and acceretionary lapilli of variable size (Fig. 2b, c). The younger tuff has no or small desiccation polygons and small uniform acceretionary lapilli (Fig. 2d, e). Consistency and thickness of the ash at the time of the footprint events varied from site to site, with some tracks comprising shallow and distinct impressions, while others include impressions up to several centimeters in depth, with signs of extrusion of wet ash about the footprint margins and between the toes.

ETHNOLOGICAL CONTEXT

Traditionally, footprints of the younger ash layer, dating to 1790, are attributed to Keoua's army. Keoua, chief of the Ka'u district, had mobilized his warriors to skirt the island and engage the rival chief, Kamehameha. They found Pele in a restless mood; the earth shook by successive earthquakes. After prayers and offerings, the first contingents of warriors, traveling with their families, passed safely beyond the Halema'uma'u crater. As the second contingent reached the flank of Kiluea, a violent phreatic eruption overtook hundreds of warriors and their families, asphyxiating them in a cloud of searing ash and poisonous gases. However, it has yet to be conclusively demonstrated that the footprints of the upper ash layer are in fact *singularly* associated with that event. It is clear that the trackways are often following established paths through the a'a lava fields. Many tracks take a definite bearing on the Halema'uma'u crater rim near the USGS observatory. In fact the observatory, situated on the crater's edge, stands upon the site of an ancient ceremonial platform. Thus, ceremonial processions from coastal settlements to the crater's edge may account for some of the footprints. Others may have been left during excursions to the inland forest for natural resources, or when gathering volcanic glass in the lava flows as material for the manufacture of tools and decorative artifacts. The presence of hundreds of archeological sites, including windbreaks, shelters, and quarries indicate an on-going presence in the desert at this time.

MATERIALS: THE FOOTPRINTS

Surveys of the distribution of footprints and archeological sites within the park boundaries were conducted by Cultural Resource Management staff (Moniz Nakamura, unpubl. report). The author examined footprints at a number of these sites and located and surveyed additional localities recorded using GPS (field notes filed with the CRM office).

The clarity and preservation of the footprints varies considerably from site to site. In some instances the footprint definition is extremely sharp (Fig. 3). Where the ash layer has been exposed to the elements, erosion of the footprints is considerable. In some instances the compaction of the ash under the footprints has differentially preserved them, while the surrounding non-compacted ash has eroded away, leaving a series of pedestalled footprints (Fig. 4). The best preserved tracks were located in areas of dunes where the overburden of sand has conserved the footprints. At the trailing edge of the dunes, new footprints are continually being uncovered by wind action, but simultaneous erosion of the interbedded sands undercuts the exposed ash layer, which then crumbles away, rapidly destroying the uncovered footprints. Footprints are often in-filled with black sand. In these cases, trial attempts to remove the sand to reveal the contact surface were of limited success. Presumably,

Figure 3. Imprint of a forefoot showing distinct impression of toes and distal metatarsals.

the action of moisture and temperature obliterated the consolidation of the contact surface. However, the outlines of these footprints are still useful for deriving structural proportions and gait measurements.

It should be borne in mind when dealing with measures of footprints that the footprint is the record of the interaction of the soft tissue and skeletal structure of the foot with the substrate over the course of a given step. There are bound to be distortions resulting from the consistency of the substrate, the depth of the impression and the dynamics of each step. The most accurate method to determine the dimensions of the foot is to consider the "minimum outline" of the footprint (Halfpenny, 1998). The minimum outline is determined by the break points—where the rounded margin of the foot turns upward. Identifying the break point can be a subjective or altogether elusive determination, especially if the footprint is in-filled with sand or matrix. In these instances the maximal outline is still useful, especially if appropriate indices are calculated.

RESULTS

Contrasts between Hawaiian and Laetoli Footprints

Casts of Laetoli G1-26, 35, 36; G2/3-26, 27 were examined as well as 2 series of stereophotographs of the better preserved southern section of the trackway, made during

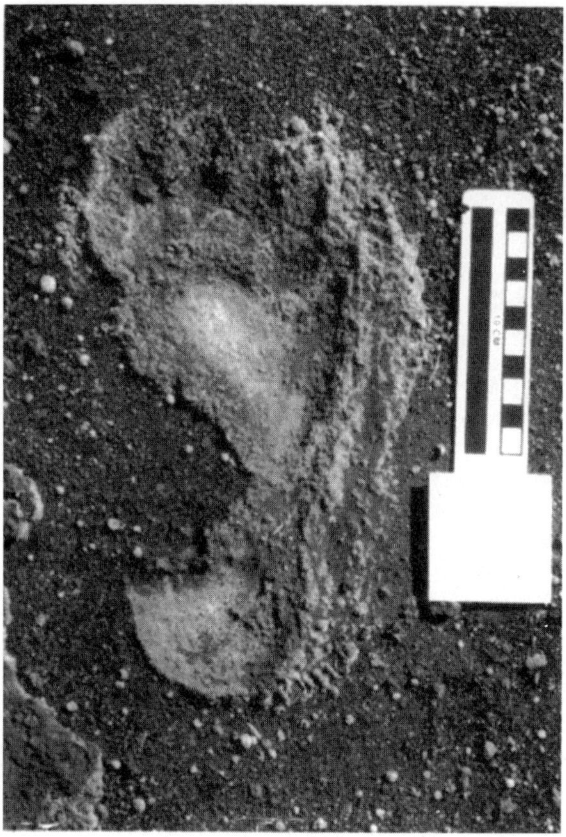

Figure 4. Compacted ash beneath footprints has resisted erosion leaving a series of pedestaled footprints.

the original 1978 and 1979 excavations (photos courtesy of Tim White). These were taken under both a.m. lighting and p.m. lighting respectively, providing contrasting illumination that proved critical for interpreting detailed features of the footprints. The condition of the initially exposed tracks evident in these photographs was noticeably superior to that documented by the subsequent re-excavation project of the Getty Conservation Institute (Demas, 1996, 2000).

As most researchers have not been afforded the opportunity to examine the trackways firsthand, but must instead rely on the few published photos or the planimetric plats, several observations should be noted. The differences between the details apparent in the tracks when lit from the east versus the west can be dramatic. As has been noted elsewhere, but which cannot be appreciated from the planimetric maps alone, there are large areas of exfoliation of the original contact surface of the individual footprints. Likewise surface irregularities observed in the photos are difficult to interpret without varied lighting and stereo-imaging. For example, in the photogrametric image of G 1-37, represented in Day and Wickens (1980, Figure 3), the contours are suggestive of a well developed medial arch, but are the result of an artifactual irregularity on the medial border of the print. Finally, there are some minor inaccuracies in the positions and orientation of the footprints as depicted in the map in Leakey and Harris (1987, Figure D-3). Therefore, care should be taken when drawing comparisons between the Laetoli footprints and contemporary unshod pedestrian footprints based on these, and similar, published sources viewed in isolation.

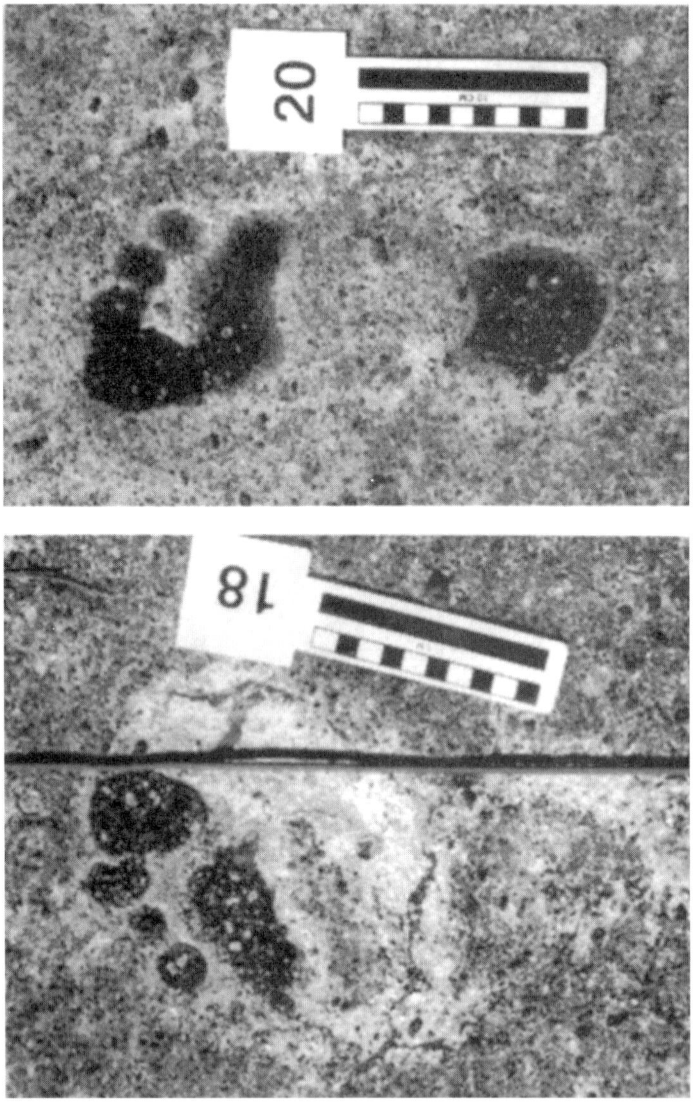

Figure 5. Two representative examples of sand-filled footprints illustrating outline, longitudinal arch development and pressure ridge (a) proximal to metatarsal heads.

Occasionally, a selective comparison between modern and Laetoli footprints may display superficial similarity to one another. However, when representative samples of Hawaiian and Laetoli footprints are examined collectively, there are clearly repeated distinguishing features. Features in the Laetoli footprints, such as the expression of an apparent medial arch, must be present consistently to support assertions of modernity in form and function.

The Hawaiian footprints display a distinctly broad and rounded heel with evidence of a well-developed heel pad, characteristic of human feet (Fig. 5). In some instances, where the longitudinal arch was exceptionally high, the pronounced heel pad impression was mistaken by one casual observer for a boot heel. This configuration of the heel pad is correlated with

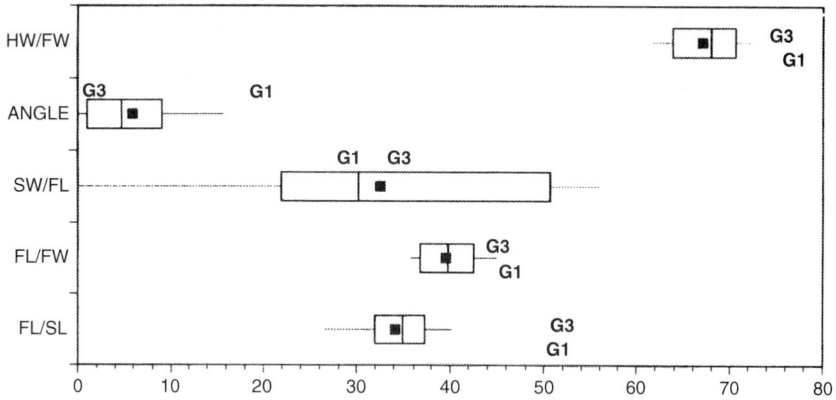

Figure 6. Selected indices of foot form and gait parameters for a sample of Hawaiian footprints. Average values for the G I and G3 Laetoli hominids are indicated separately. HW/FW = index of heel width to foot width; ANGLE = angle of gait, measured as the angle of a line through the center of the heel and the second toe, relative to the line connecting the center of the heels in the adjacent contralateral footprints; SW/FL = index of step width to foot length; FL/FW = index of foot length to foot width; FL/SL = index of foot length to step length.

the increased robusticity of the human calcaneal tubercle as compared with non-human hominoids. In contrast, the Laetoli footprints exhibit a tapered heel that is rather chimp-like in appearance. This contrast in heel outline and impression has been understated by some observers (e.g. Robbins, 1987) or overlooked by those who advocate the humanness of the footprints (e.g. Tuttle et al., 1991). Deloison (1991, 1992) noted the tapered heel characterizing the Laetoli tracks. However it is cumbersome to quantify the taper of the heel, or to express it as ratio against other dimensions of the footprint. The greatest breadth of the hindfoot distal to the onset of the taper can be expressed as a ratio of greatest overall foot width, measured across the metatarsal heads. The Laetoli footprints fall above the range of values of this ratio (HW/FW) for the Hawaiian sample (Fig. 6). This can be accounted for by the lack of development of the breadth of the ball and forefoot of the Laetoli trackmakers. The tapering heel accords with the lack of development of the lateral process in the *Australopithecus afarensis* calcaneus (Susman and Stern, 1984; Latimer and Lovejoy, 1989). This contrast has implications for the dissipation of force at heel strike, and implies a non-modern form of gait.

The taper of the heel is further accentuated by the convexity of the proximomedial edge of the foot in the Laetoli footprints, particularly evident in a number of the G1 imprints, which are generally clearer in outline. This bulge has been attributed to the prominence of the abductor hallucis muscle belly (Deloison, 1991). Tuttle (1996) dismisses it as a variable feature. In modern human feet with well developed arches, this area makes little or no contact with the substrate, except in deep impressions. However, in flat-footed individuals, even those who are habitually unshod, a similar, but more distally placed bulge may be present (Stewart, 1970). In this case, the prominence is not due to a large abductor hallucis, but due to contact of the talonavicular joint on the ground (Dee, 1989). Neither form of medial prominence was obvious in the Hawaiian footprints. Relative mobility of the hallux will be addressed below.

A well-developed longitudinal arch is present in the Hawaiian footprints. Indeed, many examples display high arches with a raised lateral margin at the midfoot (Fig. 5b). As the longitudinal arch is one of the hallmarks of the modern human foot adaptation, the presence

Figure 7. Cast of the G 1-26 footprint from Laetoli, illustrating the plastic flow of ash located proximal to the inferred position of the transverse tarsal joint.

or absence of this feature has been the focus of much attention and divergent interpretations. A number of investigators have noted the presence of a medial arch in some of the Laetoli footprints (Leakey and Hay, 1979; Robbins, 1979; White, 1980; Day and Wickens, 1980; Day, 1985; Tuttle, 1985, 1996; Latimer, pers. comm.; Schmid, this volume). However, it must be emphasized that the mere presence of a raised medial edge of the foot does not denote a fixed longitudinal arch comparable to the configuration of the typical modern human foot. As has been noted, even the relatively flat mobile foot of a chimpanzee can leave an impression that presents a raised medial border of the foot (Stern and Susman 1983; Meldrum, 1999). However, the expression of a raised medial arch in the chimp foot is variable and often inconsistent from print to print, as is the case with the Laetoli tracks. Even when a raised medial border is present in a given Laetoli footprint, the resulting outline is readily distinguished from that characteristic appearance of the modern human foot, with its rounded heel and prominent ball. This distinctive outline has been referred to as the "figure of eight" contour (Day, 1991). Instead, numerous examples of Laetoli tracks indicate a transverse axis of flexion at the midtarsus, as noted by Deloison (1991, 1992). Tuttle (1996) offers an unsatisfactory dismissal, suggesting this feature can be explained by the presence of the medial longitudinal arch in a footprint impressed in soft substrate. Earlier, I suggested that a dynamic feature present in a number of Laetoli footprints might represent a midtarsal pressure ridge, further indicating a degree of midfoot flexibility (Meldrum, 1998). After careful examination of stereophotos and casts, it cannot be conclusively demonstrated that most of these features are other than artifacts of preservation or excavation. However, given their repeated and consistent position within the footprints, it seems unlikely that they are merely random artifact. They are not attributable to insect perturbation since the burrowing termites did not traverse the areas of compacted ash beneath the footprints. Indeed, the footprint G1-26, exhibits what appears to be a plastic flow of the ash proximal to the inferred axis of midfoot flexion (Fig. 7). The appearance of the flow margin is precisely the same as the appearance of the ash that extruded between the toes in some footprints (a very clear example is seen between the first and second toes in G1-36). This flow, associated with a pressure ridge, would have resulted from the wet ash being extruded from under the midfoot, behind or proximal to the midtarsal joint axis, and beneath the elevated heel segment. No such feature is evident on any of the Hawaiian footprints examined. Instead, there are repeated examples of pressure releases situated just proximal to the metatarsal heads (Fig. 8), a feature not present in any of the Laetoli tracks.

Figure 8. Hawaiian footprint illustrating a pressure ridge proximal to the metatarsal heads.

The Hawaiian footprints consistently exhibit the presence of a well-developed ball beneath the hallucial metatarso-phalangeal joint. The rounded impression routinely projects medial to the hallux impression. This is a second hallmark of the modern human foot, which, when taken in conjunction with the longitudinal arch, marks the medial transfer of weight during the latter part of stance. Contra White and Suwa (1987), I can find no example, from the southern trackway, that clearly indicates an unambiguous impression of a human-like medial ball proximal to the hallux. The deepest impressions at the metatarsal heads, in the Laetoli tracks, are consistently associated with the lateral rays (in agreement with Schmid, this volume).

The impression of the toe pad of the hallux is well differentiated in many of the Hawaiian footprints, although there is considerable variation in clarity depending on the degree of weathering and the depth of the impression. In examples of footprint quality most comparable to the Laetoli tracks, the proximal margin of the toe pad is well delineated and separated from the sole pad. This would indicate a substantial thickness to the toe pad, differentiated from the fascia under the toe stem, and some degree of plantarflexion at the interphalangeal joint. In contrast, the hallucial impressions in the Laetoli footprints are generally continuous with the impression of the midfoot and are often shallower, suggesting less development and differentiation of the hallucial toe pad. However, in some cases, such as G2/3-27, the impression of the hallux is more distinct and deeper. Again, it is the inconsistency from step to step that suggests that neither a modern human morphology, nor the stereotyped human toe-off was fully in place.

The gap between the hallux and lateral toes is quite pronounced in the Hawaiian footprints, as is typical of habitually unshod pedestrians (Barnicot and Hardy, 1955), although the Hadza display less fanning of the forefoot and a narrower gap (Tuttle, 1997). More critical than the interdigital gap is the orientation of the long axis of the hallux. In unshod modern humans, the hallux is oriented such that the projection of the long axis of the toe intersects the center of the heel (Hoffman, 1905; cf. Fig. 1, 5). The tracing of G1-34, published by Deloison (1991) slightly exaggerates the dimensions and degree of abduction of the hallux, due to exfoliation about the margins of the hallucial impression and what appears to be drag-out artifact distolateral to the hallux. This "double-tipping" artifact was noted by Robbins (1985) and reprised by Tuttle (1996), who maintained that the long axis of the hallux is virtually aligned with the long axis of the foot. However, even a more conservative rendering of the outline of the hallux is not comparable to the orientation seen in modern barefooted pedestrians. The hallux of G 1-34 is indeed slightly abducted and the projection of the long axis intersects well lateral to the heel. This condition is even more evident in the G2/3 individuals, especially evident in G2-19 and G2-26. This observation confirms the conclusion that *Australopithecus afarensis* hallucial-tarsometatarsal joint retains a minor degree of abduction (Stern and Susman, 1983) as opposed to mere splaying of the forefoot. While acknowledging the consistent position of relative adduction of the hallux displayed in the Laetoli trackways, it must be acknowledged that even some individual chimps and gorillas walk with the hallux preferentially held in an adducted position (personal observation; Fig. 9). The superficial resemblance of these ape footprints to the Laetoli tracks, with regard to the position of the hallux, does not betray the full range of abduction of the hallux known to exist in the ape foot.

Lateral toe separation is generally distinct in the Hawaiian footprints. The span of the toes is typically the widest part of the footprint (Fig. 1, 5, 8). The individual toe pads are often conspicuously and deeply imprinted and the underlying ash compacted (Fig. 3). In

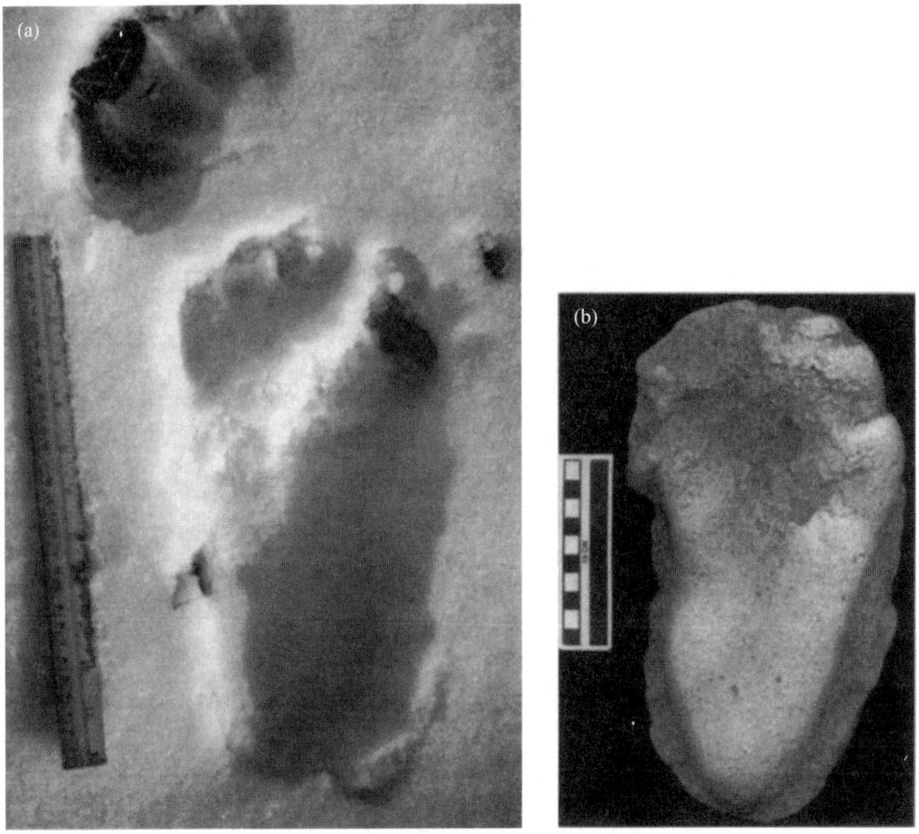

Figure 9. Examples of idosyncratic adduction of the of the hallux in the footprint of a gorilla in snow (a) and in a cast of a chimpanzee footprint in sand (b).

instances where the surrounding ash has eroded, leaving the compacted ash beneath the footprint pedestalled, the individual toe impressions are typically clearly preserved. In contrast, the Laetoli tracks lack clear impressions of the lateral toes. Tuttle (1996) has suggested that in some instances the lateral toe impressions are obscured by remaining matrix infill. I found no evidence of remaining matrix in the lateral toes. In many instances within the G1 trail the lateral toe area is exfoliated, but even in the remaining prints of the G1 and the G2/3 trails, the area of the lateral toe impressions is ill defined. This would seem to present three possibilities: first, there was some qualitative difference between the substrates at Hawaii vs. Laetoli; second, the lateral toes never made weighted contact with the Laetoli substrate; third, where the lateral toes did make clear contact there was minimal compacting of the substrate, resulting in obliteration by subsequent ash infill, or by erosional/excavational exfoliation. There is no compelling reason to attribute the differences to conjectural differences in the substrate. The latter two possibilities imply that the lateral toes made minor contribution to traction in forward progression, or propulsion in toe-off.

With the long axis of the foot defined as the line intersecting the center of the heel and the second toe, the Hawaiian footprints display either lateral toes that do not extend distal to the hallux, or the second toe is slightly longer that the hallux. This is typical of barefooted human populations (e.g. James, 1939). The relative length of the toes of the Laetoli tracks has been the subject of considerable debate. Stern and Susman (1983) have suggested that the apparent variation in the distal extent of the lateral toes may be the result of relatively long toes being curled under at times and more extended at others. It seems unnecessary to resort to complete curling under of the toes as in a chimpanzee to account for the apparent variance in lateral toe projection. The relatively longer toes of an australopithecine would leave noticeably longer impressions when fully extended versus shorter impressions when flexed. White and Suwa (1987) have argued that there is no appreciable variation in apparent toe length. Tuttle et al. (1991) attempted to determine the relative projected length of the toes of the Laetoli tracks. They defined the long axis as the line intersecting the midpoints of the heel and the ball of the foot, and concluded that the hallux is generally longer than the lateral toes. Given the conditions of exfoliation and lack of compaction, it seems unlikely that the length of the lateral toes can be accurately assessed from the footprints. The distal extent of the toes (as well as the basic outline) of footprints considered individually may be open to some interpretation. However, the simple method of superimposing planimetric tracings of successive footprints reveals consistent landmarks from imprint to imprint that can be identified with some degree of confidence, yielding a more reliable sense of the relative toe impressions. In consideration of this and even taking a liberal length for the hallux in G 1-36, it is evident that the lateral toes extend distal to it (Fig. 10).

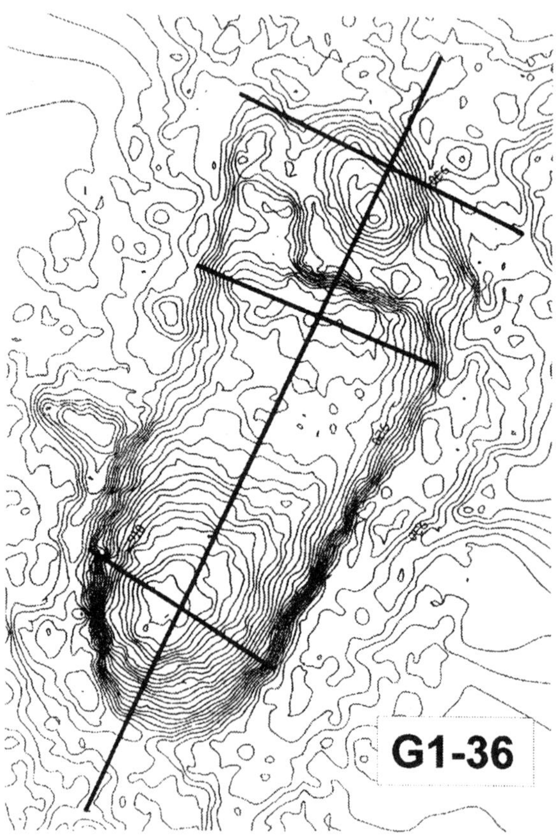

Figure 10. Contour tracing of the G1-36 footprint from Laetoli, indicating the relative lengths of the hallux and lateral toes (modified from Demas, 1996).

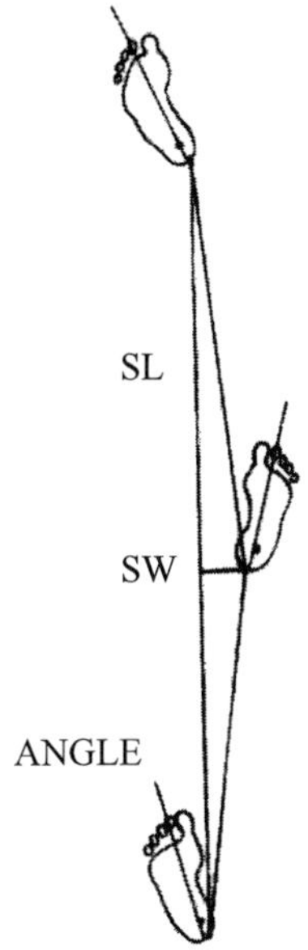

Figure 11. Method of measuring step length (SL), step width (SW), and angle of gait (ANGLE).

Gait

In some instances of Hawaiian trackways, a sufficient series of consecutive footprints are preserved to permit an assessment of the gait parameters. A sample of 20 trackways were measured. Descriptive gait parameters were documented including step length, step width or straddle, and angle of gait (Fig. 11). Stride length was measured between the most proximal points on two consecutive ipsilateral footprints. Step length was measured by dropping a perpendicular line from the most proximal point of the heel of the intervening contralateral footprint to the line representing the stride. Angle of gait was measured as the angle of the line drawn through the approximate center of the heel and the second digit, as in Hilton (1997) and Norkin and Levangie (1992). Values were consistent with results of other habitually bare-footed pedestrian populations (Tuttle et al., 1990; Hilton, 1997). Results are summarized graphically in Figure 6. The most noteworthy contrast between modern and hominid values is between the index for footlength/steplength (FL/SL). The relatively shorter step length is expected if the imprints are attributable to *Australopithecus afarensis* with its relatively shorter limb length and relatively longer foot length (Jungers, 1982). However, the short step is also to be expected, given the lack of consistent evidence for a striding extended-legged gait in the footprints of these early hominids. In light of recent fossil discoveries, the possibility must be acknowledged that some other contemporary hominid species may have been responsible for the tracks. It seems unnecessary to appeal to the exotic nature of the substrate to imply the gait was exceptionally hesitant or cautious. This work involved a chimp that had experienced only concrete and linoleum "terrestrial" surfaces prior to this experiment. His initial hesitancy to tread upon a sand trackway, until he had tested its touch and taste, was overcome by the motivation for tasty treats. Thereafter his ambulation through the sandbox appeared spontaneous and uninhibited. The unwavering course of the Laetoli trackmakers

Table 1. Summary of principal contrasts apparent between the Hawaiian footprints from Ka'u and the Laetoli hominid footprints.

Ka'u	Laetoli
• broad rounded heel	• tapered heel
• no bulge of abductor hallucis/navicular contact	• bulge of abductor hallucis/navicular contact
• medial longitudinal arch	• midtarsal flexion/pressure ridge
• prominent ball	• no ball
• pressure ridge at ball	• no pressure ridge at forefoot
• hallux adducted	• modest hallux abduction
• distinct lateral toe impressions	• indistinct lateral toe impressions
• short lateral toes	• long lateral toes

suggests they were also motivated toward some goal. The angle of gait and relative step width are quite variable in the Hawaiian footprints and encompass the values for the Laetoli trackways, with the exception of the extreme angle of gait evidenced in the G1 trail.

DISCUSSION AND CONCLUSIONS

Considerable debate continues to surround both the interpretation and significance of the Laetoli tracks. Conclusions drawn by investigators range from the position that they are indistinguishable from the footprints of modern human pedestrians to the position that little more can be ascertained beyond the observation that "something walked on two legs through the ash." As is often the case, it seems reality lies somewhere between these extremes of interpretation. The opportunity to examine spontaneous footprints of traditionally bare-footed modern humans, under relatively comparable substrate conditions, underscores the evident distinctions of the hominids' tracks from those of modern humans. While it seems apparent that the Laetoli hominids were capable of sustained bipedalism, the distinctions of the footprints seem to indicate a foot morphology and kinematics that do not imply equivalency with the modern human condition. The morphological contrasts are summarized in Table 1. The modern human foot is adapted to a striding stiff-legged gait with distinct heel-strike and toe-off. The morphology of this adaptation comprises the enlargement of the heel, heel pad, fixed longitudinal arch, padded ball and propulsive/tractional toes. In contrast, the early hominid foot had a less robust heel, a flexible midfoot, little modification of the ball, and relatively long prehensile toes. The morphology of the hallucial metatarsal head in fossil remains of *Australopithecus* and *Paranthropus* bears on this last point. The rounded dorsally tapered head of the first metatarsal would have precluded close-packing of the metatarsophalangeal joint in dorsiflexion, suggesting a lack of the human-like toe-off at the end of stance (Stern and Susman, 1983; Susman and Brain, 1988). The lack of distinct impressions of the lateral toes, combined with the suggestion of a midtarsal pressure ridge, suggest the possibility of a more compliant gait with a push-off coming more from the midfoot.

Although the midtarsal break is evident in the chimpanzee foot in both kinematic observation as well in pressure studies (Meldrum and Wunderlich, 1998) it did not produce, with any consistency, a midtarsal pressure ridge in a sandbox trackway experiment with one chimpanzee. There were a few instances in which a pressure disc was displaced proximal to the midtarsal joint. However, the typically abducted position of the chimpanzee hallux complicates comparisons to the hominid footprints. The abducted hallux is usually aligned with the direction of travel, presumably to spare the long lateral toes the bending stresses that might be imposed in a bipedal gait (Meldrum, 1991). This may alter the plantar pressure at the medial side of the midtarsus, which is the most flexible aspect of the midfoot (Meldrum and Wunderlich, 1998). Under these conditions it seems a midtarsal pressure ridge is less likely to occur. There was one footprint that clearly exhibited a primary pressure disc displaced proximal to the talonavicular joint (Fig. 12). It is roughly comparable to the ridge seen in the G1-26 footprint, allowing for the differences in hallux position. I have observed idiosyncrasies in preferential placement of the hallux, i.e. the degree of abduction/adduction, and angle of gait in both chimps and gorillas. Some individuals preferentially hold the hallux in an adducted position and exhibit a much lower angle of gait. However, the opportunity has not been realized to contrast the plantar pressures of these distinct foot postures. Clearly these contrasts need to be made.

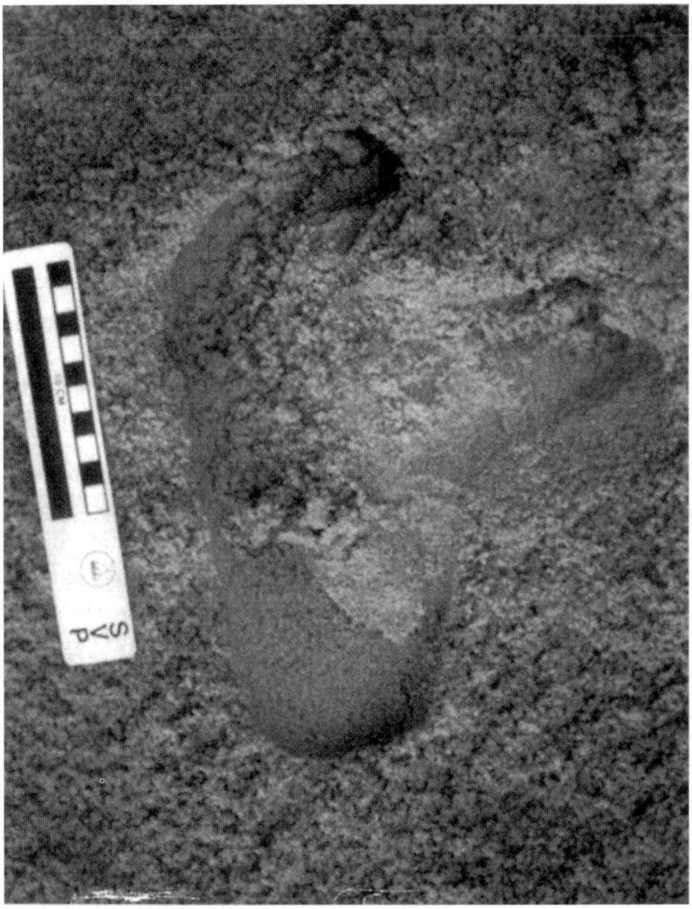

Figure 12. Chimpanzee footprint in sand that exhibits a primary pressure release disc proximal to the medial aspect of the transverse tarsal joint.

If indeed early hominids practiced a form of bipedalism that differed significantly from modern humans, the question follows, when did the modern foot form emerge? Unfortunately, there is a considerable gap in the fossil record for either the foot skeleton or footprints for the early to late middle Pleistocene hominids. Renewed analysis of the OH8 foot skeleton reaffirms the limited stability of the medial column of the foot (Susman, 1983; Kidd et al., 1996; Harcourt-Smith et al., 2002). Analyses of the StW 573 foot skeleton, attributed to *Australopithecus africanus* suggest an even more ape-like ankle complex (Harcourt-Smith et al., 2002; Deloison, this volume). Again, this is especially evident on the medial aspect of the midtarsal joint. A recent biomechanical model of the human foot indicated that 5 degrees of supination at the subtalar joint increased moments about the calcaneocuboid joint by 55% (Arangio et al., 2000). This would seem a significant selection pressure for stabilization of the lateral column of the foot in an early hominid biped.

A potential glimpse may be had into the foot form of middle Pleistocene *Homo*, in the form of the footprints at the site of Terra Amata, France (ca. 400 kya). The footprints await description; however, the only published photograph suggests a pressure ridge proximal to the midtarsal region (White, 1973). The combination of midfoot flexibilty and virtually "modern" limb proportions in the postcranial skeleton of middle Pleistocene *Homo* would seem to run counter to presumptions about the emergence of hominid locomotor adaptations. However, the footprint lacks distinct lateral toe impressions, although the hallux impression is well defined and separated from the midfoot. Furthermore, a distinct ball and longitudinal arch are also lacking. Recently, the first metatarsal, KNM-BK 63, of a 500 kya hominid from Baringo was described (Fisher and McBrearty, 2002). In most metrics, it falls within the range of overlap for neandertals and modern humans. However, it can be distinguished in having more gracile shaft dimensions. This is noteworthy since the few postcranial remains of middle Pleistocene hominids are characteristically robust by comparison. The functional implications of this finding have yet to be determined (Fisher, pers. comm.), although the attainment of hallucial robusticity has been cited as one of the landmarks of modern human foot morphology (Morton, 1926; Trinkaus, 1983). The Neandertal fossil record affords both a number of pedal fossils as well as footprints. While emphasizing the modern aspects of the Neandertal foot skeleton, including the locking of the transverse tarsal joint, Trinkaus (1983) notes the retention of some archaic features. These include slightly longer toes and longer calcanei. These features are what might be expected as pleisiomorphies from the earlier hominin foot. Increasing leverage of the plantarflexors of a foot with midfoot flexibility is accomplished by elongation of the calcaneus. The toes in a foot with midfoot flexibility would be spared selection pressure for shortened length, due to less loading in dorsiflexion. A footprint preserved in an Italian cave reflects the robustness of the skeleton, but a distally extended sole pad obscures the long toes (Day, 1991; Howell, 1970). It displays a medially projecting ball and well developed longitudinal arch.

Perhaps the evolution of modern human postcranial adaptations unfolded in more of a mosaic fashion than has previously been recognized. The possibility must be considered that not only did the Pliocene hominids walk bipedally in a manner unlike modern humans, but that bipedalism on flat flexible feet may have been a relatively stable adaptation for over 3 million years. Fully modern human walking may not have appeared until the late Pleistocene (Meldrum, 2002). Recently Haeusler and McHenry (2003) reevaluated the hindlimb morphology of OH 62 attributed to *Homo habilis* and concluded that this early hominin had human-like hindlimb proportions, and suggested that distance travel might have evolved early in human evolution. Given the morphology of the OH 8 foot, also attributed to *H. habilis*, this finding lends corroboration to the proposal that the combination of human-like hindlimb proportions and relatively flat flexible feet constituted an adaptive suite that may have persisted at least through *H. ergaster*. Clearly this hypothesis will continue to be a focus of attention pending more data and further analysis.

ACKNOWLEDGEMENTS This research was supported by grant no. 829 from the ISU Faculty Research Committee and by the LSB Leakey Foundation. Cooperation and assistance of the following: Jill Rivoli for bringing the Hawaiian footprints to my attention; the staff of the CRM and HAVO; Susan Larson and the SUNY Stony Brook animal facility staff, supported by NSF BCS980629; the hospitality and insights of Tim White; Martha Demas for providing the unpublished reports of the Getty Foundation's Laetoli Project; discussions with Bruce Latimer; and numerous discussions with contributors to this volume. This manuscript benefited from comments by Chuck Hilton, Jack Stern, Jr., Russell Tuttle and additional anonymous reviewers.

REFERENCES

Agnew, N., and Demas, M., 1998, Preserving the Laetoli footprints. *Scientific American* 278: 44–55.

Arangio, G.A., Phillippy, D.C., Xiao, D., Gu , W.K., and Salathe, E.P., 2000, Subtalar pronation relationship to the medial longitudinal arch loading in the normal foot. *Foot and Ankle International* 21:216–220.

Barnicot, N.A., and Hardy, R.H., 1955, The position of the hallux in west Africans. *Journal of Anatomy* 89:355–361.

Clarke, R.J., 1999, Discovery of complete arm and hand of the 3.3 million-year-old Australopithecus skeleton from Sterkfontein. *South African Journal of Science* 95:477–480.

Coppens, Y., and Senut, B., 1991, *Origine(s) de la Bipedie chez les Hominides*. Editions du Centre National de la Recerche Scientifique, Paris.

Day, M.H., 1991, Bipedalism and prehistoric footprints. In *Origine(s) de la Bipedie chez les Hominides*, edited by Y. Coppens and B. Senut, pp. 199–213. Editions du Centre National de la Recherche Scientifique, Paris.

Day, M.H., 1985, Hominid locomotion: from Taung to the Laetoli footprints. In *hominid evolution: Past, present, and future*, edited by P. Tobias, pp. 115–127. Alan R. Liss, New York.

Day, M.H., and Wickens, E.H., 1980, Laetoli Pliocene hominid footprints and bipedalism. *Nature* 286:385–387.

Dee, R., 1989, Flatfoot. In *Principles of Orthopaedic Practice*, edited by R. Dee, M. Enrico and L.C. Hurst, pp. 1145–1148. McGraw-Hill, New York.

Deloison, Y., 1991, Did Australopithecines walk as we do? In *Origine(s) de la Bipedie chez les Hominides*, edited by Y. Coppens and B. Senut, pp. 177–185. Editions du Centre National de la Recherche Scientifique, Paris.

Deloison, Y., 1992, Emprientes de pas a Laetoli (Tanzanie). Leur apport a une meillure connaissance de la locomotion des Hominides. *CR Academie des Sciences Paris*, Ser. II, 315–103–109.

Demas, M., 2000, Laetoli Project: Report on the 1996–1997 field seasons and the Olduvai Museum exhibition (unpublished report).

Demas, M., 1996, Laetoli Project: Report on the 1995 field season (unpublished report).

Fisher, R., and McBrearty, S., 2002, The comparative morphology of hominin postcranial remains from the Kapthurin Formation, Baringo District, Kenya. *American Journal of Physical Anthropology Supplement* 34:70.

Halfpenny, J.C., 1998, *Scats and Tracks of the Rocky Mountains*. Falcon, Helena, Montana.

Haeusler, M., and McHenry, H.M., 2003, Limb proportions of *Homo habilis* reviewed. *American Journal of Physical Anthropology Supplement* 36:107.

Harcourt-Smith, W., Higgins, P.O., and Aiello, L., 2002, From Lucy to Littlefoot: A three dimensional analysis of Plio-Pleistocene hominin tarsal remains. *American Journal of Physical Anthropology Supplement* 34:82.

Hilton, C.E., 1997, *Comparative locomotor kinesiology in two contemporary hominid groups: Sedendary Americans and mobile Venezuelan foragers*. Ph.D. dissertaion, University of New Mexico. University Microfilms, Ann Arbor.

Hoffman, P., 1905, Conclusions drawn from a comparative study of the feet of barefooted and shoe-wearing peoples. *American Journal of Orthopedic Surgery* 3:105–136.

Howell, F.C., 1970, *Early Man*. Time-Life Books, New York.

James, C.S., 1939, Footprints and feet of natives of the Solomon Islands. *Lancet* 2:1390–1393.

Jungers, W.L., 1982, Lucy's limbs: skeletal allometry and locomotion in Australopithecus afarensis. *Nature* 297:676–678.

Kidd, R.S., Higgins, P.O., and Oxnard, C.E., 1996, The OH8 foot: A reappraisal of the functional morphology of the hindfoot utilizing a multivariate analysis. *Journal of Human Evolution* 31:269–291.

Latimer, B., and Lovejoy, C.O., 1989, The calcaneus of Australopithecus afarensis and its implications for the evolution of bipedality. *American Journal of Physical Anthropology* 78:369–386.

Leakey, M.D., and Harris, J.M. (eds.), 1987, *Laetoli: A Pliocene site in northern Tanzania*. Oxford Science Publications, Oxford.

Leakey, M.D., and Hay, R.L., 1979, Pliocene footprints in the Laetoli beds at Laetoli, north Tanzania. *Nature* 278:317–323.

Lovejoy, O.C., 1988 Evolution of human walking. *Scientific American* 259:82–89.

Martin, L.G., Chritiansen, R.L., Swanson, D.A., Stauffer, P.H., and Hendley, II, J.W., 1999, Explosive Eruption at Kilauea Volcano, Hawaii? *USGS Fact Sheet*-132–98.

Meldrum, D.J., 2002, Midfoot flexibility and the evolution of bipedalism. *American Journal of Physical Anthropology Supplement* 34:111–112.

Meldrum, D.J., 2000, Footprints in the Ka'u Desert, Hawaii. *American Journal of Physical Anthropology Supplement* 30:226–227.

Meldrum, D.J., 1999, Evaluation of alleged Sasquatch footprints and inferred functional morphology. *American Journal of Physical Anthropology Supplement* 27:161.

Meldrum, D.J., 1991, The kinematics of the cercopithecine foot on arboreal and terrestrial substrates with implications for the interpretation of hominid terrestrial adaptations. *American Journal of Physical Anthropology* 84:273–289.

Meldrum, D.J., and Wunderlich, R.E., 1998, Midtarsal flexibility in ape foot dynamics, early hominid footprints and bipedalism. *American Journal of Physical Anthropology Supplement* 26:161.

Moniz Nakamura, J., "*Keonehelelei—The Falling Sands. Uncovering the origin of preserved footprints and associated Kau Desert features: A historical overview and archeological survey of the Kau Desert, Hawaii Volcanoes National Park, Island of Hawaii.*" (unpublished report).

Morton, D.J., 1926, Significant characteristics of the Neanderthal foot. *Journal of the American Museum of Natural History* 26:310–315.

Norkin, C.C., and Levangie, P.K., 1992, *Joint structure and function: A comprehensive analysis, 2nd ed.* F.A. Davis, Philadelphia. p. 459.

Robbins, L.M., 1987, Hominid footprints from site G. In *Laetoli: A Pliocene Site in Northern Tanzania*, edited by M.D. Leakey and J.M. Harris, pp. 497–502. Oxford University Press, Oxford.

Stern, J.T., and Susman, R.L., 1983, The locomotor anatomy of Australopithecus afarensis. *American Journal of Physical Anthropology* 60:279–317.

Stewart, S.F., 1970, Human gait and the human foot: An enthnological study of flat foot (Part 1). *Clinical Orthopaedics and Related Research* 70:111–123.

Susman, R.L., 1983, Evolution of the human foot: Evidence from Plio-Pleistocene hominids. *Foot and Ankle* 3:365–376.

Susman, R.L., and Brain, T.M., 1988, New first metatarsal (SKX 5017) from Swartkrans and the gait of *Paranthropus robustus*. *American Journal of Physical Anthropology* 77:7–16.

Susman, R.L., Stern, Jr., J.T., and Jungers, W.L., 1984, Arboreality and bipedality in Hadar hominids. *Folia Primatologica* 43:113–156.

Swanson, D.A., and Christiansen, R.L., 1973, Tragic base surge at Kilauea volcano. *Geology* 1:83–86.

Trinkaus, E., 1983, Functional aspects of Neandertal pedal remains. *Foot and Ankle* 3:377–390.

Tuttle, R., 1996, The Laetoli hominid G footprints: Where do they stand today? *Kaupia* 6:97–102.

Tuttle, R., 1985, Ape footprints and Laetoli impressions: A response to the SUNY claims. In *Hominid evolution: past, present, and future*, edited by P. Tobias, pp. 129–133. Alan R. Liss, New York.

Tuttle, R., Webb, C.M., and Tuttle, N., 1991, Laetoli footprint trails and the evolution of hominid bipedalism. In *Origine(s) de la Bipedie chez les Hominides*, edited by Y. Coppens and B. Senut, pp. 187–198. CNRS, Paris.

Tuttle, R., Webb, D., Weidl, E., and Baksh, M., 1990, Further progress on the Laetoli trails. *Journal of Archeological Science* 17:347–362.

Westervelt, W.D. (ed.), 1963, *Hawaiian Legends of the Volcanoes*. Charles E. Tuttle, Co., Rutland, VT.

White, E., 1973, The mystery of the first men. In *The First Men*. Editors of Time-Life Books, p. 61. Time-Life Books, New York.

White, T.D., 1980, Evolutionary implications of Pliocene hominid footprints. *Science* 208:175–176.

White, T.D., and Suwa, G., 1987, Hominid footprints at Laetoli: facts and interpretations. *American Journal of Physical Anthropology* 72:485–514.

Chapter 6

In What Manner Did They Walk on Two Legs?

An Architectural Perspective for the Functional Diagnostics of the Early Hominid Foot

GILLES BERILLON

Abstract. When Mary Leakey and collaborators discovered the Laetoli footprints (Tanzania) in 1978–1979, they exhumed the direct evidence that a species of primate was able to walk bipedally 3.75 Mya. This evidence of an early bipedalism has also been corroborated by contemporary fossil bones, especially those of *Australopithecus afarensis*. But in what manner did these early hominids walk on two legs? This point remains the subject of discussion. Our purpose is to add pedal architectural data to the debate concerning the locomotor modalities in early hominids, especially in *Australopithecus afarensis*.

Australopithecine tarsal and metatarsal bones from Hadar are compared to a sample of *Proconsul* from Rusinga and Songhor (Kenya), other early hominids from Omo (Ethiopia), Olduvai (Tanzania), Koobi Fora, Ileret and Baringo (Kenya) and Kromdraai (South Africa), Neandertals and extant species (43 *Homo*, 36 *Pan* et 38 *Gorilla*). The architecture of the foot is assessed by examining proximodistal relationships, and by measuring angulations taken on the dislocated tarsometatarsal skeleton.

The architectural analysis of the *Australopithecus afarensis* hallucal tarsometatarsal complex suggests that the hallux was slightly abducted in neutral position and of a different pattern than in African apes. Furthermore, the *Australopithecus afarensis* foot is flat in neutral position, resembling the extant African apes condition. Although foot remains are too incomplete to propose a complete architectural pattern, architectural data strongly corroborate the uniqueness of the *Australopithecus afarensis* foot pattern, in which a slightly abducted hallux and no plantar vault in neutral position are associated. Accordingly, if *Australopithecus afarensis* were the makers of the Laetoli trails, as the Laetoli fossil bones suggest, then their bipedal locomotion would have necessitate a very strong muscular action. These features corroborate the idea of a very distinctive pattern of bipedality in *Australopithecus afarensis*.

INTRODUCTION

Identifying the evidences of bipedal aptitudes in fossil primates constitutes one of the main aims of the paleoanthropological debate. Historically, the 2.5–3My-old Taung Child, the type of *Australopithecus africanus* Dart, 1925, with its anterior position of the foramen magnum, constitutes the first of these evidences (Dart, 1925). When Mary Leakey

From Biped to Strider: The Emergence of Modern Human Walking, Running, and Resource Transport, edited by D. Jeffrey Meldrum and Charles E. Hilton. Kluwer Academic/Plenum Publishers, New York, 2004.

and collaborators discovered the Laetoli footprints (Tanzania) in 1978–1979, they revealed the first direct evidence that a species of primate was able to walk bipedally 3.75 Mya (Clarke, 1979; Leakey et al., 1979). Contemporary fossil remains attributed to the species *Australopithecus afarensis* Johanson et al., 1978 also documented an early bipedalism (Johanson et al., 1978; Johanson et al., 1982; Latimer et al., 1982). While the majority of authorities acknowledge such early bipedal aptitudes, the way they walked on two legs continues to be debated. And while more fossil remains were recovered and tools of study were becoming more powerful, interpretations about the early stages of bipedality also evolved. Now the debate is much more multiparametric than it was 75 years ago, and the picture of the "walk-on-two-legs" Plio-Pleistocene primates is less clear. Since the 1970's, as already presented by Kramer et al. (1999), two main responses have been given to that question based on the postcranial anatomy. The first school argues for a modern-human pattern of bipedalism (*e.g.*, Johanson et al., 1982; Latimer et al., 1982; Latimer, 1991) while the other argues for a transitional form of locomotion, combining bipedal and arboreal abilities (*e.g.*, Senut, 1981; Jungers, 1982; Stern et al., 1983; Tardieu, 1983; Susman et al., 1984, 1991; Deloison, 1993). Recent modelings and simulations using some global parameters of functional interest such as the lower leg proportions or the musculoskeletal organization of the pelvic girdle aim to describe what this pattern could be and argue for a very unique pattern of locomotion (Berge, 1994; Crompton et al., 1998; Kramer, 1999; Kramer et al., 1999). Because it is technically difficult to integrate all the features of functional interest ("*Homo*-like" and "ape-like") in a common biomechanical modeling on the one hand and because the correlation of form to the function is often hypothetical on the other hand, it remains difficult to offer an objective perception of precisely how these early hominids walked on two legs. In that debate, each part of the locomotor skeleton has been the subject of numerous studies. The foot, as the contact area to the ground, has been the subject of numerous studies, which follow the presented dichotomy (Latimer et al., 1982; Gomberg et al., 1984; Lamy, 1986; Latimer et al., 1987; Latimer et al., 1989, 1990a, b; Langdon et al., 1991; *contra* Stern et al., 1983; Deloison, 1984; Susman et al., 1984, 1985, 1991; Deloison, 1991, 1993; Duncan et al., 1994). Most paleoanthropological studies focus on the restitution of the movement, mainly because of the fragmentary nature of the fossil postcrania (*e.g.*, Day & Wood, 1968; Oxnard, 1979; Lewis, 1980; Stern et al., 1983; Susman et al., 1984; Deloison, 1986; Lamy, 1986; Latimer et al., 1989; Deloison, 1993), while architectural approaches remained mainly limited to well preserved and previously rearticulated specimens, such as OH 8, the *Homo habilis* specimen from Olduvai, (*e.g.*, Day et al., 1964; Oxnard et al., 1980; Deloison, 1993) or Stw 573 from Sterkfontein (Clarke & Tobias, 1995).

Our purpose is to add pedal architectural data to the debate concerning the locomotor modalities in early hominids, especially in *Australopithecus afarensis*, through a better understanding of its foot functioning during bipedal locomotion. We have developed a quantitative approach using disarticulated skeletons that allows estimating certain architectural parameters of isolated foot elements. In this issue, we present the results concerning the organization of the hallucal column and the longitudinal structure of the *Australopithecus afarensis* foot.

MATERIAL

The *Australopithecus afarensis* specimens (Hadar, Ethiopia) are compared to the sample of fossil specimens given in the Table 1. The sample of extant species consists of 43 *Homo sapiens* from the Laboratoire d'Anthropologie Biologique of the Muséum

Table 1. Inventory of the fossil sample.

Proconsul (*P.*), from the Miocene sites of Rusinga and Songhor (Kenya)	*P. nyanzae*: KNM-RU-1656 (medial cuneiform), KNM-RU-1743 (left talus), KNM-RU-1755 (right calcaneus). *P. heseloni*: KNM-RU-1745 (right talus), KNM-RU-2036 (left intermediate cuneiform and first metatarsal). *P. africanus*: KNM-SO-1402 (right talus). *P. major*: KNM-SO-389/390 (right talus and calcaneus).
Australopithecus afarensis, from Hadar (Ethiopia)	A.L. 333-13 (left fifth metatarsal), A.L. 333-28 (right medial cuneiform), A.L. 333-36 (right navicular), A.L. 333-47 (right navicular), A.L. 333-54 (left first metatarsal), A.L. 333-78 (left fifth metatarsal), A.L. 288-1-as (right talus).
Other early hominids	Omo 33-74-895 (right calcaneus), Omo 323-76-898 (right talus) from the Shungura Formation (Ethiopia); Tm 1517 (right talus) from Kroomdrai (South Africa); KNM-ER-1464 (right talus), KNM-ER-1476A (left talus), KNM-ER-813A (right talus) from the Koobi Fora Formation (Kenya); BK 63 (right first metatarsal) from Baringo (Kenya), OH 8 (partial left foot) from Olduvai (Tanzania).
Neandertals	La Chapelle-aux-Saints (left talus and calcaneus), La Ferrassie I (right and left feet), La Ferrassie II (left foot), Regourdou 1 (right talus and calcaneus) from France; Tabun CI (right and left feet), Amud 1 (left talus) from Israel; Subalyuk 1 (left fifth metatarsal, left second and third metatarsals) from Hungaria; Kiik-Koba I (right and left feet) from Ukraine; Neanderthal 1 (left talus) from Germany.

National d'Histoire Naturelle (Paris), 28 *Pan troglodytes*, 8 *Pan paniscus* and 38 *Gorilla gorilla* from the Laboratoire d'Anatomie Comparée of the Muséum National d'Histoire Naturelle (Paris), the Natural History Museum, Mammals Group (London) and the Musée Royal de l'Afrique Centrale (Tervuren).

METHOD

The foot architecture is described in terms of proximodistal geometric relationships between tarsal and metatarsal bones, each articular column being considered separately (Berillon, 2000). This architectural analysis rests on the principle of optimal congruency or neutral position: considering an articular column and its joints, there is a direct relationship between the orientation of the joint and the architecture of the column when both areas of each joint are maintained parallel. For a given isolated bone, each relative joint orientation is measured in degrees by three perpendicular angular components, horizontally, longitudinally, and transversally (Table 2, Fig. 1). The combinations of these angulations are expressions of architectural parameters. Finally, the structure of the foot, as a static geometric object made of different components, is described according to the orientation of their contact areas.

A variance analysis for each angulation was undertaken to compare the distributions of the individual values in each group of extant species and the performed "F" values as well as angular statistics and fossil values are given in Tables 3 and 4. Architectural parameters are illustrated graphically (Fig. 2).

RESULTS

Architecture of the Hallucal Ray and the Medial Column

The longitudinal component as an expression of the degree of medial arching—As seen on Fig. 2 and Table 3, both *Australopithecus afarensis* navicular specimens have a high

Table 2. Definitions of the measured angles used in the analysis of the longitudinal architecture of the tarsometatarsal complex (see Figure 1 for illustrations).

a—Talus—The talus rests on its trochlear plane.
TalHT: Talar head torsion: in distal view, angle between the mediolateral axis of the joint area for the navicular (aa') and the trochlear plan (TrP) (Martin et al., 1928).

b—Calcaneus—The long axis of the calcaneus (CaLA) is horizontally holded and the dorsoplantar axis of its distal joint area (cc') rests in a vertical plane.
CalCuD: Deviation of the calcaneal joint surface for the cuboid: in dorsal view, horizontally, angle between the mediolateral axis of the joint surface for the cuboid and the long axis of the calcaneus.
CalCuI: Inclination of the calcaneal joint surface for the cuboid: in medial view, vertically, angle between the dorso-plantar axis of the joint surface for the cuboid and the long axis of the calcaneus.

c—Cuboid—The cuboid rests on its calcaneal area, its mediolateral (aa') and dorsoplantar (bb') axis both horizontally.
Cub4D-Cub5D: Deviation of the cuboid joint surfaces for the 4th and 5th metatarsals: vertically, angle between the mediolateral axis of the joint surfaces for the 4th (cc') and 5th (dd') metatarsals respectively and the vertical plane.
Cub4I-Cub5I: Inclination of the cuboid joint surfaces for the 4th and 5th metatarsals: vertically, angle between the dorso-plantar axis of the joint surfaces for the 4th (cc') and 5th (dd') metatarsals respectively and the vertical plane.

d—Navicular—The navicular rests on its talar joint area, its mediolateral (aa') and dorsoplantar (bb') axis both horizontally.
NavmD: Navicular medial deviation of the joint area for the medial cuneiform: vertically, angle between the mediolateral axis of joint area for the cuneiform (cc') and the vertical plane.
NavmT: Navicular medial torsion: horizontally, angle between the dorsoplantar axis of joint area for the cuneiform (dd') and the mediolateral axis of the joint area for the talus.
NavmI-NavII-NavlI: Navicular medial, intermediate and lateral inclinations: vertically, angle between the dorsoplantar axis of joint area for the cuneiform (dd', ff' and hh' respectively) and the dorsoplantar axis of the joint area for the talus.

e—Cuneiforms—The medial cuneiform rests on its navicular joint area, its mediolateral (aa') and dorsoplantar (bb') axis both horizontally.
Cf1dD: Dorsal divergence of the medial cuneiform joint area for the first metatarsal: in dorsal view, horizontally, angle between the dorsal mediolateral axis (cc') of the medial cuneiform joint area for the first metatarsal and the vertical plane (Berillon, 1999).
Cf1pD: Plantar divergence of the medial cuneiform joint area for the first metatarsal: in plantar view, horizontally, angle between the plantar mediolateral axis (dd') of the medial cuneiform joint area for the first metatarsal and the vertical plane (Berillon, 1999).
Cf1mD: Calculated medium divergence of the medial cuneiform joint area for the first metatarsal: (Cf1dD+Cf1pD)/2 (Berillon, 1999).
Cf1T: Proximodistal torsion of the medial cuneiform: in distal view, horizontally, angle between the dorsoplantar axis of the joint areas for the metatarsal (ee') and the navicular one (bb').
Cf1I: Inclination of the medial cuneiform joint area for the first metatarsal: in medial view, vertically, angle between the dorsoplantar axis of the metatarsal joint area and the vertical plane.

f&g—Metatarsals—The long axis of the metatarsal (MtiLA) is horizontally holded and the dorso-plantar axis of the proximal joint area (cc') rests in a vertical plane.
Mt1dD: Dorsal divergence of the first metatarsal joint area for the medial cuneiform: in dorsal view, horizontally, angle between the dorsal mediolateral axis (aa') of the first metatarsal joint area for the medial cuneiform and the long axis of the bone (Berillon, 1999).
Mt1pD: Plantar divergence of the first metatarsal joint area for the medial cuneiform: in plantar view, horizontally, angle between the plantar mediolateral axis (bb') of the first metatarsal joint area for the medial cuneiform and the long axis of the bone (Berillon, 1999).
Mt1mD: Calculated medium divergence of the first metatarsal joint area for the medial cuneiform: (Mt1dD+Mt1pD)/2 (Berillon, 1999).
Mt5D: Divergence of the fifth metatarsal joint area for the cuboid: in dorsal view, horizontally, angle between the mediolateral axis (aa') of the metatarsal joint area for the cuboid and the long axis of the bone.
Mt1T: Proximodistal metatarsal torsion: in distal view and vertically, angle between the dorsoplantar axis of the head (dd') and the dorsoplantar axis of the joint area for the corresponding tarsal bone (cc').
Mt1I-Mt5I: Metatarsal inclination: in medial view, vertically, angle between the long axis of the bone and the dorso-plantar axis of the joint area for the corresponding tarsal bone.

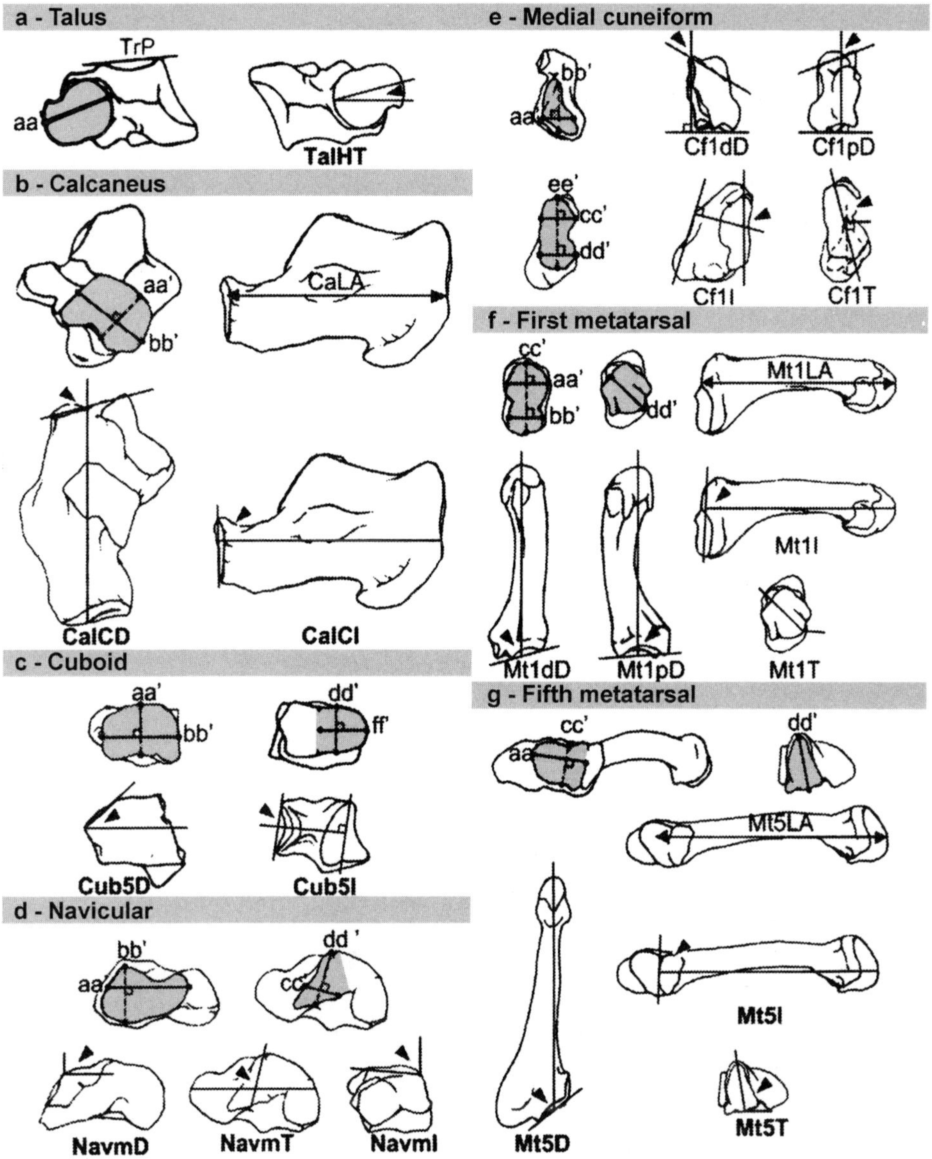

Figure 1. Definitions of the points and measured angles used in the architectural analysis of the tarsometatarsal complex. They are illustrated on the *Pan troglodytes* left bones (see Table 2 for texts).

dorsal obliquity of the navicular joint areas for their corresponding cuneiforms (δ NavmI, δNaviI, δNavlI). The medial cuneiform presents a very low dorsal obliquity of its distal area (δCf1I), while the first metatarsal shows a low dorsal obliquity of its long axis (δMt1I). In the extant species sample, the obliquity of the navicular joint areas for the cuneiforms is strongly indicative of the degree of flexion of the medial foot relatively to the tarsal transverse joint. Interspecific variations are highly significant. *Pan* and *Gorilla* have high obliquities; such values highly contribute to the dorsiflexion of the medial foot relative to the transverse tarsal joint (AC1I, AC2I and AC3I are a much higher than 0°) and thus to

Table 3. Angular statistics (mean, standard deviation and range, in degrees) corresponding to the hallucal column in a sample of 43 *Homo*, 35 *Pan* and 38 *Gorilla* and values in *Australopithecus afarensis*, other early hominids, neandertals and *Proconsul*. Results of the variance analysis (F with 3 DF) are given.

	TalHT	NavmD	NavmT	NavmI	NaviI	NavlI	Cf1dD	Cf1pD	Cf1T	Cf1I	Cf1mD	Mt1I	Mt1dD	Mt1pD	Mt1T	Mt1mD
Homo sapiens (43)																
Mean	30.6	90	35.8	90.77	75.8	68.8	85.3	95.9	92.9	97.6	90.6	90.5	90.5	76.4	97.7	83.4
SD	7	4.1	5.3	6.13	5.7	5.6	9	8.2	5.4	4.8	8	2.4	4.7	4.1	4.8	3.3
Min/Max	14/53	97/82	26/45	73/101	64/87	55/83	65/110	73/112	84/108	85/107	69/109.5	84/95	81/105	67/83	88/107	75.5/92.5
Pan troglodytes (27)	**															
Mean	17	89.4	27.5	63.59	48.7	34.3	54	66.7	97.5	107.4	60.4	97	98.8	74.4	115.7	86.6
SD	6.3	4.5	5.2	5.51	5.8	7.0	8.4	8.9	4.9	5.3	7.3	2	4	5.7	9.2	3.4
Min/Max	4/27	81/101	16/37	56/78	37/61	23/50	40/74	54/87	90/107	98/119	47.5/76.5	94/102	88/105	65/87	91/128	80/93
Pan paniscus (8)																
Mean	17.5	83.6	26.8	70.0	57.4	35.3	60	62.3	97.4	105.1	61.1	94.1	97.6	78.1	121.4	87.9
SD	3.5	3.8	5.5	6.32	7.3	4.4	7.5	11.4	5.9	7.3	7.1	2	5	6	6.9	4.7
Min/Max	13/22	77/88	20/36	62/78	45/67	28/43	51/71	48/81	87/109	94/118	53/74	92/98	90/104	72/89	113/134	81/93.5
Gorilla gorilla (38)		*	*	*	*	*										
Mean	14.4	93.1	36.9	65.11	49.9	35.7	50.2	69.8	105	95.8	60	96	100.4	74.4	107.3	87.4
SD	5.8	7.7	8.9	6.38	7.2	7.2	9.2	7.6	8	6.8	7.5	2.6	3.7	5.7	8.8	3.9
Min/Max	4/25	80/115	18/63	49/79	32/67	19/50	28/67	55/84	88/122	78/112	44/72	90/101	93/110	64/90	94/126	81.5/95.5
Neandertals																
Mean	41.4	86.3	123.3	84.5	71.5	61.0	88.8	95.3	90.8	90	92	91.4	88.6	77.6	105.2	83.1
SD	5.8	3.9	13.1	8.5	13.2	5.9	10.4	5.7	7.5	2.4	7.2	3.3	3	6.2	9.2	3.2
Min/Max	31/51	81/90	109/140	78/96	59/90	54/68	78/103	87/100	80/96	87/93	82.5/100	87/96	85/93	69/86	98/118	78/85.5

OH 8	27	88	127	83	70	62	99	87	83	98	93	92	86	77		81.5
BK 63												88	94	86	99	90
A.L. 288–1-as	28															
A.L. 333–36		83	114	75	57	47										
A.L. 333–47		87	107	73	60	34										
A.L. 333–28							87	103	94	93	95					
A.L. 333–54												85	87	66	>98	76.5
Omo 323-76-898	22															
Tm 1517	28															
KNM-ER-1476A	34															
KNM-ER-1464	23															
KNM-ER-813A	30															
KNM-RU-2036												96	96	56	93	76
KNM-RU-1656							49	77	163	102	63					
KNM-RU-1743	3															
KNM-RU-1745	4															
KNM-SO-1402	5															
KNM-SO-390	3															
F (DF=3)	50.3	6.98	14.04	171.31	162.04	270.41	102.56	85.44	25.05	23.57		54.07	42.24	2.25	43.11	

* N-1. ** N-2

Table 4. Angular statistics (mean, standard deviation and range, in degrees) corresponding to the fifth articular column of the foot in a sample of 43 *Homo*, 35 *Pan* and 38 *Gorilla* and values in *Australopithecus afarensis*, other early hominids, neandertals and *Proconsul*. Results of the variance analysis (F with 3 DF) are given.

	CalCD	CalCI	Cub4D	Cub5D	Cub4I	Cub5I	Mt5I	Mt5D	Mt5T
Homo sapiens (43)									
Mean	103.1	107.2	23.6	41.4	90.1	89.8	96.3	54.1	111.7
SD	5.2	3.8	5	6.8	5.4	5.6	4.4	7.4	12.3
Min/Max	90/116	99/117	13/39	27/55	78/103	78/103	87/105	33/69	91/135
Pan troglodytes (27)	*	*							
Mean	91.8	80.4	17.6	50.9	82.7	82.7	91.2	48.8	100.2
SD	3.5	4.4	5.2	8.8	6.5	5.9	5.9	6.4	9.3
Min/Max	85/99	72/89	5/25	34/69	71/98	73/98	81/104	35/63	85/122
Pan paniscus (8)									
Mean	95.4	83.9	23	52	85.1	84.4	90.4	50	91.8
SD	2.7	4.3	4.6	5	6.5	6.6	5.9	5.7	6.3
Min/Max	92/100	77/88	18/30	46/61	77/95	74/94	83/102	39/58	82/99
Gorilla gorilla (38)	*	*							
Mean	96.3	89.4	19.4	48.7	85.5	88	92.1	48.7	91.4
SD	4.1	4.9	5.2	6.3	7	7	6.3	7.2	8.7
Min/Max	88/106	77/99	6/29	27/60	73/98	73/101	80/110	33/63	73/109
Neandertals									
Mean	101.8	106	28	42.5	89.5	84.5	90.8	55.8	117
SD	4	4.4	10.6	9.3	5.3	7.6	5.3	5.7	2.6
Min/Max	97/106	100/112	14/38	35/55	84/94	77/94	85/98	48/63	115/120
OH 8	96	105	25	55	92	87	99	52	
A.L. 333–13							83	40	
A.L. 333–78							84	34	
Omo 33-74-895	96	105							
KNM-ER 803F							86	40	
KNM-RU 1755	90	79							
KNM-SO 969	93	81							
KNM-SO 389	98	57							
F (DF = 3)	40.89	216.03	9.24	13.44	8.83	7.98	7.02	4.63	29.19

* N-1

the flattening of their medial foot. In modern humans, neandertals and OH 8 (although at a lower degree) the obliquity of the navicular joint areas is low, reducing the dorsal flexion of the medial foot. In that feature, *Australopithecus afarensis* closely resembles the African apes average condition and is very distinct from those of OH 8, neandertals and modern humans. Moreover, the dorsal obliquity of the *Australopithecus afarensis* first metatarsal proximal joint area (δMt1I) contributes slightly to increase the dorsiflexion of the first articular column. To conclude, in *Australopithecus afarensis*, the longitudinal architectural features of the hallucal column produce a flat medial foot in neutral position, as it can be observed in extant African apes.

The horizontal component as an expression of the degree of hallucal abduction— The tarsal and metatarsal elements of the *Australopithecus afarensis* first pedal column are characterized by a medial cuneiform joint area for the first metatarsal that is in a frontal position (δCf1mD $\sim$ 0°) and a medial obliquity of the proximal joint area of the

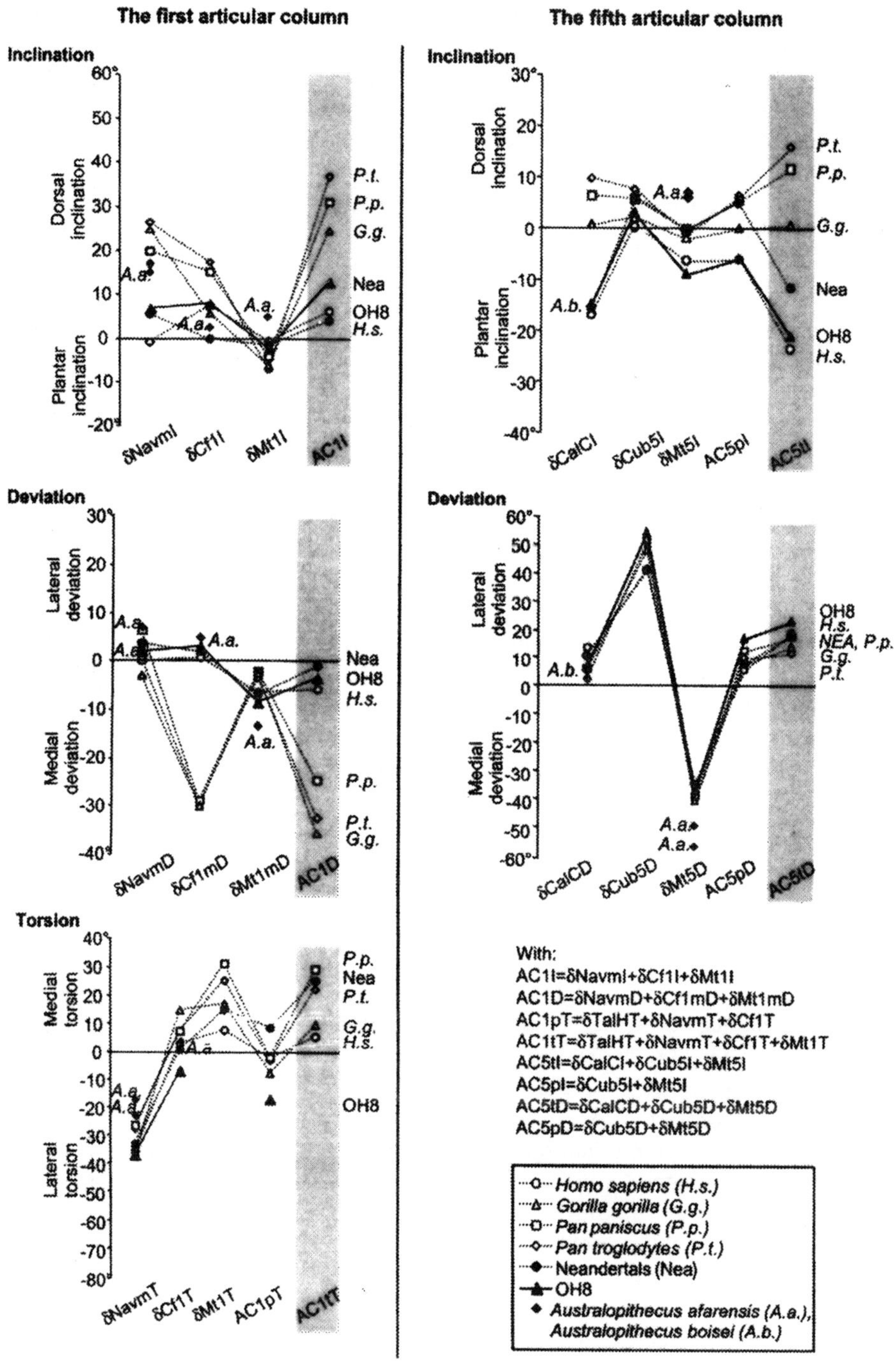

Figure 2. Architectural patterns of the first and fifth articular columns. Hadar specimens values are compared to those of *Proconsul* and other plio-pleistocene specimens, and to the average patterns of the neandertals, modern humans and african apes samples. Values are explained in terms of contributions to a dorsal/plantar inclination or torsion and medial/lateral deviation. For each articular column, the total degree of flexion, deviation or inclination (ACiI) are the sum of elementary bones contributions (δ).

first metatarsal (δMt1mD $\sim -15^\circ$). They resemble the *Homo* condition in the first feature and the *Proconsul heseloni* specimen in the second one. Together, such conditions would correspond to a slightly abducted hallux in neutral position. This abduction is much lower and of a different pattern than those of African apes, in which the abduction is mainly due to the strong medial obliquity of the distal joint area of the medial cuneiform (δCf1D). Both in *Australopithecus afarensis* and in *Homo*, the abduction, although low, is due to the obliquity of the proximal joint area of the first metatarsal.

The transverse component—The *Australopithecus afarensis* talar head torsion (δTalHT), like those of all the studied Plio-Pleistocene specimens, is more pronounced than those of African apes. In *Australopithecus afarensis*, it is associated with a moderate torsion of the navicular bone (δNavmT) and a low torsion of the medial cuneiform (δCf1T), being ape-like in the first feature and human-like in the second one. These joint orientations tend to rest the dorsoplantar axis of the first cuneometatarsal joint axis in a vertical plane. Such orientation of the cuneometatarsal joint dorsoplantar axis is also described both in extant African great apes and modern humans (AC1pT) although their patterns are different.

Although, a certain degree of medial torsion within the shaft of the first metatarsal is observed (8°), this torsion is too low to place the functional axis of the hallucal metatarsophalangeal joint in a high valgus position (internal rotation) such as the one of *Pan* ($25^\circ <$ AC1tT $< 30^\circ$).

In summary, the architectural pattern of the *Australopithecus afarensis* hallucal tarsometatarsal complex is unique, the joint orientations arguing for a flat medial foot and a slightly abducted hallux of which the cuneometatarsal joint is maintained in a vertical and frontal plane. It differs from the *Homo* condition by a higher degree of abduction, and from those of African apes by an abduction that is due to the obliquity of the metatarsal proximal joint area. The hallucal metatarsophalangeal joint seems to have been in a very low valgus position.

The Structure of the Lateral Foot

Figure 2 illustrates that both the obliquity of the calcaneal joint area for the cuboid (δCalcI) and the degree of inclination of the fourth and fifth metatarsals' long axes in relation to the proximal joint area (δMt4I, δMt5I respectively) contribute substantially to the degree of flexion of the lateral skeleton (AC4I, AC5I respectively). In humans, they are less than 0° (or CalcI, Mt4I, Mt5I $> 90^\circ$) and thus contribute to the plantar flexion of the lateral skeleton. In *Gorilla*, the average condition is nil (CalcI, Mt4I, Mt5I #90°) corresponding to a flat lateral foot. Concerning *Pan*, it is slightly dorsally flexed (CalcI, Mt4I, Mt5I $\leq 90^\circ$).

Both 5th metatarsal bones of *Australopithecus afarensis* are very similar with a proximal joint area which is dorsally oblique relative to their long axes (δMt5I $> 0^\circ$), although their distal portion are incomplete. In extant species, such values are only observed in *Pan* and correspond to a dorsal flexion of the lateral skeleton (AC5tI $> 0^\circ$) and thus to the absence of a longitudinal arch. On the contrary, the Omo calcaneus (δCalcI $> 0^\circ$) as well as the OH 8 foot, Neandertals (Nea) and modern humans (*H.s.*) structures correspond to a plantarflexed 5th articular column (AC5I $< 0^\circ$). The high medial deviation of the *Australopithecus afarensis* 5th metatarsal long axis (horizontal component—Mt5D), would also contribute to the flattening of its lateral skeleton.

DISCUSSION AND CONCLUSIONS

Within the limitations of the fragmentary state of the fossil material, an architectural approach to the analysis of foot remains has been developed. As proposed in this study, considering the foot in the context of neutral position, the joint orientations define the architecture of the foot. Transposing this principle to isolated bones, the relative orientations of the joint areas define the architecture of the foot. This study contributes to that point in particular regarding to the longitudinal structure of the foot and the degree of hallucal abduction. This way of studying the pedal complex does not take into account the movement, but it offers a reproducible methodology by which the movement can be secondarily inferred.

Although each bone of the foot contributes to the whole pedal structure through the orientation of its joint areas, some of these joint orientations sufficiently distinguish humans from African apes to characterize the general pedal architectural parameters. Concerning the longitudinal structure of the foot, the dorsoplantar obliquity of the calcaneal joint area for the cuboid was already used as a marker of the plantar arch (*e.g.*, Deloison, 1993; Sarmiento, 1994). This angular analysis proposes two other features of architectural interest, which are (1) the dorsoplantar orientation of the navicular joint areas for the cuneiforms and (2) the dorsoplantar orientation of the fifth and fourth metatarsal joint areas for the cuboid. Concerning the hallucal complex, Schultz's studies of articulated skeletons demonstrated that the mediolateral orientation of the medial cuneiform area for the first metatarsal was a good marker of the degree of hallucal abduction (Schultz, 1930). They were followed by those of Latimer and Lovejoy (1990a) who considered it through the study of the isolated medial cuneiform. While agreeing with their conclusions, the analysis of the whole hallucal tarsometatarsal complex demonstrates that the obliquity of the first metatarsal area for the medial cuneiform as well should be considered as a indicator of the hallucal architectural pattern. From that point of view, the African ape pattern of hallucal abduction could represent a derived condition. The primitive pattern of hallucal abduction from which the african apes, early hominids and *Homo* patterns would have been constructed would have been much more like that of *Proconsul* (Berillon, 1999).

Considering the *Australopithecus afarensis* foot, both medial and lateral skeletons indicate that *Australopithecus afarensis* lacked a plantar longitudinal arch in neutral position. Given the principle of optimal coaptation, this does not indicate the lack of plantar vault during the full locomotor cycle. Langdon et al. (1991) argued that the *Australopithecus afarensis* foot presented a human-like mechanism of medial arching, with the association of an inverted subtalar joint during the later phase of the gait and a close-packed transverse joint due to a certain amount of talar head torsion. According to the contact area to the ground, some normal humans appear to have no effective arch in a relaxed state but create one as the subtalar joint inverts during the later part of the stance phase in walking. On the other hand, X-rays analysis of normal flat feet showed that, although the whole plantar face contacts the ground while standing upright, the skeleton remains arched (*e.g.*, Chabeuf et al., 1955; Wangermez et al., 1972). Then, in a normal flat foot, the skeleton is arched. Thus, although the inversion of the subtalar joint and the close-packing of the transverse tarsal joint occur during the gait, in human both mechanisms are not necessary to create an arched skeleton because of the orientation of its joints. In *Australopithecus afarensis*, the joints orientation describes a flat foot in neutral position. Thus it is likely to envisage that, in this species, a plantar arch would need a complementary mechanism to be created such as the plantar flexion of the tarsal joints under the action of muscles.

In addition, the hallucal architectural pattern of *Australopithecus afarensis* is unique in associating: (1) a high dorsal flexion, (2) a slightly abducted hallux of which the cuneometatarsal joint is maintained in a frontal plane, and (3) a cuneometatarsal joint of which the dorsoplantar axis is vertical. It differs from the *Homo* condition by a higher degree of abduction and from those of African apes by a lower abduction, due to the obliquity of the metatarsal proximal joint area. Given the frontal position of the hallucal cuneometatarsal joint, the torsion within the shaft of the first metatarsal would have placed the metatarsophalangeal joint in a very low valgus condition; thus, likely, this joint would have worked in a nearly vertical plane, which is consistent with Latimer and Lovejoy (1990b) conclusions, and consequently, would have not been involved in the amount of hallucal abduction. Yet, given that a certain hallucal mobility would have been allowed, due to the shape of the hallucal cuneometatarsal joint and given that the joint orientation indicates that, in neutral state, the first metatarsal is clearly medialward oriented compare to the cuneometatarsal joint, it seems likely to conclude that the *Australopithecus afarensis* hallux could be variably abducted, from a low to a moderate degree following a pattern which clearly differs from those of *Gorilla*, *Pan*, and *Homo*.

What do these observations imply for the understanding of the Laetoli footprints and the early hominid's way of walking on two legs? The debate concerning the Laetoli trails has been constructed around two linked issues: Who made these footprints? How did they walk? Three positions have been proposed. Two of them conclude that australopithecines (*Australopithecus afarensis*) were the makers of the trails—the first one proposing that australopithecines, especially *Australopithecus afarensis*, were fully bipedal with an adducted hallux and a plantar arch (Clarke, 1979; Leakey et al., 1979; Day et al., 1980; White, 1980; Charteris et al., 1981, 1982; Day, 1985; Tuttle, 1985; Robbins, 1987; White et al., 1987; Day, 1991), the other one identifying some arboreal abilities (Susman et al., 1984; Deloison, 1991, 1992a, b, 1993). The third position argues that *Australopithecus afarensis* is too arboreally adapted, especially in its foot organization, to be the maker of the Laetoli footprints, which they considered as being of a modern bipedal form (Tuttle, 1987; Tuttle et al., 1991, 1992; Tuttle, 1996). What does the present architectural analysis of the *Australopithecus afarensis* foot add to this debate? Although a few authors have proposed, on the basis of postcranial evidences, that two species are represented in the Hadar material (*e.g.*, Senut et al., 1985), we assume, like the majority of researchers that *Australopithecus afarensis* made the Laetoli trails, for two reasons: because of the exclusive presence of *Australopithecus afarensis* remains in the contemporary Laetoli levels and because our angular data do not allow us to demonstrate the contrary. Thus, what can we answer to the question "in what manner did they walk on two legs?" Among the characters commonly observed, the Laetoli footprints present a proximodistal arching. Two explanations were proposed: (1) the presence of a plantar arch, advanced by White, Tuttle and their respective collaborators in particular, (2) a plantar flexion of the pedal transverse joints under the action of the muscles, proposed by Deloison (this volume) and Meldrum (this volume). The present architectural analysis of the *Australopithecus afarensis* foot remains concludes in an absence of a plantar arch in neutral position. Thus, in addition to a possible inversion of the subtalar joint and a certain amount of close-packing of the transverse tarsal joint, a plantar arching would have necessitated a strong and coordinated muscular action to be created. This interpretation is consistent with previous observations, although variably interpreted, concerning the huge areas of insertion of the peroneus longus, the peroneus brevis, the tibialis anterior and the tibialis posterior muscles and their localization (Latimer et al.,

1982; Stern et al., 1983; Gomberg et al., 1984; Susman et al., 1984; Langdon, 1985; Lamy, 1986; Latimer et al., 1989; Latimer et al., 1990a; Deloison, 1993). This implies that the amount of close-packing of the transverse tarsal joint in *Australopithecus afarensis* would have been lower than in modern human, as suggested by the moderated talar head torsion of the early hominids talus. Such functional pattern is distinct from those of human and African apes.

In the literature, the big toe of the Laetoli prints is variably interpreted. It is adducted on the print G-1/37 (Tuttle et al., 1991) or abducted and associated with a strong print of the hallucal abductor muscle on the print G-1/34 (Deloison, 1991, 1992a, 1993). Others interpret this abduction as the trace left by the hallux at the time of the toe-off and should not be consequently considered as a real hallucal abduction (White et al., 1987). The angular analysis of the *Australopithecus afarensis* hallucal tarsometatarsal complex highlights a moderated degree of abduction in neutral position. Such an organization would have allowed both the adduction of the hallux under the action of adductor muscles and a moderate abduction under the action of the hallucal abductor muscle, which is compatible with previous morphofunctional observations (Latimer et al., 1990 and Deloison, 1992, 1993, respectively).

CONCLUSION

Measuring the joint area orientations allows us to quantify certain architectural parameters of the fossil foot. Although foot remains are too rare to allow us to propose a complete architectural pattern, the architectural data corroborate the uniqueness of the *Australopithecus afarensis* foot, in which a unique pattern of first ray and no plantar arch in neutral position are associated. This analysis considers the foot skeleton in neutral position and thus in static conditions. Since muscles could also participate in the organization of the foot, an integrative approach both morphofunctional and architectural must be undertaken to better understand the functioning of the early hominid foot and thus to the knowledge of the first steps to bipedalism in primates. Assuming that *Australopithecus afarensis* is the maker of the Laetoli footprints, we have to admit that the plantar arching described in the footprints as well as the variable degree of hallucal abduction were the results of a muscular action of both intrinsic and extrinsic muscles of the foot. Thus, the *Australopithecus afarensis* foot functioning would have been different from those of later hominids including modern humans in which the arching would not have necessitated a strong additional muscular action to be created. These features corroborate the idea of a unique pattern of bipedality in *Australopithecus afarensis*.

ACKNOWLEDGEMENTS I am very grateful to Pr. H. de Lumley, Pr. A. Langaney, Pr. D. Robineau, Pr. C. Stringer, Dr. W. Van Neer, Dr. P. Jenkins who allowed me to study specimens in their care. I thank Dr. M.A. de Lumley, Dr. D. Grimaud-Hervé, Dr. Y. Deloison, A. Dambricourt-Malassé, Dr. F. Valentin and Dr. S. Condemi for their scientific support in this research. Thanks to J. Langdon and J. Meldrum and anonymous reviewers for reviewing this manuscript and for their helpful comments. This research was supported by a *Bourse de Doctorat pour Ingénieur* (CNRS, France).

REFERENCES

Berge, C., 1994, How did the australopithecines walk? A biomechanical study of the hip and thigh of Australopithecus afarensis. *Journal of Human Evolution* 26: 259–273.

Berillon, G., 1999, Geometric pattern of the hominoid hallucal tarsometatarsal complex. Quantifying the degree of hallux abduction in early hominids. *Comptes Rendus de l'Académie des Sciences, Paris* Série IIa, 328: 627–633.

Berillon, G., 2000, *Le pied des hominoïdes miocènes et des hominidés fossiles. Architecture, locomotion, évolution. Cahiers de Paléoanthropologie*, CNRS Éditions, Paris, 166 p.

Chabeuf, M., and Fontaine, M., 1955, La voûte plantaire des noirs du Soudan. *Bulletins de la Société d'Anthropologie de Paris*, 10ème série, 6: 174–180.

Charteris, J., Walls, J.C., and Nottrodt, J.W., 1981, Functional reconstruction of gait from the Pliocene hominid footprints at Laetoli, Northern Kenya. *Nature* 290: 496–498.

Charteris, J., Walls, J.C., and Nottrodt, J.W., 1982, Pliocene hominid gait: new interpretations based on available footprint data from Laetoli. *American Journal of Physical Anthropology* 58: 133–144.

Clarke, R.J., 1979, Early hominid footprints from Tanzania. *South African Journal of Science* 75: 148–149.

Clarke, R.J., and Tobias P.V., 1995, Sterkfontein Member 2 foot bones of the oldest south african hominid. *Science* 269: 521–524.

Crompton, R.H., Yu, L., Weijie, W., Gunther, M., and Savage, R., 1998, The mechanical effectiveness of erect and "bent-hip, bent-knee" bipedal walking in *Australopithecus afarensis*. *Journal of Human Evolution* 35: 55–74.

Dart, R., 1925, *Australopithecus africanus*, the man-ape of South Africa. *Nature* 115: 195–199.

Day, M.H., 1985, Hominid locomotion: from Taung to Laetoli footprints. In *Hominid evolution: Past, Present and future*, edited by P.V. Tobias, pp. 115–127, Alan R. Liss, New York.

Day, M.H., 1991, Bipedalism and prehistoric footprints. In *Origine(s) de la bipédie chez les Hominidés*, edited by B. Senut and Y. Coppens, pp. 199–213, *Cahiers de Paléoanthropologie*, CNRS Éditions, Paris.

Day, M.H., and Napier, J.R., 1964, Hominid fossils from Bed I, Olduvai Gorge, Tanganyika. Fossil foot bones. *Nature* 201: 969–970.

Day, M.H., and Wood, B.A., 1968, Functional affinities of the Olduvai Hominid 8 talus. *Man* 3: 440–454.

Day, M.H., and Wickens, E.H., 1980, Laetoli Pliocene hominid footprints and bipedalism. *Nature* 286: 385–387.

Deloison, Y., 1984, Etude comparative de calcaneums de primates et les rapports chimpanzé-australopithèque-homme. *Comptes Rendus de l'Académie des Sciences, Paris* Série III, 299, 5: 115–118.

Deloison, Y., 1986, Description d'un calcaneum fossile de primate et sa comparaison avec des calcaneums de Pongidés, d'Australopithèques et d'*Homo*. *Comptes Rendus de l'Académie des Sciences, Paris* Série III, 302, 7: 257–262.

Deloison, Y., 1991, Les Australopithèques marchaient-ils comme nous? In *Origine(s) de la bipédie chez les Hominidés*, edited by B. Senut and Y. Coppens, pp. 177–186, *Cahiers de Paléoanthropologie*, CNRS Éditions, Paris.

Deloison, Y., 1992a, Empreintes de pas de Laetoli (Tanzanie). Leur apport à une meilleure connaissance de la locomotion des Hominidés fossiles. *Comptes Rendus de l'Académie des Sciences, Paris* Série III, 315: 103–109.

Deloison, Y., 1992b, Sur les traces de pas de Laetoli en Tanzanie. In *Cinq millions d'années, l'aventure humaine*, edited by M. Toussaint, pp. 63–72, Liège.

Deloison, Y., 1993, *Etude des restes fossiles des pieds des premiers hominidés*: Australopithecus *et* Homo habilis. *Essai d'interprétation de leur mode de locomotion.* Thèse de Doctorat d'Etat es Sciences, Université René Descartes, Paris, 783 p.

Duncan, A.S., Kapellman, J., and Shapiro, L.J., 1994, Metatarsophalangeal joint function and positional behavior in *Australopithecus afarensis*. *American Journal of Physical Anthropology* 93: 67–81.

Gomberg, D.N., and Latimer, B., 1984, Observations on the transverse tarsal joint of *Australopithecus afarensis*, and some comments on the interpretation of behaviour from morphology. *American Journal of Physical Anthropology* 63: 164.

Johanson, D.C., White, T.D., and Coppens, Y., 1978, A new species of the genus *Australopithecus* (Primates: Hominidae) from the Pliocene of Eastern Africa. *Kirtlandia* 28: 1–14.

Johanson, D.C., Lovejoy, O.C., Kimbel, W.H., White, T.D., Ward, S.C., Bush, M.E., Latimer, B.M., and Coppens, Y., 1982, Morphology of the Pliocene partial hominid skeleton (AL 288–1) from the Hadar formation, Ethiopia. *American Journal of Physical Anthropology* 57: 403–451.

Jungers, W.L., 1982, Lucy's limbs, skeletal allometry and locomotion in *Australopithecus afarensis*. *Nature* 297: 676–678.

Kramer, P.A., 1999, Modelling the locomotor energetics of extinct hominids. *Journal of Experimental Biology* 202: 2807–2818.
Kramer, P.A., and Eck, G.G., 1999, Locomotor energetics and leg length in hominid bipedality. *Journal of Human Evolution* 38: 651–666.
Lamy, P., 1986, The settlement of the longitudinal plantar arch of some African Plio-Pleistocene Hominids: a morphological study. *Journal of Human Evolution* 15: 31–46.
Langdon J.H., 1985, Fossils and the origin of bipedalism. *Journal of Human Evolution*, 14: 615–635.
Langdon, J.H., Bruckner, J., and Baker, H.H., 1991, Pedal mechanics and bipedalism in early hominids. In *Origine(s) de la bipédie chez les Hominidés*, edited by B. Senut and Y. Coppens, pp. 159–167, *Cahiers de Paléoanthropologie*, CNRS Éditions, Paris.
Latimer, B., 1991, Locomotor adaptations in *Australopithecus afarensis*: the issue of arboreality. In *Origine(s) de la bipédie chez les Hominidés*, edited by B. Senut and Y. Coppens, pp. 169–176, *Cahiers de Paléoanthropologie*, CNRS Éditions, Paris.
Latimer, B., and Lovejoy, O., 1989, The calcaneus of *Australopithecus afarensis* and its implications for the evolution of bipedality. *American Journal of Physical Anthropology* 78: 369–386.
Latimer, B., and Lovejoy, O., 1990a, Hallucal tarsometatarsal joint in *Australopithecus afarensis*. *American Journal of Physical Anthropology* 82: 125–133.
Latimer, B., and Lovejoy, O., 1990b, Metatarsophalangeal joints of *Australopithecus afarensis*. *American Journal of Physical Anthropology* 83: 13–23.
Latimer, B., Lovejoy, O., Johanson, D.C., and Coppens, Y., 1982, Hominid tarsal, metatarsal and phalangeal bones recovered from the Hadar formation: 1974–1977 collections. *American Journal of Physical Anthropology* 57: 701–719.
Latimer, B., Ohman, J.C., and Lovejoy, O., 1987, Talocrural joint in African hominoids: implications for *Australopithecus afarensis*. *American Journal of Physical Anthropology* 74: 155–175.
Leakey, M.D., and Hay, R.L., 1979, Pliocene footprints in the Laetoli beds at Laetoli, Northern Tanzania. *Nature* 278: 317–323.
Lewis, O.J., 1980, The joint of the evolving foot. Part III. The fossil evidence. *Journal of Anatomy* 131: 275–298.
Oxnard, C.E., 1979, Convention and controversy in human evolution. *Homo* 30: 225–246.
Oxnard, C.E., and Lisowski, F.P., 1980, Functional articulation of some hominoid foot bones: implications for the Olduvai (Hominid 8) foot. *American Journal of Physical Anthropology* 52: 107–117.
Robbins, L.M., 1987, Hominid footprints from site G. In *Laetoli, a Pliocene site in Northern Tanzania*, edited by M.D. Leakey and J.M. Harris, pp. 497–502, Oxford University Press, Oxford.
Sarmiento, E.E., 1994, *Terrestrial traits in the hands and feet of Gorillas. American Museum Novitates* 3091, 56 p.
Schultz, A.H., 1930, The skeleton of the trunk and limbs of Higher Primates. *Human Biology* 2: 303–409.
Senut, B., 1981, *L'humérus et ses articulations chez les Hominidés Plio-Pléistocènes. Cahiers de Paléoanthropologie*, CNRS Éditions, Paris, 141 p.
Senut, B., and Tardieu, C., 1985, Functional aspects of Plio-Pleistocene hominid limb bones: implications for taxonomy and phylogeny. In *Ancestors: the hard evidence*, edited by E. Delson, pp. 193–201, Alan R. Liss, New York.
Stern, J.T., and Susman, R.L., 1983, The locomotor anatomy of *Australopithecus afarensis*. *American Journal of Physical Anthropology* 60: 279–318.
Susman, R.L., Stern, J.T., and Jungers, W.L., 1984, Arboreality and bipedality in the Hadar hominids. *Folia Primatologia* 43: 113–156.
Susman, R.L., Stern, J.T., and Jungers, W.L., 1985, Locomotor adaptations in the Hadar Hominids. In *Ancestors: the hard evidence*, edited by E. Delson, pp. 184–192, Alan R. Liss, New York.
Susman, R.L., and Stern, J.T., 1991, Locomotor behavior of early hominids: epistemology and fossil evidence. In *Origine(s) de la bipédie chez les Hominidés*, edited by B. Senut and Y. Coppens, pp. 121–131, *Cahiers de Paléoanthropologie*, CNRS Éditions, Paris.
Tardieu, C., 1983, *L'articulation du genou. Analyse morphofonctionnelle chez les Primates et les Hommes fossiles. Cahiers de Paléoanthropologie*, CNRS Éditions, Paris, 108 p.
Tuttle, R.H., 1985, Ape footprints and Laetoli impressions: a response to the suny claims In *Hominid evolution: Past, Present and future*, edited by P.V. Tobias, pp. 129–133, Alan R. Liss, New York.
Tuttle, R.H., 1987, Kinesiological inferences and evolutionary implications from Laetoli bipedal trails G-1, G-2/3, and A. In *Laetoli, a Pliocene site in Northern Tanzania*, edited by M.D. Leakey and J.M. Harris, pp. 503–523, Oxford University Press, Oxford.
Tuttle, R.H., 1996, The Laetoli Hominid G footprints, where do they stand today? *Darmsädter Berträge zur Naturgeschichte* 6: 97–102.

Tuttle, R.H., Webb, D.M., and Tuttle, N.I., 1991, Laetoli footprints and the evolution of hominid bipedalism. In *Origine(s) de la bipédie chez les Hominidés*, edited by B. Senut and Y. Coppens, pp. 187–198, *Cahiers de Paléoanthropologie*, CNRS Éditions, Paris.

Tuttle, R.H., Webb, D.M., Tuttle, N.I., and Baksh, M., 1992, Footprints and gaits of bipedal apes, bears, and barefoot people: Perspective on Pliocene tracks. In *Topics in primatology, Vol. 3, Evolutionary biology, reproductive endocrinology and virology*, edited by S. Matano, R.H. Tuttle, H. Ishida and M. Goodman, pp. 221–242, University of Tokyo Press, Tokyo.

Wangermez, J., and Courouge, F., 1972, Etude radiologique de la platypodie chez les Vili et Lari du Congo. *Bulletin de la Société Anthropologique du Sud-Ouest*, 8(2): 16–23.

White, T.D., 1980, Evolutionary implications of Pliocene hominid footprints. *Science* 208: 175–176.

White, T.D., and Suwa, G., 1987, Hominid footprints at Laetoli: facts and interpretations. *American Journal of Physical Anthropology* 72: 485–514.

Chapter 7

The Behavioral Ecology of Locomotion

PATRICIA ANN KRAMER

Abstract. One of the cornerstones of modern evolutionary biology is the insight that an individual that is more tailored to his environment should have a higher survival and reproductive potential than his less adapted neighbors. Behavioral ecology attempts to explain behaviors as adaptive, or not, only in particular ecological contexts. And life history theory highlights that the developmental stage of an individual is an integral part of the environment and that development is an inherently sequential process. But how do these seminal ideas help us understand the evolution of hominid bipedality?

Bipedality can be thought of as a suite of possible styles, with different forms optimizing diverse traits. The locomotor form of an individual can be optimized for many characteristics, including 1) efficient use of energetic, thermal, temporal and water budgets; 2) range; 3) velocity or acceleration; 4) endurance, fatigue and injury protection; and 5) burden carrying. All could be important in different ecological circumstances and, indeed, many may act in concert with others, but which ones? when? how? and how can we tell from the fossil record?

While female australopithecines appear to have excelled at slow speeds and short distances, the locomotor form of *Homo* seems to be oriented toward relatively long distance travel and endurance walking and running. Understanding these aspects of the locomotor regime does not, however, inform about the other elements of the behavioral repertoire, nor does it elucidate why the transition took place and what the functional implications of it are.

The task of understanding bipedality as a fundamental adaptation of Hominidae can only be a truly fruitful endeavor if we keep in mind that males and females may have different ecological perspectives; children, adolescents, and adults may interact with the environment in different ways; and the environment is a complex set of interacting, dynamic factors.

INTRODUCTION

One of the cornerstones of modern evolutionary biology is the insight that an individual that is more tailored to his environment should have a higher survival and reproductive potential than his less customized neighbors and that these adaptive characteristics should be differentially expressed in subsequent generations (Darwin, 1872). Behavioral (or evolutionary) ecology expands the notion of "survival of the fittest" beyond physical characteristics and attempts to explain behaviors as adaptive—or not—within particular ecological contexts (Smith, 1983; Smith and Winterhalder, 1992). Among the most important behaviors in a primate's life is the ability to move through its environment (Zihlman, 1992) and, for

From Biped to Strider: The Emergence of Modern Human Walking, Running, and Resource Transport, edited by D. Jeffrey Meldrum and Charles E. Hilton. Kluwer Academic/Plenum Publishers, New York, 2004.

Hominidae, bipedality is the locomotive method of choice. Life history theory requires that the developmental stage of an individual be considered an integral part of their environment and highlights that fitness-related tradeoffs shape the developmental process and schedule (Stearns, 1992). Development is an inherently sequential undertaking and the effect of traits beneficial at one maturational stage must be considered for other stages (Stearns, 1992).

While the relevance of these notions is apparent, it is easy to lose sight of them when analyzing obscure anatomical traits. In this report, I will step back from morphology and question: how do these seminal ideas help us understand the role that locomotion has played in shaping the hominid family? I provide two brief examples of mechanical analysis that contribute to understanding the hominid form of bipedality from the perspective of behavioral ecology.

Evolution, Behavioral Ecology, and Life History Theory

Axiomatic to this research is the concept that a creature that minimizes its energy expenditure on locomotion while still accomplishing the tasks necessary for survival and reproduction will leave more descendents than one that "squanders" energy in inefficient locomotion (Foley, 1992). In more general terms, the constraints of resource allocation are the mediators of survival and successful reproduction of an individual. A resource is a limited commodity and energy to survive and procreate is one of the most essential resources (Fisher, 1958) because it is both non-renewable and finite (Borgerhoff-Mulder, 1992). An adaptation that allows a creature to devote less energy to a necessary task will experience tremendous selective advantage because that energy can then be apportioned to another requirement.

All phases of the life cycle and both sexes must be considered in any complete evaluation of the constraints of resource allocation (Stearns, 1992). Adult female primates may be particularly energetically constrained, however, because they must provide for offspring that are highly energetically demanding for long periods of development (Kaplan, 1996). Gestation and lactation demand significant energetic expenditures (Kaplan, 1996) as can infant transport (Kramer, 1998a). Equally as important to the energetic constraints of adult females are those of immature individuals. Juveniles, with their still developing physiological processes and social skills, may also be particularly vulnerable to resource stress due the twin demands of development and maintenance (Bogin, 1999). An explanatory framework that posits an understanding that an "adaptation" may be more or less advantageous to each sex and at various points in an individual's life is an appropriate paradigm from which to view changes in locomotor anatomy and their energetic consequences (Zihlman, 1992) because "[l]ocomotion is fundamental to foraging and feeding and to predator avoidance... [and] to an individual's ability to survive, to mate and to ensure the survival of its offspring (p. 315)."

Bipedality as a Multifaceted Behavior

Although the "the principal cost of any locomotor system is energy (p. 139)" and other behaviors are secondary (Foley, 1992), energetic efficiency[1] can take many forms. Bipedality can be thought of as a suite of possible styles, with different forms optimizing diverse traits (Kramer, 1998b; Kramer, 1999; Kramer and Eck, 2000). The locomotor form of an individual can be optimized for many characteristics, including 1) efficient use of energetic, thermal, temporal and water budgets; 2) range; 3) burden transport; 4) endurance, fatigue and injury protection; and 5) velocity or acceleration. All could be important in different ecological circumstances and act in concert with others.

Optimization for one form of bipedality does not, however, prohibit an individual from participating in other locomotor regimes, though it may do so at increased energetic (Elton et al., 1998) or other cost. Modern humans, despite exhibiting feet and upper bodies "designed" for other tasks, can climb a tree, given the proper inducement. Other fully effective bipeds could also have exploited this ability, and done so without compromising their bipedality. It is seductive, but inaccurate, to dichotomize primates into bipeds or climbers, brachiators or quadrupeds. Certainly, adapting the foot, for instance, to bipedality probably makes it less useful for climbing (Latimer et al., 1987) and increases locomotive effort (Elton et al., 1998), but it doesn't prohibit an individual from climbing. Chimpanzees, bonobos and gorillas are knuckle-walking quadrupeds, yet all exhibit other locomotive tendencies. Indeed, some individuals spend a significant portion of their activity budget on "other" forms of locomotion (Doran, 1993; Hunt, 1992; Remis, 1995). The possible forms of bipedality, which may be optimized for any combination of characteristics, must remain a part of the functional complex of that individual. "Secondary behaviors" can be of tremendous selective importance to the individual. Indeed, no form of bipedality will long survive if it prevents the individual who exhibits it from acquiring food or other resources or having and raising offspring.

Burden transport is an integral part of the behavioral complex of primates. Dependent young are carried until they are capable and efficient at locomoting on their own (Altmann and Samuels, 1992; Kramer, 1998a). For instance, baboon infants are always carried by their mothers during infancy, but the amount of carrying time decreases as the infant ages. Additionally, older offspring are carried more frequently when their mothers are fast moving than when they move slowly (Altmann and Samuels, 1992). This observation lead Altmann and Samuels to the conclusion that energetics underlies the decision to carry: Mothers carry offspring when it is less energetically costly to the mother for her to do so. Unfortunately, they had no way to directly calculate energetic expenditure, but had to rely on general equations developed by Taylor and colleagues (Taylor et al., 1982), which do not control for the inefficiencies associated with juvenility nor allow for the addition of burdens (other than as an increase in total body mass). Similarly, human children in some environmental contexts are conveyed by relatives, usually their mothers, for 4 or more years (Bentley, 1984; personal observation). Constraints other than efficient use of energy undoubtedly have a role in deciding when to carry offspring, but energetic considerations are important (Kramer, 1998a).

Although non-human primates infrequently transport burdens other than offspring, modern humans routinely must carry resources, including food, water, firewood and raw materials (Bentley, 1984; Hilton and Greaves, 2000; Maloiy et al., 1986). These tasks are cross-culturally ubiquitous and have an energetic price. Frequently, women bear the majority of transport tasks (e.g., Bentley, 1984; Maloiy et al., 1986), incurring this energetic drain, even though female reproduction is more strongly linked to energy supply than male reproduction (Ellison, 2001). Theories of human evolution that advance complex tool manufacture, food sharing, and central-place foraging as defining characteristics of *Homo* (e.g., Dahlberg (1981), Isaac (1978), Lee and DeVore (1968), Potts (1988), and van Schaik et al. (1999)), all assume the transport of burdens other than offspring. The locomotor anatomy of *Homo* should, therefore, have experienced selective pressure for efficiency at burden transport.

Bipedality among Hominids

Two schools of thought have traditionally been put forth to describe the bipedality of early hominids (Duncan et al., 1994; McHenry, 1983): that australopithecines were transitional forms, using a form of bipedality that was compromised by a continued reliance on arboreal activities (Jungers, 1982; Robinson, 1972; Stern and Susman, 1983; Zihlman, 1978; Zihlman and Brunker, 1979), or that australopithecines exhibited a form of bipedality identical to that of modern humans (Latimer and Lovejoy, 1989; Latimer et al., 1987; Lovejoy, 1988; Lovejoy et al., 1973; McHenry, 1975; White, 1980). Evidence for both schools has been found in the same fossils, especially in AL 288–1 (Johanson et al., 1982), and sometimes within the same character (e.g., the pelvis AL 288–1ap). Rak (1991) proposed a third scenario: australopithecines were fully, habitually bipedal, but in a manner unique to their morphology.

From an energetic perspective, I have proposed that female australopithecines, with their short legs and wide pelves, appear to have excelled at slow speeds and short distances (Kramer, 1998b; Kramer, 1999; Kramer and Eck, 2000), while the locomotor form of *Homo*[2] seems to be oriented toward relatively long distance travel (Isbell et al., 1998) and endurance walking and running. This generalization for australopithecines is based on the morphology of the pelvis and leg of AL 288–1, which is similar to other known australopithecine specimens (e.g., STS 14 (Abitbol, 1995) and KNM-ER 1500 (Grausz et al., 1988)). The configurations representing AL 288–1 were found to be more efficient than that of a composite modern woman (Kramer, 1998b; Kramer, 1999; Kramer and Eck, 2000). This energetic advantage, however, came at the price of decreased maximum velocity and, possibly, decreased daily range compared to *Homo*. Energetic efficiency is optimized by the australopithecine configurations while range and/or velocity appears less important.

Others have proposed that the relatively long legs of *Homo* are an adaptation to allow for increased maximal walking velocity and increased daily range (Isbell et al., 1998). Long legs are also advantageous for increasing the dispersion of heat (Wheeler, 1991; Wheeler, 1993), which can build up to unacceptably high levels in fast velocities. The configuration of *Homo*, then, may optimize range, velocity and heat dispersion at the cost of energetic efficiency.

This relatively simple picture of the forms of bipedality exhibited in Hominidae—australopithecines adapted to slow speeds and short distances, *Homo* adapted to higher speeds and longer distances—may well become more complex as additional fossil skeletons emerge. For instance, McHenry and Berger (1998) have indicated that a south African form exhibited characteristics that suggest more reliance on arboreality than AL 288–1. Some fossils attributed broadly to *Homo habilis* may also indicate this morphology (e.g., OH 62 and KNM ER 3735 (McHenry and Berger, 1998)). This possible plethora of morphological forms, presumably adapted to various ecological niches, serves to reinforce Rak's initial proposal of a kind of bipedality unique to a particular morphology.

While the future may bring other fossil skeletons to light, today AL 288–1 remains the only australopithecine with an associated pelvis and leg, the minimum necessary ingredients to perform a mechanical analysis to evaluate energetic expenditure. Although other characteristics may well be important, the emphasis herein is on energetic constraints.

Consequently, this report expands on the existing analysis (Kramer, 1998b; Kramer, 1999; Kramer and Eck, 2000) by exploring two additional issues (burden transport and the difference between male and female modern humans).

METHODS AND MATERIALS

A detailed description of the modeling techniques used herein is given elsewhere (Kramer, 1998b; Kramer, 1999; Kramer and Eck, 2000), but a brief explanation follows. Hominid bipedality has been idealized as an inverted pendulum with energy exchanged between potential and kinetic forms (Cavagna et al., 1977); this transferal is not perfect, however, and some energy must be added during each step. A simple inverted pendulum does not capture many of the important features of bipedal walking (like knee flexion), so a more complex idealization was created. This model is symmetrical, with rigid links representing the body segments that compose the locomotor anatomy: right and left foot, calf, thigh and pelvis. A lumped mass at the centerline of pelvis represents the head, arms and trunk. This idealization allows a standard equation of dynamics (equation 1) (Meriam, 1978) to be used to calculate the energy state of each segment at various temporal points.

$$E = 0.5mv^2 + 0.5Im\omega^2 + \text{mgh}$$

where E is the instantaneous energy of a segment (*J*), m mass (kg), v translational velocity (m/s), I mass moment of inertia (kgm^2), ω angular velocity (radians/s) and h distance from the centroid of a segment to a reference point, typically the ground (m). The additional energy needed to move the body and legs through a stride (E_{sum}, J) is the difference in energy between successive temporal increments summed over all increments. Mass-specific mechanical power (energy used per unit time per unit body mass, $E^{\text{Y}}_{\text{sum}}$/m, W/kg) and cost of transport (energy used per unit distance per unit body mass, CoT/m, W/kgm) are determined from $\dot{E}_{\text{sum}}$.

Equation 1 requires that two types of input be available for each link: movement profiles and segmental parameters. Movement profiles are the angular excursions that each body segment experiences and are developed from Inman et al. (1981) and Winter (1987). Three equivalent velocities are used which represent a slow, normal and fast walk. Angular velocity is calculated directly from angular excursions while translational velocity is determined from the excursions and the length of each segment. The movement profiles used herein are identical to those used previously (Kramer, 1998b; Kramer, 1999; Kramer and Eck, 2000).

To calculate the unburdened segmental parameters, I utilize four configurations, three from previous work (modern human female (mh), AL 288–1 with human-like mass distribution (aah), and AL 288–1 with chimpanzee-like mass distribution (aac)) and a new one (modern human male (mhm)). The segmental data needed to solve equation (1) includes segmental mass, length and mass moment of inertia, which are listed in Table 1. Segmental length for the modern humans is available from the literature (Cavagna et al., 1977; NASA, 1978; Tague, 1992), while that of the australopithecine is determined from long bone lengths. Segmental mass for the modern humans is also available from the literature (NASA, 1978; Zihlman, 1984) and that for the australopithecines is estimated from data for humans and for bonobos and common chimpanzees (Zihlman, 1984; Zihlman and

Table 1. Segmental parameters of foot, calf and thigh. Data for modern human female and AL 288–1 from Kramer and Eck (2000).

	Modern human female			AL 288–1 (human-like)			AL 288–1 (chimpanzee-like)			Modern human male		
	Foot	Calf	Thigh	Foot	Calf	Thigh	Foot	Calf	Thigh	Foot	Calf	Thigh
Length	.110	.390	.373	.080	.265	.252	.080	.265	.252	.121	.429	.41
Mass	.84	3.00	6.00	0.46	1.65	3.30	0.56	0.89	2.24	0.91	2.88	7.11
Relative mass	1.4	5.0	10.0	1.4	5.0	10.0	1.7	2.7	6.8	1.3	4.12	10.15
I	.003	.035	.110	.003	.019	.053	.003	.019	.067	.003	.037	.128

Cramer, 1978). Segmental mass moments of inertia are calculated from whole body and segmental values using available equations (Crompton et al., 1996; NASA, 1978). Segmental properties for the modern woman and the two australopithecine configurations are identical to those used in previous analyses (Kramer, 1998b; Kramer, 1999; Kramer and Eck, 2000), while that for the modern man is new.

The unburdened configurations are modified in the current analysis by the addition of a burden to the back of the head-arms-trunk (HAT) segment. Each burden is idealized as a vertical cylinder located symmetrically about the sagittal plane with a density of 1000 kg/m^3 (approximately that of the body segments (1060 kg/m^3)) and with appropriate length and diameter combinations (Table 2). Mass moments of inertia of the burdened HATs (Table 2) are calculated using the parallel axis theorem, which allows the inertia of the burden to be added to that of the unburdened HAT (Meriam, 1978). The segmental parameters for the foot, calf and thigh do not change.

Table 2. Mass moments of inertia of the HAT. Unburdened values for mh, aah, and aac from Kramer and Eck (2000).

	Burden					
Configuration	Mass (kg)	Length (m)	Diameter (m)	Ixx (kgm^2)	Iyy (kgm^2)	Ixy (kgm^2)
mh	0.0			8.90	0.43	0.95
	2.5	0.25	0.031	8.06991	0.98235	0.39903
	10	0.5	0.125	8.71982	1.29167	0.65568
	20	0.5	0.25	9.69705	1.70736	0.99788
aah	0.0			2.34	0.18	0.43
	2.5	0.25	0.031	4.00748	0.42825	0.16438
	10	0.5	0.125	4.56331	0.66989	0.30481
	20	0.5	0.25	5.4151	0.99534	0.49206
aac	0.0			2.71	0.21	0.49
	2.5	0.25	0.031	4.0822	0.51129	0.21079
	10	0.5	0.125	4.54605	0.76123	0.35232
	20	0.5	0.25	5.2752	1.09774	0.54103
mhm	0.0			12.00	1.35	0.39
	2.5	0.25	0.031	12.2572	1.39648	0.48243
	10	0.5	0.125	13.0395	1.7058	0.76474
	20	0.5	0.25	14.1934	2.12149	1.14116

Each configuration is analyzed in the unburdened state and "carrying" burdens of 2.5, 10 20 kg and, for the larger modern configurations, 30 and 40 kg, walking at slow, normal and fast velocities. Increase over the unburdened state is calculated as the difference between the value of the burdened and unburdened (baseline) states divided by the baseline value. A comparison between the modern configurations is also made. As a check to the validity of the model, mass-specific power and cost of transport are calculated for the modern configurations using a standard empirically-derived equation (Pandolf et al., 1977) and compared to the values derived from the model.

RESULTS

As the mass of the burden or the velocity increases, so too does mass-specific mechanical power (Figure 1) and cost of transport (Figure 2). This expected phenomenon is seen in all configurations. The energetic expenditures of the two australopithecine configurations are similar in magnitude and less than that of either modern configuration. The modern man requires more mechanical power than the modern woman to move at slow and natural velocities, but, interestingly, uses less at fast walking speeds. Cost of transport for the australopithecine configurations is less than the modern humans at lower burdens, but rapidly becomes greater than modern humans as the mass of the burden increases.

When the absolute numbers are converted to increases over the baseline (unburdened) state (Figure 3), two trends are apparent. First, as velocity increases, the relative increase over baseline decreases in all cases. This occurs because the portion of the expended energy required to move the burden becomes smaller relative to the energy required to move the body. Second, the australopithecine configurations require a relatively larger increase over their unburdened state than does that of the modern humans.

The increases in baseline mechanical power and cost of transport calculated from empirically-based equations (Pandolf et al., 1977) are presented in Table 4. The model

Table 3. Mass-specific mechanical power ($\dot{E}^{Y}_{sum}$/mass) and Cost of transport (CoT/mass).

		$\dot{E}^{Y}_{sum}$/mass (J/kgs or W/kg)				CoT/mass (J/kgm)			
Velocity	Load (kg)	mh	aah	aac	mhm	mh	aah	aac	mhm
Slow	0.00	1.08	0.86	0.84	1.18	1.08	1.05	1.02	1.12
	2.50	1.13	0.92	0.90	1.22	1.13	1.12	1.09	1.16
	10.00	1.25	1.09	1.07	1.33	1.25	1.33	1.31	1.27
	20.00	1.41	1.33	1.31	1.49	1.41	1.62	1.60	1.42
Natural	0.00	1.88	1.48	1.38	2.06	1.38	1.32	1.23	1.44
	2.50	1.94	1.56	1.47	2.12	1.43	1.40	1.31	1.48
	10.00	2.12	1.82	1.72	2.29	1.56	1.62	1.54	1.60
	20.00	2.36	2.16	2.06	2.53	1.73	1.92	1.84	1.77
Fast	0.00	4.94	4.03	3.67	4.90	2.47	2.44	2.22	2.33
	2.50	5.09	4.24	3.88	5.02	2.54	2.57	2.35	2.39
	10.00	5.52	4.88	4.52	5.40	2.76	2.96	2.74	2.57
	20.00	6.09	5.72	5.37	5.91	3.05	3.47	3.25	2.81

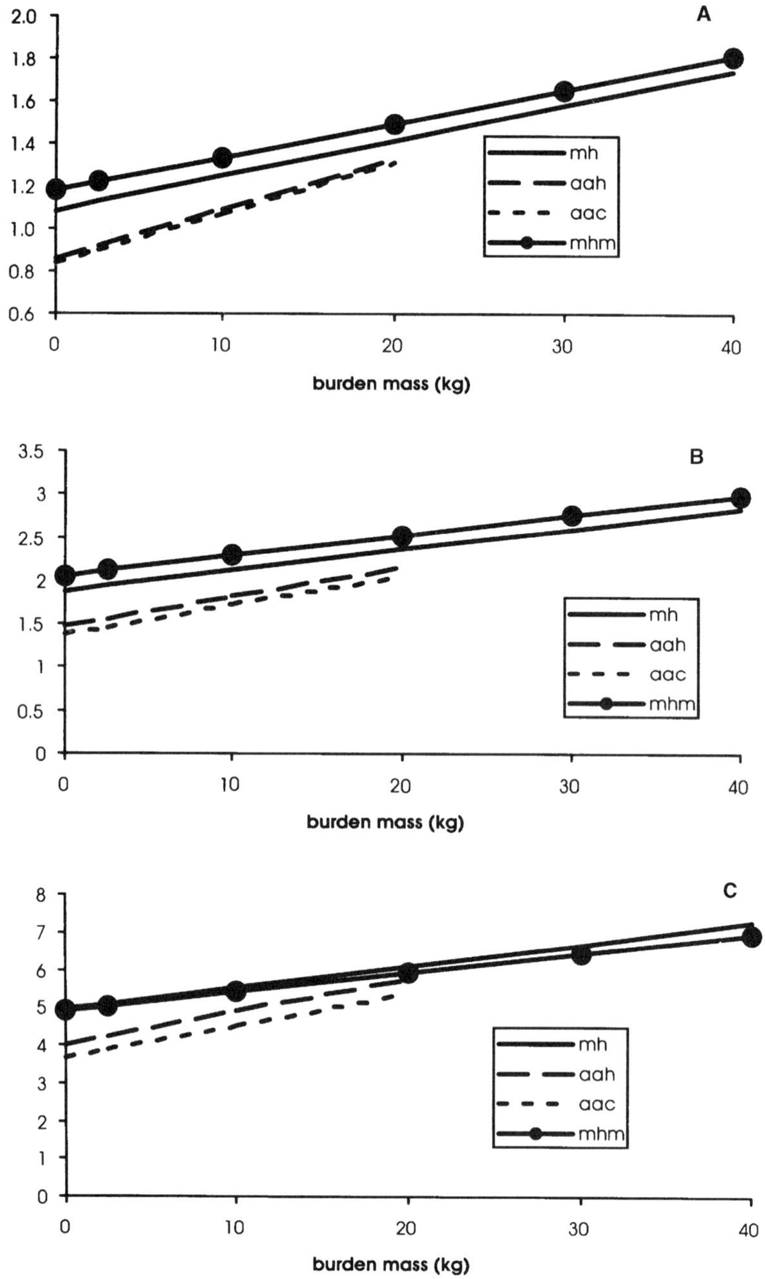

Figure 1. Mass-specific mechanical power. a) slow; b) natural; c) fast

presented herein and the empirical model yield results that indicate a similar trend, although the empirical model tends to predict higher increase than the theoretical one, especially with higher burdens. Also, the difference between women and men is not apparent in the empirical model because it is based solely on body mass and sex-based differences in body configuration are ignored.

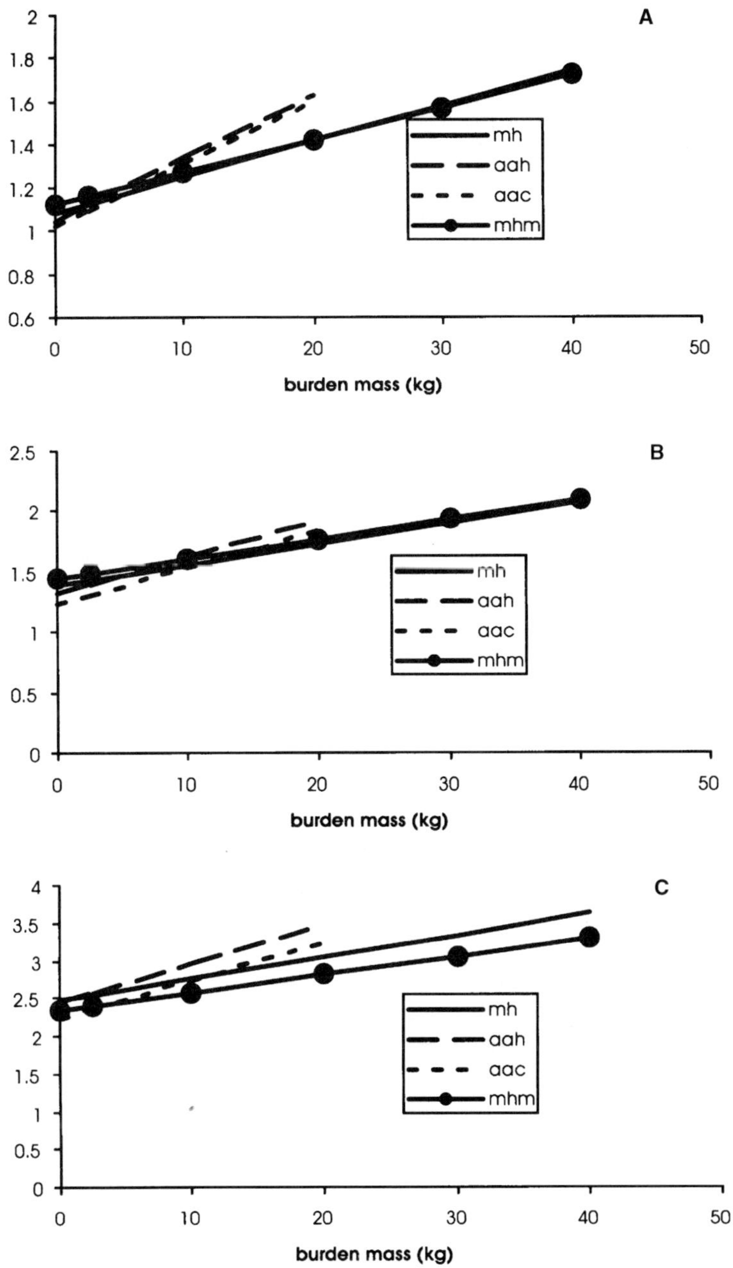

Figure 2. Cost of transport (J/kgm). a) slow; b) natural; c) fast

DISCUSSION

The transport of burdens—whether those burdens include an antelope, a tree branch or an infant—is a seemingly ubiquitous primate behavior. For modern humans, burden transport is not a trivial undertaking. Modern foragers routinely carry 40–50% of their total

Table 4. Increase over baseline, unburdened state calculated using an empirical method (Pandolf et al., 1977).

Velocity	Load (kg)	mh	aah	aac	mhm
Slow	2.50	0.04	0.08	0.08	0.04
	10.00	0.17	0.35	0.35	0.15
	20.00	0.35	0.64	0.64	0.29
Natural	2.50	0.04	0.08	0.08	0.04
	10.00	0.16	0.30	0.30	0.14
	20.00	0.31	0.55	0.55	0.26
Fast	2.50	0.04	0.07	0.07	0.03
	10.00	0.15	0.27	0.27	0.13
	20.00	0.28	0.47	0.47	0.24

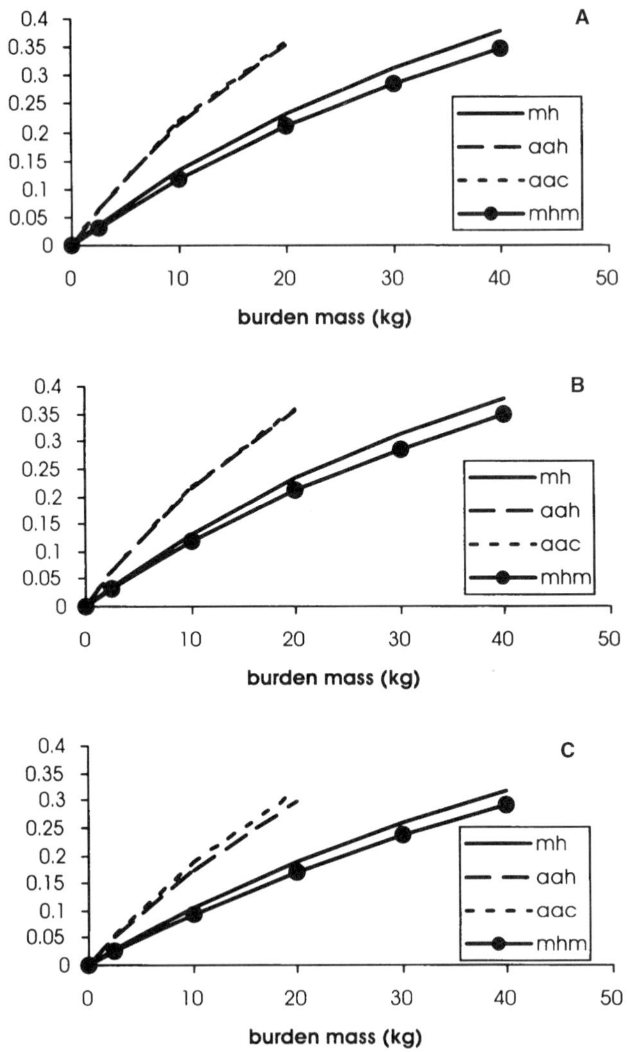

Figure 3. Increase in mass-specific mechanical power or cost of transport over baseline (J/kgm). a) slow; b) natural; c) fast

Table 5. Increase over baseline, unburdened state.

Velocity	Load (kg)	mh	aah	aac	mhm
Slow	2.50	0.04	0.06	0.07	0.03
	10.00	0.13	0.21	0.22	0.12
	20.00	0.23	0.35	0.36	0.21
Natural	2.50	0.03	0.05	0.06	0.03
	10.00	0.11	0.19	0.20	0.10
	20.00	0.20	0.31	0.33	0.19
Fast	2.50	0.03	0.05	0.05	0.03
	10.00	0.10	0.17	0.19	0.09
	20.00	0.19	0.30	0.32	0.17

body mass (Hilton and Greaves, 2000) and occasionally carry up to 70% (Lee, 1979 cited in Bentley, 1984; Maloiy et al., 1986). Firewood, water, housing materials, food, children—all must be carried, sometimes for substantial distances. Empirical evaluation of humans indicates that the efficiency of carrying a particular burden is related to the individual's total body mass. Larger individuals can carry heavy burdens more efficiently than smaller individuals (Kramer, 1998a; Pandolf et al., 1977), although women frequently carry heavier burdens than males (Hilton and Greaves, 2000).

Interestingly, the general trend toward increasing energy expenditure with increasing burden mass was not seen in a study of East African women who habitually carried very heavy loads (70% of total body mass) (Maloiy et al., 1986). These women were able to carry up to 20% of their body mass without an increase in the rate of energy consumption. The mechanism by which this phenomenon could occur remained obscured until recently when it was revealed that these African women change their gait in subtle ways that allow them to carry loads without increasing their rate of energy use (Pohl, 2002). The details of this gait transformation are not yet available, but an interesting extension of the analysis presented herein would be to use these techniques to model the different gait characteristics of this group of African women. The applicability of this variation in gait to other populations of modern humans or to other species is currently unexplored. Nonetheless, it is clear that most humans do not routinely use this accommodation and do increase their energetic expenditure when they carry burdens.

The australopithecine configurations are effective at bipedal walking (Kramer, 1998b; Kramer, 1999; Kramer and Eck, 2000) and require less mass-specific mechanical power to move their bodies in all velocity-burden configurations. They also have a lower cost of transport. While the short legs and a wide pelvis do not hinder the ability of AL 288–1 to walk bipedally, her small total body does inhibit her ability to carry heavy burdens efficiently. Although energetic data do not exist to evaluate burden carrying in species other than modern humans, the modeling detailed herein suggests that the increases over the unburdened condition for the australopithecine configurations are higher than for the modern ones. If increase over a baseline is taken as a method to measure "efficiency," then the australopithecine configurations appear to be less efficient at burden transport than the modern ones. From this, it is reasonable to suggest that australopithecines inhabited a slow speed environment that did not require more than minimal burden transport. The configuration of *Homo* appears to be fashioned for the conveyance of heavy burdens at faster velocities.

The comparison of mass-specific mechanical power between the configurations of the modern man and woman is particularly intriguing in that it indicates a cross-over point: the male configuration used more mechanical power than the female at low velocities, but slightly less power at fast walking speeds. If the same comparison is made for cost of transport, the difference at high velocity is more pronounced. Modern males are, on average, 10% larger than females of similar environmental background and ethnicities in such gross measures as stature, body mass and leg length (NASA, 1978), a basic size difference incorporated into the model configurations. Sexual dimorphism is more pronounced in other measures such as body mass distribution, where males carry a higher percentage of their total body mass in their arms and torso than women, and pelvic width where women are substantially wider (NASA, 1978). Despite these recognized differences, no empirical study has shown a significant difference in energetic efficiency between the two sexes, possibly because the difference between the sexes is small in comparison to within sex variance. Although the distribution of mass among the body segments varies between the male and female modern configuration, the principle difference between the two configurations is in the length of the link representing the pelvis: the woman has a 7% wider pelvis than the man.

I suggest that this wider pelvis allows modern woman to walk more effectively at slower speeds than the more narrow-hipped man, but hinders their effectiveness at the fastest walking velocities. The effect of the configuration differences on running effectiveness remains to be determined. Of course, the shape of the female locomotor anatomy is influenced by at least one other primary constraint—that of an adequately sized birth canal (Rosenberg, 2000). Selective forces acting for a large birth canal in modern woman are at odds with those selecting for effective fast walking ability. This comparison points to a scenario where females walk at slower speeds while males walk faster.

Unfortunately, it is impossible to make a similar comparison between adult male and female australopithecines because intact male pelves and lower limbs remain elusive. It is clear from both east and South African fossil assemblages (Reed et al., 1993), however, that considerable size dimorphism exists, but the character of that dimorphism remains uncertain.

Although not analytically addressed herein, accommodation to immaturity is an important aspect of locomotor adaptation. Human children require more power to move a kg of body mass than do adults (Waters et al., 1983; Waters et al., 1988), but the causes for this inefficiency are not completely understood. The younger the child, the more costly is locomotion (Waters et al., 1983). In many instances, it is more efficient for a mother to carry her dependent youngster rather than to require it to walk independently (Kramer, 1998a). The transport of dependent youngsters is not limited to humans, of course, but little empirical evidence exists to detail the locomotor efficiency of non-human primate juveniles. Nonetheless, the decision to carry dependent young by non-human primate mothers has an energetic component (Altmann and Samuels, 1992). Immature australopithecines were also almost certainly carried, but understanding the details of the energetic calculation remains a topic for future work.

From these basic components, we can begin to build a picture of an "australopithecine adaptation." Female australopithecines appear to be adapted to a form of relatively slow speed foraging that required little burden carrying beyond infant transport. They probably had relatively limited daily ranges and did not need to disperse large amounts of excess heat to the environment. The evidence does not yet exist to firmly establish whether males and females exploited different ecological niches or locomotor profiles and the locomotor potential of juveniles is equally unknown. It is clear, however, that they employed an energetically efficient form of bipedality, suggesting that they utilized an environment that was different from that of modern humans.

The osteological morphology of *Homo* presents a different picture. The locomotor anatomy of women appears to be optimized for longer distance and/or higher speed travel, carrying heavier burdens, than their australopithecine cousins. Men appear to be more adept at fast walking than women, suggesting that the sexes exploit different ecological niches or utilized different foraging strategies, scenarios common in modern hunter-gatherers. Both sexes can disperse heat effectively (Wheeler, 1991; Wheeler, 1993). Children are inefficient and must frequently be carried or left at central places. The "human adaptation," then, appears to revolve around longer distance/higher velocity travel with burdens and with sex-based differential resource utilization.

CONCLUSION

The configuration of female australopithecines, as indicated by AL 288–1, is less efficient at burden transport than that of modern humans. Men are less efficient than women at slow walking velocities, but more efficient at higher velocities. The locomotor anatomy of female australopithecines appears to be adapted to a form of relatively slow speed foraging that required little burden carrying beyond infant transport and resulted in relatively limited daily ranges. That of *Homo* appears to revolve around longer distance/higher velocity travel with burdens and with sex-based differential resource utilization.

ENDNOTES

1 "Efficiency" and its cognates are used in this report in their common definition—productive without waste—and not in their formal mechanical meaning—power produced by a system divided by power required by it.

2 By "locomotor form of *Homo*" I mean relatively longer legs and narrower pelves than are present in australopithecines. This characterization may or may not be warranted for some pre-*erectus* forms.

REFERENCES

Abitbol, M. M., 1995, Reconstruction of the STS 14 (*Australopithecus africanus*) Pelvis. *American Journal of Physical Anthropology* 96:143–58.

Altmann, J., and Samuels, A., 1992, Costs of Maternal Care: Infant-Carrying in Baboons. *Behavioral Ecology and Sociobiology* 29:391–398.

Bentley, G., 1984, Hunter-Gatherer Energetics and Fertility: A Reassessment of the !Kung San. *Human Ecology* 13:79–109.

Bogin, B., 1999, *Patterns in Human Growth.* Cambridge University Press, Cambridge.

Borgerhoff-Mulder, A., 1992, Reproductive Decisions. In *Evolutionary Ecology and Human Behavior*, edited by E. A. Smith and B. Winterhalder, pp. 339–374. Aldine De Gruyter, New York.

Cavagna, G., Heglund, N., and Taylor, C., 1977, Mechanical Work in Terrestrial Locomotion, Two Basic Mechanisms for Minimizing Energy Expenditure. *American Journal of Physiology* 233:R243–R261.

Crompton, C., Yu, L., Alexander, R., Wang, W., and Gunther, M., 1996, Segmental Inertial Properties of Primates: New Techniques for Laboratory and Field Studies of Locomotion. *American Journal of Physical Anthropology* 99:547–570.

Dahlberg, F., 1981, *Woman the Gatherer.* Yale University Press, New Haven.

Darwin, C., 1872, *On the Origin of Species by Means of Natural Selection or the Preservation of Favored Races in the Struggle for Life.* Random House, New York.

Doran, D. M., 1993, Comparative Locomotor Behavior of Chimpanzees and Bonobos: the Influence of Morphology on Locomotion. *American Journal of Physical Anthropology* 91:83–98.
Duncan, A., Kappelman, J., and Shapiro, L., 1994, Metatarsalphalangeal Joint Function and Positional Behavior in *Australopithecus afarensis*. *American Journal of Physical Anthropology* 93:67–81.
Ellison, P., 2001, *On Fertile Ground*. Harvard University Press, Cambridge.
Elton, S., Foley, R., and Ulijaszek, S. J., 1998, Habitual Energy Expenditure of Human Climbing and Clambering. *Annals of Human Biology* 25:523–31.
Fisher, R. A., 1958, *The Genetical Theory of Natural Selection*. Dover Press, New York.
Foley, R. A., 1992, Evolutionary Ecology of Fossil Hominids. In *Evolutionary Ecology and Human Behavior*, edited by E. A. Smith and B. Winterhalder, pp. 131–164. Aldine De Gruyter, New York.
Grausz, H., Leakey, R., Walker, A., and Ward, C., 1988, Associated Cranial and Postcranial Bones of *Australopithecus boisei*. In *Evolutionary History of the "Robust" Australopithecines*, edited by F. Grine, pp. 127–132. Aldine de Gruyter, New York.
Hilton, C. E., and Greaves, R. D., 2000, Age, Sex, and Resource Transport in Venezuelan Foragers (Abstract). *American Journal of Physical Anthropology*, Supplement 30:179.
Hunt, K. D., 1992, Positional Behavior of *Pan troglodytes* in the Mahale Mountains and Gombe Stream National Parks, Tanzania. *American Journal of Physical Anthropology* 87:83–105.
Inman, V., Ralston, H., and Todd, F., 1981, *Human Walking*. Williams & Wilkins, Baltimore.
Isaac, G., 1978, The Food-Sharing Behavior of Protohuman Hominids. *Scientific American* 238:90–108.
Isbell, L. A., Pruetz, J. D., Lewis, M., and Young, T. P., 1998, Locomotor Activity Differences Between Sympatric Patas Monkeys (*Erythroecebus patas*) and Vervet Monkeys (*Cercopithecus* a *ethiiops*): Implications for the Evolution of Long Hindlimb Length in *Homo*. *American Journal of Physical Anthropology* 105:199–207.
Johanson, D., Taieb, M., and Coppens, Y., 1982, Pliocene Hominids from the Hadar Formation, Ethiopia (1973–1977): Stratigraphic, Chronological, and Paleoenvironmental Contexts, with Notes on Hominid Morphology and Systematics. *American Journal of Physical Anthropology* 57:373–402.
Jungers, W., 1982, Lucy's Limbs: Skeletal Allometry and Locomotion in *Australopithecus afarensis*. *Nature* 297:676–678.
Kaplan, H., 1996, A theory of Fertility and Parental Investment in Traditional and Modern Societies. *American Journal of Physical Anthropology* 39:91–135.
Kramer, P. A., 1998, The Costs of Human Locomotion: Maternal Investment in Child Transport. *American Journal of Physical Anthropology* 107:71–85.
Kramer, P. A., 1998, *Locomotor Energetics and Leg Length in Hominid Evolution*. Ph.D. dissertation, University of Washington. University Microfilms, Ann Arbor.
Kramer, P. A., 1999, Modeling the Locomotor Energetics of Extinct Homnids. *Journal of Experimental Biology* 202:2807–2818.
Kramer, P. A., and Eck, G. G., 2000, Locomotor Energetics and Leg Length in Hominid Evolution. *Journal of Human Evolution* 38:651–666.
Latimer, B., and Lovejoy, C., 1989, The Calcaneous of *Australopithecus afarensis* and Its Implications for the Evolution of Bipedality. *American Journal of Physical Anthropology* 78:369–386.
Latimer, B., Ohman, J., and Lovejoy, C., 1987, Talocrural Joint in African Hominoids: Implications for *Australopithecus afarensis*. *American Journal of Physical Anthropology* 74:155–75.
Lee, R., 1979, *The !Kung San: Men, Women and Work in a Foraging Society*. Cambridge University Press, Cambridge.
Lee, R., and DeVore, I., 1968, *Man the Hunter*. Aldine, Chicago.
Lovejoy, C., 1988, Evolution of Human Walking. *Scientific American* 259:118–135.
Lovejoy, C., Heiple, K., and Burnstein, A., 1973, The Gait of *Australopithecus*. *American Journal of Physical Anthropology* 38:757–780.
Maloiy, G. M., Heglund, N. C., Prager, L. M., Cavagna, G. A., and Taylor, C. R., 1986, Energetic Cost of Carrying Loads: Have African Women Discovered an Economic Way? *Nature* 319:668–9.
McHenry, H., 1975, The Ischium and Hip Extensor Mechanism in Human Evolution. *American Journal of Physical Anthropology* 43:39–46.
McHenry, H., 1983, The Capitate of *Australopithecus afarensis* and *A. africanus*. *American Journal of Physical Anthropology* 62:187–198.
McHenry, H. M., and Berger, L. R., 1998, Body Proportions of *Australopithecus afarensis* and *A. africanus* and the Origin of the Genus *Homo*. *Journal of Human Evolution* 35:1–22.
Meriam, J., 1978, *Dynamics*. John Wiley and Sons, New York.
NASA, 1978, *Anthropometric Source Book*. NASA, Houston.

Pandolf, K., Givoni, B., and Goldman, R., 1977, Predicting Energy Expenditure with Loads While Standing or Walking Very Slowly. *Journal of Applied Physiology* 43:577–581.

Pohl, O., 2002, Improving the Way Humans Walk the Walk, *Times*. New York, p. 3.

Potts, R., 1988, *Early Hominid Activities at Olduvai*. Aldine de Gruyter, New York.

Rak, Y., 1991, Lucy's Pelvic Anatomy: Its Role in Bipedal Gait. *Journal of Human Evolution* 20:283–290.

Reed, K. E., Kitching, J. W., Grine, F. E., Jungers, W. L., and Sokoloff, L., 1993, Proximal Femur of *Australopithecus africanus* from Member 4, Makapansgat, South Africa. *American Journal of Physical Anthropology* 92:1–15.

Remis, M., 1995, Effects of Body Size and Social Context on the Arboreal Activities of Lowland Gorillas in the Central African Republic. *American Journal of Physical Anthropology* 97:413–33.

Robinson, J., 1972, *Early Hominid Posture and Locomotion*. University of Chicago Press, Chicago.

Rosenberg, K., 2000, *Walking, Having Babies, and Keeping Warm: Multiple Selective Pressures On Human Pelvic Morphology. American Journal of Physical Anthropology*, Supplement 30:264.

Smith, E. A., 1983, Anthropological Applications of Optimal Foraging Theory: A Critical Review. *Current Anthropology* 24:625–651.

Smith, E. A., and Winterhalder, B., 1992, Natural Selection and Decision Making: Some Fundamental Principles. In *Evolutionary Ecology and Human Behavior*, edited by E. A. Smith and B. Winterhalder, pp. 25–60. Aldine de Gruyter, New York.

Stearns, S. C., 1992, *The Evolution of Life Histories*. Oxford University Press, New York.

Stern, J., and Susman, R., 1983, The Locomotor Anatomy of *Australopithecus afarensis*. *American Journal of Physical Anthropology* 60:279–317.

Tague, P., 1992, Sexual Dimorphism in the Human Bony Pelvis, with a Consideration of the Neandertal Pelvis from Kebara. *American Journal of Physical Anthropology* 88:1–21.

Taylor, C., Heglund, N., and Maloiy, G., 1982, Energetics and Mechanics of Terrestrial Locomotion. I. Metabolic Energy Consumption as a Function of Speed and Body Size in Birds and Mammals. *Journal of Experimental Biology* 97:1–21.

van Schaik, C., Deaner, R., and Merrill, M., 1999, The Condition for Tool Use in Primates: Implications for the Evolution of Material Culture. *Journal of Human Evolution* 36:719–41.

Waters, R. L., Hislop, H. J., Thomas, L., and Campbell, J., 1983, Energy Cost of Walking in Normal Children and Teenagers. *Dev. Med. Child Nerol.* 25:184–188.

Waters, R. L., Lunsford, B. R., Perry, J., and Byrd, R., 1988, Energy-Speed Relationship of Walking: Standard Tables. *Journal of Orthopaedic Research* 6:215–222.

Wheeler, P. E., 1991, The thermoregulatory Advantages of Hominid Bipedalism in Open Equatorial Environments: The Contribution of Increased Convective Heat Loss and Cutaneous Evaporative Cooling. *Journal of Human Evolution* 21:107–115.

Wheeler, P. E., 1993, The Influence of Stature and Body Form on Hominid Energy and Water Budgets: A Comparison of *Australopithecus* and *Homo* Physiques. *Journal of Human Evolution* 24:13–28.

White, T., 1980, Additional Hominid Fossils from Laetoli Tanzania. *American Journal of Physical Anthropology* 53:487–504.

Winter, D., 1987, *The Biomechanics and Motor Control of Human Gait*. University of Waterloo Press, Ontario.

Zihlman, A., 1978, Interpretations of Early Hominid Locomotion. In *Early Hominids of Africa*, edited by C. J. Jolly, pp. 361–377 Duckworth, London.

Zihlman, A., and Brunker, L., 1979, Hominid Bipedalism. *Yearbook of Physical Anthropology* 22:132–162.

Zihlman, A. L., 1984, Body Mass and Tissue Composition in *Pan paniscus* and *Pan troglodytes*, With Comparisons to Other Hominoids. In *The Pygmy Chimpanzee Evolutionary Biology and Behavior*, edited by R. Susman, pp. 179–200 Plenum, New York.

Zihlman, A. L., 1992, Locomotion as a Life History Character: The Contribution of Anatomy. *Journal of Human Evolution* 22:315–25.

Zihlman, A. L., and Cramer, D., 1978, Skeletal Differences Between Pygmy (*Pan paniscus*) and Common Chimpanzees (*Pan troglodytes*). *Folia Primatologia* 29:86–94.

Chapter 8

Bipedalism in *Homo ergaster*: An Experimental Study of the Effects of Tibial Proportions on Locomotor Biomechanics

LAURA TOBIAS GRUSS AND DANIEL SCHMITT

Abstract. A full understanding of the evolution of bipedalism requires consideration of the effects of body form on locomotor biomechanics. This is particularly important given the great range of body proportions found in fossil hominids. For example, although *Homo ergaster* had an essentially modern body form, it differed from most modern humans in having very long limbs, and particularly long distal limb elements, for its size. This study experimentally investigated the effects of crural index (tibia length relative to femur length) on the kinematics and kinetics of normal human walking. We hypothesized that a relatively longer tibia would lead to higher bending moments along the leg. We found, however, that modern humans with higher crural indices employ a series of postural adjustments that keep bending moments at comparable levels. An examination of the relevant fossil material suggests that early hominids with long tibiae, particularly *Homo ergaster*, likewise did not experience particularly high bending moments along the leg during locomotion. This implies that early hominids may have modified their locomotion behaviorally to moderate the bending stresses incurred during walking. Although an essentially modern human body form had evolved by the early Pleistocene, the form of walking that characterizes most modern humans may not have evolved until significantly later.

INTRODUCTION

The modern human body form seems to have appeared suddenly and almost fully-blown in the early Pleistocene. This is clearly illustrated by the 1.55 million-year-old *Homo ergaster* skeleton from Nariokotome, Kenya. The proportions of this almost-complete specimen stand in stark contrast to comparatively small earlier hominids such as *Australopithecus afarensis* and *Homo habilis*, with their relatively short legs and long arms (Ruff and Walker, 1993; Wolpoff, 1999). Measurements of KNM-WT 15000 suggest that *H. ergaster* had a height and weight within or above the range for most modern humans, and had human-like

From Biped to Strider: The Emergence of Modern Human Walking, Running, and Resource Transport, edited by D. Jeffrey Meldrum and Charles E. Hilton. Kluwer Academic/Plenum Publishers, New York, 2004.

Table 1. Intramembral indices of the KNM-WT 15000 *H. ergaster* specimen compared with means for Neandertals and modern human populations. Brachial index = (radius length/humerus length)*100; crural index = (tibia length/femur length)*100. Ranges given for living humans are for sample means, not individual observations, and crural indices are calculated from maximum bone lengths rather than external landmarks, as in our study (see text for details). Data from Ruff and Walker (1993), except Neandertal data (Trinkaus, 1981).

Specimen/Sample	Brachial index	Crural index
KNM-WT 15000	79.9	88.0
Living Africans	76.4–78.9	82.8–85.8
Living Europeans	72.9–74.0	78.4–83.1
Neandertals	74.4	78.9

interlimb (upper limb/lower limb) proportions (Brown et al., 1985; Leakey and Walker, 1989; Ruff and Walker, 1993).

Although in *Homo ergaster* we see for the first time a body shape that is clearly human, this species differed in several important ways from the majority of modern humans, notably in intralimb proportions. The Nariokotome youth's body proportions have been described as "hyper-tropical," in reference to his narrow trunk, extremely long arms and legs, and very high brachial and crural indices (i.e. long distal elements of the upper and lower limbs, respectively, relative to the proximal elements; Ruff and Walker, 1993). Compared with modern humans, WT 15000 is most similar in body shape to Nilotic Africans. However, the specimen lies at the upper end of the range for even this tall, slender, long-limbed population, which has the highest intramembral indices among modern humans (Table 1; Ruff and Walker, 1993; Ruff, 1994). This individual may have suffered from axial dysplasia, causing abnormal reduction of trunk size; however, the limb bones seem to have been unaffected (Latimer and Ohman, 2001). As a result, Ruff and Walker's (1993) height estimate, which was based on limb bone lengths, may be somewhat high *for this individual*, but should be representative of non-pathological *H. ergaster*. Inferences regarding inter- and intramembral indices should be valid.

The opposite extreme in human body proportions is found among fossil hominids as well. Late Pleistocene Neandertals from glacial Europe are "hyper-arctic," with wide trunks, short arms and legs, and very low brachial and crural indices, like those of modern humans inhabiting arctic regions (Trinkaus, 1981). The mean Neandertal crural index, for example, is equal to that of Lapps, who have the relatively shortest tibiae among modern *Homo sapiens* populations (Trinkaus, 1981). The cause of the extreme contrasts in body shape among fossil hominids has been the subject of considerable discussion (e.g. Holliday, 1997b; Isbell et al., 1998; Kramer, 2000; Rosenberg, 2000; Ruff, 1994; Trinkaus, 1981). The earliest and most widely accepted hypothesis that explains differences in proportions among fossil and modern hominids is that body shape varies for thermoregulatory reasons in accordance with Bergmann's and Allen's Rules (Fig. 1; Allen, 1877; Bergmann, 1847; see Biasutti, 1951; Crognier, 1981; Hiernaux and Froment, 1976; Holliday, 1997a; Newman, 1953; Roberts, 1978; Ruff and Walker, 1993; Ruff, 1994; Trinkaus, 1981). An alternative explanation, discussed below, is that intralimb proportions are under selection for locomotor attributes such as speed and efficiency.

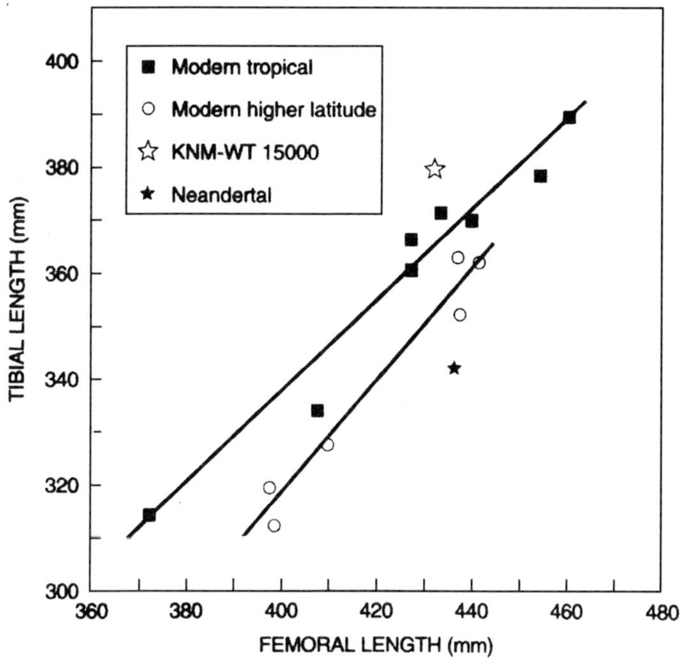

Figure 1. Crural index and climate in modern and fossil hominids. From Ruff and Walker (1993); reproduced with permission from Harvard University Press.

Our research focuses on a different issue, one that has rarely been addressed: what are the mechanical consequences of these proportional differences? We argue that whatever the evolutionary reasons for differences in crural index, relative length of the tibia has important effects on the biomechanics of human walking: in theory, a long tibia could increase flexion/extension moments at the knee and bending stresses along the tibia. The ways that modern humans might mediate these differing stresses (hereafter referred to as bending forces for the sake of convenience) caused by varying limb proportions may provide insight into the bipedal mechanics of fossil hominids, whose proportions varied far more widely.

Changes in body proportions, especially in the limbs, have important effects on locomotion; for example, limb configuration may strongly influence an animal's endurance and speed. Some researchers have argued that differences in body proportions, especially in the lower limb, among extinct hominids may have provided advantages in locomotion (e.g. see Holliday, 1999). The high inter- and intramembral indices of the Nariokotome skeleton may both have acted to maximize the efficiency of bipedal walking, by increasing stride length and decreasing the energy required to swing the limb, respectively.

An animal's energetic cost of locomotion depends largely on its stride length and stride frequency (Alexander and Ker, 1990; Heglund and Taylor, 1988; Kram and Taylor, 1990; Strang and Steudel, 1990). Energy cost per step is equivalent in animals of different sizes, but locomotion is more costly in smaller animals with shorter legs, which must take more steps more quickly to cover a given distance in a given time. For this reason, it appears that longer limbs can increase locomotor efficiency by increasing stride length without increasing the energy cost per step.

Variation in intralimb proportions (in this case, crural index) may also influence the work performed during locomotion, by affecting the amount of energy required to accelerate and decelerate the limbs. Theoretical and experimental studies have shown that a more proximal placement of mass in a limb (that is, longer, lighter distal elements for a given limb length) reduces its moment of inertia, decreasing the energy required to swing it (Myers and Steudel, 1985; Steudel, 1990; Taylor et al., 1974; Witte et al., 1991). These conclusions are borne out by observations of cursorial animals, which, like the Nariokotome *H. ergaster* skeleton, tend to have long limbs and high intramembral indices (Carrier, 2000; Taylor et al., 1974). In the context of a long limb, then, increasing crural index may be an effective way to increase locomotor efficiency. This configuration could lower the portion of a hominid's total energy expenditure that is devoted to locomotion.

Whatever the cause of their contrasting body shapes, the Nariokotome skeleton and those of European Neandertals show limb proportions that are matched only by the most extreme (and opposite) modern human populations. It is apparent from these fossil hominids that at least some pre-modern *Homo* groups differed significantly in aspects of body shape from the majority of extant *Homo sapiens*. What effects might these proportional differences, particularly in the lower limb, have had on the locomotion of earlier *Homo*?

Changes in relative length of limb segments should have a significant effect on the forces and movements involved in walking. Specifically, differences in crural index may influence the magnitude of bending moments acting on the knee and tibial shaft. Bending forces affect the bones of the lower limb in two different ways, acting differently on diaphyses and joints (Currey, 1984). Joints will tend to be collapsed or extended by flexion or extension moments, and these tendencies must be resisted by muscular effort and modifications in the osteoligamentous structure of the joint. In contrast, bending forces put bending stresses (tensile stresses along one surface and compressive stresses along the other) on long bone diaphyses. These forces are resisted by muscular effort and cross-sectional reinforcement of the shaft (through cortical bone remodeling).

The bending moments that act on the knee and the tibial shaft are generated by a ground reaction force and the force of muscles acting to resist these bending moments. The force that a person applies to the ground during walking results from acceleration and deceleration of the body's center of mass as a function of gravity. The ground reaction force, an equal and opposite force returned to the body, acts in three planes (Fig. 2; for detailed discussions of the kinetics of human walking see Ayyappa, 1997; Suzuki, 1985, and references therein): 1) the vertical plane (Fig. 2a); 2) the antero-posterior plane (often referred to as braking and propulsive forces) (Fig. 2b); and 3) the medio-lateral plane (Fig. 2c). The components of the ground reaction force can be resolved into a ground reaction resultant (GRR), which can be viewed as originating at the center of pressure, the point on the foot at which force is applied to the ground.

During the support phase of walking, the GRR changes in magnitude and direction as the components of the force applied change (Fig. 3). For example, when braking forces are high, the GRR is oriented posteriorly and generally passes behind the tibial shaft in the sagittal plane (as at braking peak, Fig.3b). As the braking force goes to zero at midsupport (Fig. 3c), the GRR becomes vertical. Finally, as the propulsive force increases toward the end of support phase, the GRR becomes oriented anteriorly (as at propulsive peak, Fig. 3d). As a result, the GRR may sometimes run along the long axis of the tibia, sometimes behind it, and at other times in front of it. In addition, the center of pressure moves forward along the foot during stance phase, and this movement shifts the GRR forward and affects its distance from the leg (Carrier et al., 1994; see Fig. 3). The magnitude of the bending moments acting

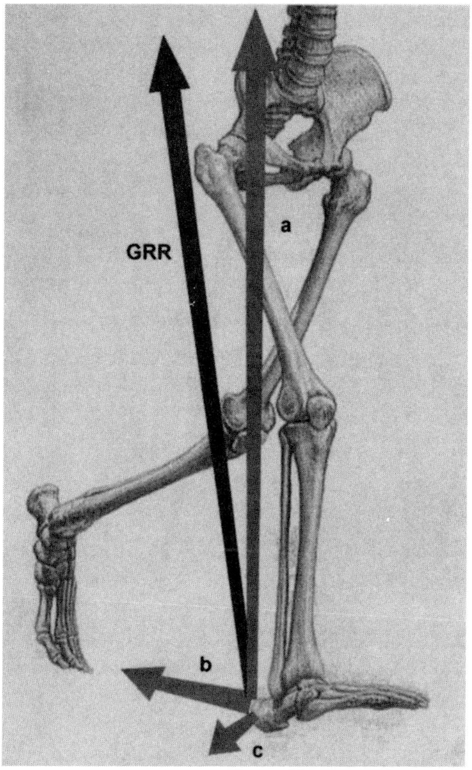

Figure 2. The ground reaction resultant (**GRR**) is composed of (**a**) vertical, (**b**) antero-posterior, and (**c**) medio-lateral forces. Modified from Gore (1997).

on any point along the leg is the product of the magnitude of the GRR and its perpendicular distance from that point (its moment arm; Fig. 4).

Given two people with similar body mass but different tibial lengths, the person with the longer tibia will experience higher bending moments at the knee and along the shaft of the tibia during the stance phase of walking. This is because in two people with different

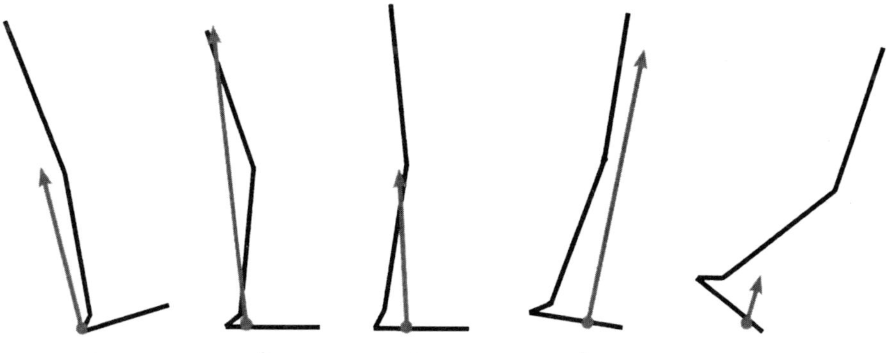

Figure 3. The movement of the GRR relative to the lower limb during stance phase. (**a**) heel strike, (**b**) braking peak, (**c**) midsupport, (**d**) propulsive peak, (**e**) toe-off. Also note the anterior movement of the center of pressure along the foot throughout stance phase. These figures were created by Motus, using kinetic and kinematic data from one of our experimental trials.

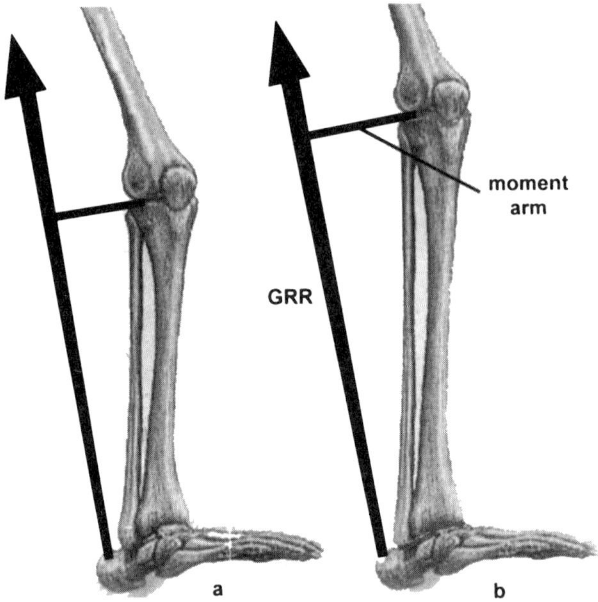

Figure 4. The effects of crural index on bending moments at the knee. Because individual **b**'s tibia is longer than individual **a**'s, the moment arm of the GRR at the knee is also longer. If the magnitude of the GRR is the same, this will result in higher bending moments at the knee in individual **b** (bending moment = force magnitude x moment arm). Modified from Gore (1997).

crural indices, the moment arm of the GRR at any homologous point along the leg (i.e. the knee or tibial midshaft) will be longer in the individual with a longer tibia. This results in larger bending moments about that point (Fig. 4).

Here we test the hypothesis that modern humans with higher crural indices (relatively long tibiae) will experience higher antero-posterior bending moments at the knee during the stance phase of walking, presumably requiring greater muscle and joint reaction forces to resist them. Alternatively, it is possible that people with higher crural indices will modify the mechanics of their walking in order to maintain dynamic similarity (Biewener, 1989; Rubin and Lanyon, 1984). For example, by keeping the knee in a more extended position throughout the stride, the moment arm of the GRR at the knee could be shortened, thereby reducing bending moments. Biewener (1989) has found that large animals, faced with large locomotor stresses because of the disproportional increase of body mass relative to bone cross-sectional area, use a similar postural mechanism to keep locomotor stresses at more tolerable levels. It is also possible that people with relatively long tibiae somehow reduce the magnitude of forces applied to the ground by decreasing vertical stiffness of their legs, which would also reduce bending moments (Farley et al., 1993; Farley and Gonzalez, 1996; Milgrom et al., 2000; McMahon et al., 1987; Schmitt, 1994).

We also test the hypothesis that differences in tibial length would have a similar effect on the medio-lateral bending moments acting on the knee. During support phase, when a medial force is applied to the ground, the GRR angles slightly outward, lateral to the knee. As with the A-P forces, the GRR should have a longer moment arm at the knee in an individual with a longer tibia, resulting in higher M-L bending moments for a given GRR magnitude. The same predictions hold: individuals with higher crural indices must experience higher

M-L bending forces on their knees unless they employ behavioral modifications to keep those bending moments lower.

This model has interesting implications for the evolution of modern bipedal walking. Fossil hominid populations with very high crural indices, such as *Homo ergaster*, either must have walked in a way that is different from the majority of modern humans, or their knees and tibial shafts must have been subjected to higher bending forces during walking than those of most people today. These hypotheses are testable against the fossil record because increased bending stresses should have osteological correlates such as larger joint surfaces, greater size and rugosity of muscle markings, and possibly increased shaft strength.

MATERIALS AND METHODS

This study investigated the effects of differences in crural index on the kinematics and kinetics of walking in 12 human subjects (6 males and 6 females). Crural index is useful in the context of this study because it is a size-free measure; it compensates for the fact that taller (and generally heavier) people also have longer tibiae. Using crural index essentially allows us to factor out total limb length and compare individuals with varying intralimb proportions (as discussed below, we also account for body mass). More importantly, the use of the crural index allows us to model more closely the condition in early *Homo*.

Subjects' crural indices (tibia length/femur length* 100) were determined by palpation of the greater trochanter and lateral epicondyle of the femur, and the medial condyle and medial malleolus of the tibia. The mean crural index in our sample was 87.3 with a standard deviation of 2.8 (Table 2). Our values differ somewhat from those given by Ruff and Walker (1993; see Table 1) because ours are based on external landmarks rather than maximum bone lengths. Calculated by our method, the crural index of the Nariokotome skeleton would be approximately 97.2, as opposed to 88.0 as calculated by Ruff and Walker (Casts were measured to calculate WT 15000's crural index by our method. Maximum bone lengths as measured on the casts were indentical to published measurements of the fossil (Ruff and Walker, 1993), ensuring that differential shrinking had not occurred during the casting process).

During experiments, subjects walked back and forth across a 6.2-meter long wooden runway in the Animal Locomotion Laboratory at Duke University. Subjects were continuously video-recorded in frontal and lateral views at 60 fields per second using Panasonic CCD cameras (Panasonic DV-W5100) and an industrial quality VCR (Panasonic Ag-7350). Both cameras were placed as far away from the subject as possible (4.25 meters for the

Table 2. Crural index summary statistics for subjects in this study. N = 12. Tibia and femur lengths are given for the subjects with the highest and lowest crural indices (C.I.) in this study. Bone lengths are in mm. Note that these measurements are based on external landmarks, unlike those given in Table 1, which are based on osteometric measurements from skeletal material. For this reason the crural indices listed here are not directly comparable to those in Table 1.

	Tibia length	Femur length	Crural index
Mean ± 1 S.D.	38.88 ± 2.83	44.25 ± 2.54	87.31 ± 2.80
Minimum C.I.	34.5	43	80.2
Maximum C.I.	40	43.5	92.0

Figure 5. Lateral and frontal views of a subject walking across a force plate in the Animal Locomotion Laboratory at Duke University. Bony landmarks are indicated with white tape. This image was taken from videotape that was later used in kinematic analysis.

lateral camera and 4 m for the frontal camera) to avoid the effect of parallax. The optical axis of the lateral camera was perpendicular to the subjects' path of travel and level with the knee. The frontal camera was collinear with the subjects' path of travel and level with the knee. Subjects wore black socks and tights, and bony landmarks of the pelvis, legs, and feet were marked with reflective white tape so that joint angles could be calculated from the video recordings (Fig. 5).

As the subjects walked, vertical, antero-posterior, and medio-lateral reaction force magnitudes and angles in stance phase were measured using a Kistler (Kistler Instruments, Buffalo, NY 14228) 9281B force platform (40 cm × 60 cm) mounted in the center of the runway. Four columns of piezoelectric quartz crystal discs are embedded at the four corners of the force plate, allowing vertical, antero-posterior, and medio-lateral forces to be recorded accurately, independent of the point of application. The platform was made flush with the runway surface by mounting a 3.75 cm thick piece of hardwood on the slotted top surface of the platform, using 13 mm screws and T-bolts designed for this purpose. The platform cover was separated from the rest of the runway by a 1 cm gap on all sides so that foot contacts on the force plate were not disturbed by contacts on other parts of the runway.

Signals from the piezoelectric crystals in the force plate were converted to proportional voltages by charge and summing amplifiers (Kistler 9856EY8). Analog signals were converted to digital signals using an A/D board, and passed to a personal computer. Digital signals were processed and converted to newtons using the Motus 2000 analog module (Peak Performance Technologies, Englewood, CO 80112). Precise coordination between video and force plate was achieved through an Event and Video Control Unit that attached a digital signal to force plate output and video recordings.

During experiments, subjects were asked to begin walking along the runway at a slow pace, and then gradually increase their speed with each successive crossing of the platform until they could go no faster without breaking into a run. Each time subjects stepped on the force plate, kinetic data was recorded. Fifty-four steps total were analyzed. For most subjects, five steps, representing the full range of speeds employed, were selected for analysis. Only steps where the subject contacted the plate completely and where the subject was walking in a straight path with no obvious acceleration or deceleration were selected for analysis.

Kinematic and kinetic variables were calculated from force plate and video data. We were able to calculate knee and ankle angles, and orientation of the tibia and GRR relative to vertical in the sagittal and coronal planes. We were also able to identify the position of the center of pressure on the foot. The GRR was anchored to the center of pressure in all of our analyses. From these data, and anthropometric data on the subjects, we were able to calculate the moment arms and bending moments at the knee and ankle. We collected and calculated all these data for four events during support phase (see Fig. 6): the first peak of vertical force, the peak of braking force, the second peak of vertical force, and the peak of propulsive force. These events represent critical points in the force curves, when peak values and orientations are most extreme. GRR magnitudes and bending moments about the knee and ankle were taken as a percentage of body mass, since people that are more massive exert a greater force on the ground and have higher absolute GRR magnitudes and greater bending moments.

The kinematic and kinetic data described above were tested for associations with crural index and tibia length. To control for any speed differences, we also used a standard ANCOVA to produce speed-adjusted means (Schmitt, 1999; Sokal and Rohlf, 1995). These means were then tested for correlations with crural index. We calculated Pearson correlation coefficients, using SAS version 6.12 software (SAS Institute, Cary, NC 27513).

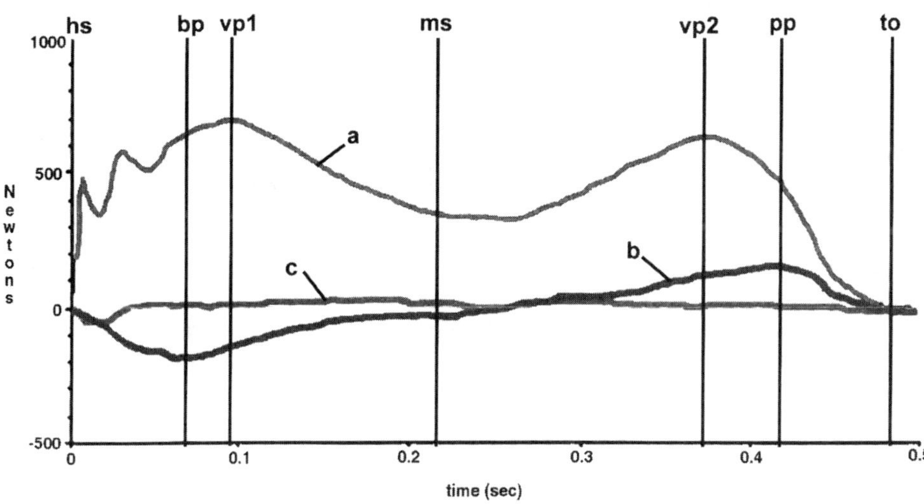

Figure 6. Kinetic output from Motus Peak Performance, showing typical values in newtons) for (**a**) vertical, (**b**) antero-posterior, and (**c**) medio-lateral forces throughout stance phase. The recording begins at heel strike (hs) and ends at toe-off (to). Other important events include peak of braking force (bp), first peak of vertical force (vp1), midsupport (ms), second peak of vertical force (vp2), and peak of propulsive force (pp).

RESULTS

Below we present the associations between crural index and kinetic and kinematic measures (Table 3). Correlations between tibia length and these variables yielded similar results. Our most important finding is that bending moments at the knee, as a percentage of body weight, are not significantly correlated with crural index during any part of stance phase. According to our model, this means that people with relatively long tibiae must be modifying their locomotion somehow to avoid high bending moments at the knee. This could be done by reducing either the magnitude of the GRR or the length of its moment arm. Both of these possibilities could be accomplished by manipulating either kinetic variables, such as the angle of the GRR to the ground, or kinematic (positional) variables, such as knee angle.

Neither the magnitude of the GRR, its angle to the ground, nor its moment arm at the knee is significantly correlated with crural index during stance phase. The lack of correlation between crural index and bending moments at the knee, therefore, is not explained by variation in these kinetic variables. Knee angle is negatively correlated with crural index throughout stance phase, most significantly in the second half of the step. Contrary to our predictions (and the mechanism for maintaining dynamic similarity described by Biewener (1989)), these correlations demonstrate that people with higher crural indices actually keep their knees in a more flexed position throughout stance phase.

The question remains; how do people with relatively long tibiae moderate bending moments at the knee? We have one possible explanation, involving the position of the center of pressure along the foot. This variable has been recognized as important during human running (Carrier et al., 1994), and may play a role during walking as well. Our results suggest that center of pressure is kept farther forward along the foot in people with longer tibiae at comparable points in stance phase. This relationship is not significant but it is consistent, suggesting that people with long tibiae may keep the point of application of force farther forward on the foot during these parts of the step than do individuals with shorter tibiae (Table 4). Since the center of pressure is the origin of the GRR, this shift effectively moves the entire GRR forward, bringing the force vector closer to the knee and

Table 3. Spearman correlation coefficients. For each variable, significant correlations with crural index are indicated. All variables are standardized for speed. No asterisk = $P \leq .05$; * = $P \leq .01$; ** = $P \leq .001$; *ns* = non-significant correlation. Events analyzed are: peak of braking force (bp), first peak of vertical force (vp1), second peak of vertical force (vp2), and peak of propulsive force (pp). Variables analyzed are: knee angle (kneeang), angle of tibia to ground (tibia/x), angle of GRR to ground (GRR/x), GRR magnitude as % body weight (GRR/bw), position of center of pressure (cop), length of moment arm at knee (kmom arm), bending moments at knee as % body weight (kmom/bw), length of moment arm at ankle (amom arm), and bending moments at ankle as % body weight (amom/bw).

Event	Knee ang	Ankle ang	Tibia/x	GRR/x	GRR/bw	Cop	Kmom arm	Kmom/bw	Amom arm	Amom/bw
bp	−.678	*ns*	−.722	*ns*	*ns*	*ns*	*ns*	*ns*	*ns*	*ns*
vp 1	−.767	*ns*	*ns*	*ns*	*ns*	*ns*	*ns*	*ns*	*ns*	*ns*
vp 2	−.901**	*ns*	*ns*	*ns*	*ns*	*ns*	*ns*	*ns*	*ns*	*ns*
pp	−.877*	*ns*	*ns*	*ns*	*ns*	*ns*	*ns*	*ns*	*ns*	*ns*

Table 4. Position of the center of pressure along the foot. Position of the center of pressure along the foot (in cm, as measured from the heel forward) at four events during stance phase for the subjects with the highest and lowest crural indices in this study. The center of pressure is consistently farther forward along the foot in the individual with the higher crural index. Center of pressure position is measured here in absolute terms rather than as a percentage of foot length because an equivalent point (e.g. 50% of foot length) on a long foot is a greater distance forward than it would be on a short foot. This difference affects the absolute length of the bending moment arm at the knee. Events are abbreviated as in Table 3.

Crural index	bp	vp1	vp2	pp
92.0	7.4	8.0	18.2	18.4
80.2	2.9	5.8	16.0	16.3

keeping its moment arm (and thus bending moments) at the knee equivalent in people with varying crural indices (Fig. 7).

There were no significant or consistent relationships between crural index and A-P bending moments or moment arms at the ankle. In addition, crural index does not appear to be related to M-L moment arms or bending moments at the knee or ankle, or indeed to any kinematic or kinetic variable in the M-L plane.

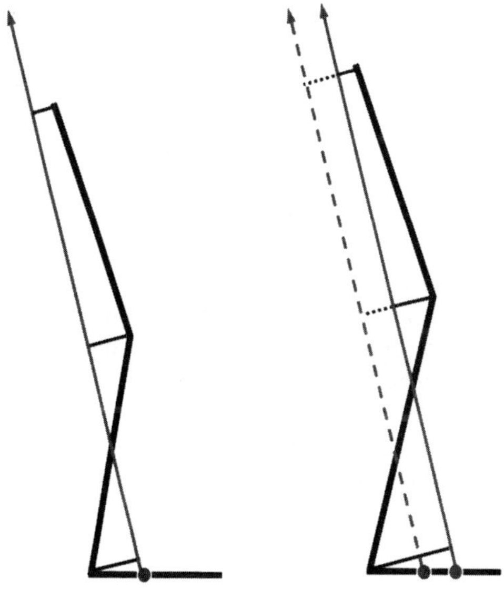

Figure 7. The effects of center of pressure position on moment arms. By moving the center of pressure farther forward on the foot, the individual with a higher crural index (**b**) can shorten the moment arm of the GRR at the knee. Assuming the magnitude of the GRR is the same, this will result in equivalent bending moments at the knee in the two individuals. The GRR's moment arm at the hip may also be shortened in individual **b** by this shift in center of pressure position. Although it appears that the moment arm at the ankle in individual **b** would be lengthened by this process, the difference is not significant.

DISCUSSION

Locomotor Mechanics and Intralimb Proportions in Modern Humans

This experimental study shows that antero-posterior bending moments about the knee do not rise with increases in crural index in modern humans. Instead, people with relatively long tibiae seem to employ a suite of postural adjustments that moderates bending stresses at the knee during locomotion. Equivalent bending moments are not achieved by changing the angle of the GRR or by keeping the knee more extended, as we predicted. In fact, in people with longer tibiae, the knee is actually kept in a more flexed position throughout stance phase. This differs from compliant walking, in which there is a high degree of yield at the knee, i.e. the joint acts as a shock-absorbing spring, flexing markedly after heel strike and straightening again before toe-off. Our subjects with higher crural indices did not exhibit higher degrees of yield, but had consistently more flexed knees. Instead, we find that people with long tibiae reduce bending forces by changing the position of the origin of the GRR. By moving the center of pressure more anteriorly along the foot, people with higher crural indices maintained moment arms and A-P bending moments about the knee similar to those in people with lower crural indices.

One other possible way of moderating bending stresses along the tibia and at the knee would be through altering the mechanics of the foot. A foot with more flexible tarsal and metatarsal joints could absorb forces before they are transmitted up the leg, reducing the forces experienced by the knee and tibial shaft, although we have no reason to suppose that people with higher crural indices also have more flexible feet. Another possibility is that changes in foot posture at equivalent points in the step—e.g. differences in degree of dorsiflexion at the metatarso-phalangeal joint—could help shift the position of the center of pressure or change the position of the knee relative to the GRR. We did not investigate foot mechanics in this study, but this area of research will be crucial in future efforts to refine our understanding of how crural index affects the mechanics of walking.

Although hip moment arms and bending moments were not analyzed in this study, our results suggest that these variables may also be moderated in individuals with relatively long tibiae. In the same way that it reduces the GRR's moment arm at the knee, the anterior shift of the center of pressure probably shortens the GRR's moment arm at the hip as well, thereby decreasing bending moments at that joint (Fig. 7). And although the increased knee flexion we observed in individuals with longer tibiae does nothing to mediate bending stresses at the knee, it may also help to decrease bending stresses at the hip. By flexing the knee without dorsiflexing the ankle, the hip can be moved backward and closer to the GRR, shortening the GRR's moment arm at the hip (Fig. 8). Alexander (1991) has shown that many animals keep the GRR close to the hip in locomotion and argues that this adjustment increases efficiency:

> It may be argued that it is more important to minimize the moments at the joints, and so the forces required of the muscles, than to minimize work... both the work of the muscles and the total of the moments at the joints are minimized by keeping the ground force in line with the hip. This will presumably also minimize the metabolic cost of locomotion. (1991:61)

However, there may also be an energetic cost to keeping the knees more highly flexed: more muscular effort may be required to support the body's weight on a more bent knee than on a straighter one. Studies comparing oxygen consumption during walking in people with

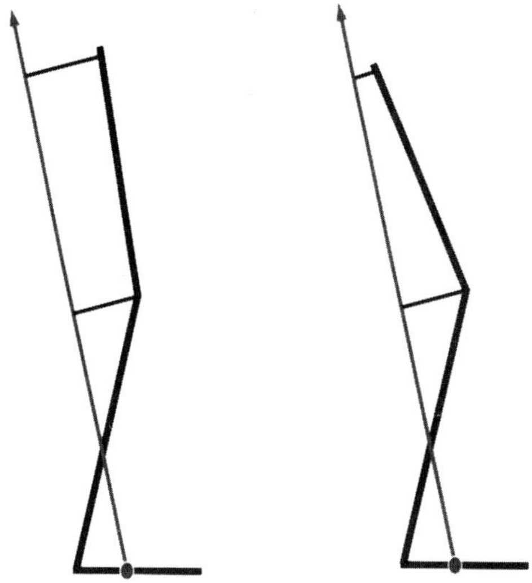

Figure 8. The effect of knee flexion on the GRR's moment arm at the hip. Note that although the moment arms at the knee are the same in individuals **a** and **b**, the moment arm at the hip is shorter in individual **b** because of the increased flexion of the knee.

varying crural indices are necessary to determine conclusively the energetic effects of this mode of locomotion.

Crural index does not seem to influence bending moments at the ankle. Theoretically, we would expect increased moment arms at the ankle in the A-P plane in our subjects with high crural indices, because the origin of the GRR is moved farther forward, away from the ankle (Fig. 7). However, we found no significant relationship between moment arm length at the ankle and tibia length. This may be because moment arms at the ankle are so short; the very small magnitude of differences in them among people of varying crural indices may make any such relationship difficult to detect and of minimal functional impact.

Although differences in crural index were related to differences in locomotor biomechanics at the knee in the A-P plane, biomechanical variables in the M-L plane seemed to be unaffected. We found no significant correlations between crural index and any kinetic or kinematic variable in the M-L plane. One explanation for this finding is that people with longer tibiae manage to shift their centers of pressure more medially on their feet to reduce moment arms about the knee, similar to what is done in the A-P plane. However, we think it is unlikely that they could move the center of pressure far enough medially to compensate. We think it is more likely that no consistent patterns exist because of the very low magnitude of the M-L forces (generally less than 5% of body weight), and the small angle that the GRR makes with the tibia (about 3° on average). If any patterns do exist, they may be obscured because the degree of variation of these factors is as great within individuals as between them.

Locomotor Mechanics and Intralimb Proportions in Extinct Hominids

If early hominids with high crural indices, such as *Homo ergaster*, relied on similar postural mechanisms to reduce antero-posterior bending moments in their lower limbs as the modern humans in our study did, then we would anticipate little or no difference in stabilizing mechanisms on the tibia between *H. ergaster* and later humans with lower crural indices. On the other hand, if these hominids were not compensating posturally for the stresses imposed by their body proportions, we would expect indications of elevated resistance to A-P bending loads at the knee. Osteological indications of higher bending loads at the knee might include a larger tibio-femoral articular surface and a robust tibial tuberosity, which is the attachment (via the patella and patellar ligament) for *m. quadriceps femoris*, the only muscle that acts to resist flexion at the knee. Antero-posterior reinforcement of the proximal diaphysis could also be informative, although the relationship between locomotion and diaphyseal robusticity is complex and incompletely understood (e.g. see Churchill and Schmitt, in press; Lanyon, 1990; Lieberman, 1997).

We reviewed descriptions in the literature of fossil hominid tibiae as they relate to musculoskeletal stabilization of the knee and diaphysis. Based on the intralimb proportions of KNM-WT 15000, it appears that *Homo ergaster* had the highest crural index of any hominid known (Ruff and Walker, 1993; Ruff, 1994). However, few well-preserved *H. ergaster* tibiae exist. Those that do preserve some of the relevant morphology (KNM-ER 741, (Leakey et al., 1972); KNM-ER 813b, (Leakey and Wood, 1973); KNM-ER 803b, (Day et al., 1974); KNM-WT 15000, (Walker and Leakey, 1993)) do not provide conclusive answers to our questions. Leakey et al. (1972) report that while *H. ergaster* tibia KNM-ER 741 has a strong tibial tuberosity, the markings for the plantar flexors of the ankle are weakly developed. This specimen, therefore, is somewhat suggestive of high bending moments at the knee relative to the ankle. The Nariokotome skeleton is gracile overall, with generally weak muscle markings, but this is probably due more to its juvenile status than to any mechanical factors (Walker and Leakey, 1993). More importantly, the proximal tibial articular surfaces that are preserved do not show any marked increase in size or osteoligamentous reinforcement relative to modern humans (Day et al., 1974; Walker and Leakey, 1993). Although the sample of lower limb bones attributable to *H. ergaster* is small, the tibiae that are available show little evidence of increased resistance to bending loads, suggesting behavioral modifications like those described in this study.

In order to determine whether the tibiae of *H. ergaster* show any patterns of structural reinforcement of the knee related to their "hypertropical" intralimb proportions, it is necessary to compare them to those of hominids from similar and contrasting climates. There are few other early *Homo* tibiae that preserve the relevant morphology (the proximal articular surface and tibial tuberosity); they include the Middle Pleistocene tibiae from Broken Hill, Zambia, and Sima de los Huesos, Atapuerca, Spain, as well as tibiae of Late Pleistocene Neandertals from Europe.

Like the *H. ergaster* tibiae from Lake Turkana, the Broken Hill tibia belonged to a tropically proportioned individual. Pycraft et al.'s (1928) estimates of the maximum lengths of the tibia and associated femur yield a crural index of 87.2, compared to 88.0 in WT 15000. Pycraft et al. (1928) note a number of peculiarities of the proximal articular surface of the Broken Hill tibia, some of which may be related to increased ligamentous support of the knee joint. However, the tibial tuberosity does not appear to be especially large, pronounced, or rugose, and there are indications (a relatively large distal articular surface and pronounced soleal line) that the ankle may have been structurally reinforced. This morphology does not clearly suggest that the knee of this tropically proportioned individual was experiencing high bending stresses.

A brief survey of fossil hominid tibiae from higher latitudes results in similar conclusions. Despite coming from much colder environments (Bischoff et al., 1997; Trinkaus,

1981), the tibiae of *H. heidelbergensis* from Atapuerca and European Neandertals do not display reduced musculoskeletal reinforcement of the knee and shaft in the A-P plane relative to tropical early *Homo*. In fact, in some cases these tibiae actually appear to be distinctly stronger at the joint and in the diaphysis, the opposite of what we would expect in individuals with short tibiae (Arsuaga et al., 1991; Lovejoy and Trinkaus, 1980; Trinkaus and Rhoads, 1999; Wolpoff, 1999). However, a number of factors caution against interpreting this morphology strictly in terms of locomotor biomechanics. Other possible influences include the allometric effects of body size and shape (Trinkaus et al., 1998; Trinkaus and Rhoads, 1999; Trinkaus and Ruff, 1999) and, all else being equal, differences in habitual activities other than normal locomotion that involve the lower limb (Black, 1999; Trinkaus and Rhoads, 1999).

We are able to find no clear indication in the hominid fossil record that relatively longer tibiae result in higher bending moments about the knee. The lack of increased structural reinforcement of the knee in extinct hominids with high crural indices suggests that, like our modern human subjects, early members of our genus may have used postural adjustments to limit A-P bending moments. However, it is also possible that the difference in stresses on the knee caused by variation in intralimb proportions is not great enough to result in morphological differences in either fossil or modern hominids. Further research on well-preserved modern human bones has the potential to answer this question.

We performed a preliminary study of muscle attachment sites on tibiae of recent American males (N = 77) from the Terry Collection, housed at the Smithsonian Institution. This work suggests that the size and rugosity of the tibial tuberosity is not related to crural index (Gruss, unpublished data), a finding consistent with expectations based on the locomotor pattern of our experimental subjects. A future study, using a larger skeletal sample and more precise methods of quantifying muscle scar size and rugosity, will search for more subtle morphological patterns. Likewise, a more detailed study of the size and shape of the tibio-femoral articular surface could bring to light osteological correlates of high crural index.

The research reported here is concerned with the effects of high crural index on walking, but it seems likely that variation in intralimb proportions would have biomechanical consequences for running as well. This possibility is especially important in considering the locomotion of *Homo ergaster*, which has been portrayed as the first hominid for which sustained running was possible and, in fact, adaptively critical (e.g. Bramble, 2000; Kramer, 2000). This issue will be addressed in future studies.

CONCLUSIONS

Two important conclusions can be drawn from this study. First, our experimental work shows that modern humans with long tibiae do not experience higher bending moments at the knee or ankle during walking. These findings suggest a previously unknown way in which humans (and possibly other animals as well) may regulate bending moments at joints or along long bone shafts. In order to limit bending moments at the knee joint, human subjects with long tibiae seem to shorten the moment arm of the ground reaction resultant at the knee through dynamic shifts that move the point of application of the force of body mass (the center of pressure) farther anteriorly on the foot. This moves the entire trajectory of the GRR closer to the knee, keeping the GRR's moment arm at the knee (and probably at the hip) equivalent to that in individuals with lower crural indices.

Second, we find no evidence that fossil hominids with very high crural indices, such as *Homo ergaster*, experienced especially high bending moments at the knee. Instead, early humans with varying limb proportions may have modified their locomotion behaviorally to keep bending moments at appropriate levels. It is not clear whether fossil hominids with high crural indices employed a mechanism for regulating bending moments at the knee similar to that of our modern human subjects. If they did not perform mechanical adjustments to keep these bending moments low (whether by moving the center of pressure or by some other, unknown method), hominids with long tibiae would have been forced to expend more energy during locomotion, either to stabilize the joint through muscular effort, or to move structurally reinforced (and thus more massive) bones. Although preliminary studies of modern and fossil humans have not revealed osteological correlates of the biomechanical effects of high crural indices, our experimental work suggests that variation in forces engendered at the knee, as well as along the tibial shaft, during walking should be considered when the functional morphology of fossil tibiae is interpreted.

A basically modern body form and mode of bipedalism had evolved in the genus *Homo* by the beginning of the Pleistocene. However, our work suggests the possibility that the locomotion of *Homo ergaster*, with its very long tibiae, may have differed in important ways from that of most modern humans. If the biomechanical trends apparent in our modern sample held true for earlier *Homo* as well, differences in intralimb proportions (whether the result of selection for thermoregulation or locomotor efficiency) may have caused aspects of bipedalism to vary significantly among fossil hominids such as *H. ergaster* and the Neandertals. It appears that there were important transitional steps between the advent of bipedalism in the earliest australopithecines and the striding walk of modern humans.

ACKNOWLEDGEMENTS We are very grateful to Jeff Meldrum and Chuck Hilton for inviting us to participate in this volume and the symposium on which it is based. LTG thanks David Hunt of the Smithsonian institution for permission to study the Terry Collection. We also wish to thank Michael Black, Matt Cartmill, Rich Gruss, Jandy Hanna, Brigitte Holt, Bill Hylander, Chris Kirk, Patrick Lewis, Henry McHenry, Carel van Schaik, Chris Vinyard, Patricia Vinyard, Chris Wall, and Todd Yokley for helpful discussions and comments on earlier drafts of this paper, and all of our experimental subjects for their time and patience. We are especially grateful to Steve Churchill, Pierre Lemelin, and Chris Ruff for their generous input. Three anonymous reviewers improved the quality of this manuscript.

REFERENCES

Alexander, R.M., 1991, Energy-Saving Mechanisms in Walking and Running. *Journal of Experimental Biology* 160:55–69.

Alexander, R.M., and Ker, R.F., 1990, Running Is Priced by the Step. *Nature* 346:220–221.

Allen, J.A., 1877, The Influence of Physical Conditions in the Genesis of Species. *Radical Review* 1:108–140.

Arsuaga, J.L., Carretero, J.M., Martinez, I., and Garcia, A., 1991, Cranial Remains and Long Bones from Atapuerca/Ibeas (Spain). *Journal of Human Evolution* 20:191–230.

Ayyappa, E., 1997, Normal Human Locomotion, Part 2: Motion, Ground Reaction Force and Muscle Activity. *Journal of Prosthetics and Orthotics* 9:42–57.

Bergmann, C., 1847, Uber Die Verhaltniesse Der Warmeokonomie Der Thiere Zu Ihrer Grosse. *Gottingen Studien* 1:595–708.

Biasutti, R., 1951, Radio-Humeral and Tibio-Femoral Indexes in Fossil and Living Man. *Homo* 3:97–99.

Biewener, A., 1989, Scaling Body Support in Mammals: Limb Posture and Muscle Mechanics. *Science* 245:45–48.

Bischoff, J.L., Fitzpatrick, J.A., Leon, L., Arsuaga, J.L., Falgueres, C., Bahain, J.J., and Bullen, T., 1997, Geology and Preliminary Dating of the Hominid-Bearing Sedimentary Fill of the Sima de los Huesos Chamber, Cueva Mayor of the Sierra de Atapuerca, Burgos, Spain. *Journal of Human Evolution* 33:129–154.

Black, M.T., 1999, The "Trunk Torsion Hypothesis" and Neandertal Superior Pubic Ramal Morphology. *Journal of Human Evolution* 36:A2.

Bramble, D.M., 2000, Head Stabilization in Human Running: Implications for Hominid Evolution. *American Journal of Physical Anthropology* Suppl. 30:111.

Brown, F.H., Harris, J., Leakey, R.E.F., and Walker, A., 1985, Early *Homo erectus* Skeleton From West Lake Turkana, Kenya. *Nature* 316:788–792.

Carrier, D.R., 2000, The Running-Fighting Dichotomy and Hominid Evolution. *American Journal of Physical Anthropology* Suppl. 30:120.

Carrier, D.R., Heglund, N.C., and Earls, K.D., 1994, Variable Gearing During Locomotion in the Human Musculoskeletal System. *Science* 265:651–653.

Churchill, S.E., and Schmitt, D., 2001, Biomechanics in Paleoanthropology: Engineering and Experimental Approaches to the Investigation of Behavioral Evolution in the Genus *Homo*. In *New Perspectives in Primate Evolution and Behavior*, edited by C. Harcourt and R. Crompton, pp. 59–90. Linnean Society, London.

Crognier, E., 1981, Climate and Anthropometric Variations in Europe and the Mediterranean Area. *Annals of Human Biology* 8:99–107.

Currey, J., 1984, *The Mechanical Adaptations of Bones*. Princeton University Press, Princeton.

Day, M.H., and Leakey R.E.F., 1974, New Evidence of the Genus *Homo* From East Rudolf, Kenya (III). *American Journal of Physical Anthropology* 41:367–380.

Day, M.H., Leakey, R.E.F., Walker, A.C., and Wood, B.A., 1974, New Hominids From East Rudolf, Kenya (I). *American Journal of Physical Anthropology* 42:461–476.

Farley, C.T., and Gonzalez, O., 1996, Leg Stiffness and Stride Frequency in Human Running. *Journal of Biomechanics* 29:181–186.

Farley, C.T., Glasheen, J., and McMahon, T.A., 1993, Running Springs: Speed and Animal Size. *Journal of Experimental Biology* 185:71–86.

Gore, R., 1997, The Dawn of Humans Expanding Worlds. *National Geographic* 191(5): 84–109.

Heglund, N.C., and Taylor, C.R., 1988, Speed, Stride Frequency and Energy Cost per Stride: How Do They Change with Body Size and Gait? *Journal of Experimental Biology* 138:301–318.

Hiernaux, J., and Froment, A., 1976, The Correlations between Anthrobiological and Climatic Variables in Sub-Saharan Africa: Revised Estimates. *Human Biology* 48:757–767.

Holliday, T.W., 1997a, Body Proportions in Late Pleistocene Europe and Modern Human Origins. *Journal of Human Evolution* 32:423–447.

Holliday, T.W., 1997b, Postcranial Evidence of Cold Adaptation in European Neandertals. *American Journal of Physical Anthropology* 104:245–258.

Holliday, T.W., 1999, Brachial and Crural Indices of European Late Upper Paleolithic and Mesolithic Humans. *Journal of Human Evolution* 36:549–566.

Isbell, L.A., Preutz, J.D., Lewis, M., and Young, T.P., 1998, Locomotor Activity Differences Between Sympatric Patas Monkeys (*Erythrocebus patas*) and Vervet Monkeys (*Cercopithecus aethiops*): Implications for the Evolution of Long Hindlimb Length in *Homo*. *American Journal of Physical Anthropology* 105: 199–207.

Kram, R., and Taylor, R., 1990, Energetics of Running: A New Perspective. *Nature* 346:265–267.

Kramer, P.A., 2000, Exploring the Locomotor System of a Biped Through a Behavioral Ecology and Life History Perspective. *American Journal of Physical Anthropology* Suppl. 30:201–202.

Lanyon, L.E., 1990, The Relationship Between Functional Loading and Bone Architecture. In *Primate Life History and Evolution*, edited by C.J. Derousseau, pp. 269–284. Wiley-Liss, New York.

Latimer, B., and Ohman, J.C., 2001, Axial Dysplasia in *Homo erectus*. *Journal of Human Evolution* 40:A12.

Leakey, R.E.F., and Walker, A., 1989, Early *Homo erectus* from West Lake Turkana, Kenya. In *Hominidae: Proceedings of the Second International Congress of Human Paleontology*, edited by G. Giacobini, pp. 209–215. Jaca, Milan.

Leakey, R.E.F., and Wood, B.A., 1973, New Evidence of the Genus *Homo* from East Rudolf, Kenya (II). *American Journal of Physical Anthropology* 39:355–368.

Leakey, R.E.F., Mungai, J.M., and Walker, A.C., 1972, New Australopithecines from East Rudolf, Kenya (II). *American Journal of Physical Anthropology* 36:235–252.

Lieberman, D.E., 1997, Making Behavioral and Phylogenetic Inferences from Hominid Fossils: Considering the Developmental Influence of Mechanical Forces. *Annual Review of Anthropology* 26:185–210.

Lovejoy, C.O., and Trinkaus, E., 1980, Strength and Robusticity of the Neandertal Tibia. *American Journal of Physical Anthropology* 53:465–470.
McMahon, T.A., Valiant, G., and Frederick, E.C., 1987, Groucho Running. *Journal of Applied Physiology* 62:2326–2337.
Milgrom, C., Finestone, A., Levi, Y., Simkin, A., Ekenman, I., Mendelson, S., Millgram, M., Nyska, M., Benjuya, N., and Burr, D., 2000, Do High Impact Exercises Produce Higher Tibial Strains than Running? *British Journal of Sports Medicine* 34:195–199.
Myers, M.J., and Steudel, K., 1985, Effect of Limb Mass and Its Distribution On the Energetic Cost of Running. *Journal of Experimental Biology* 116:363–373.
Newman, M.T., 1953, The Application of Ecological Rules to the Racial Anthropology of the Aboriginal New World. *American Anthropologist* 55:311–327.
Pycraft, W.P., Smith, G.E., Yearsley, M., Carter, J.T., Smith, R.A., Hopwood, A.T., Bate, D.M.A., Swinton, W.E., and Bather, F.A., 1928, *Rhodesian Man and Associated Remains*. British Museum of Natural History, London.
Roberts, D.F., 1978, *Climate and Human Variability*. Cummings, Menlo Park.
Rosenberg, K., 2000, Walking, Having Babies and Keeping Warm: Multiple Selective Pressures On Human Pelvic Morphology. *American Journal of Physical Anthropology* Suppl. 30:264–265.
Rubin, C.T., and Lanyon, L.E., 1984, Dynamic Strain Similarity in Vertebrates: An Alternative To Allometric Limb Bone Scaling. *Journal of Theoretical Biology* 107:321–327.
Ruff, C.B., 1994, Morphological Adaptation to Climate in Modern and Fossil Hominids. *Yearbook of Physical Anthropology* 37:65–107.
Ruff, C.B., and Walker, A., 1993, Body Size and Body Shape. In *The Nariokotome Homo erectus Skeleton*, edited by A. Walker and R.E.F. Leakey, pp. 234–265. Harvard University Press, Cambridge.
Schmitt, D., 1999, Compliant Walking in Primates. *Journal of Zoology, London* 248:149–160.
Schmitt, D., 1994, Forelimb Mechanics as a Function of Substrate Type During Quadrupedalism in Two Anthropoid Primates. *Journal of Human Evolution* 26:441–457.
Sokal, R.R., and Rohlf, F.J., 1995, *Biometry: the Principles and Practice of Statistics in Biological Research*. W.H. Freeman and Company, New York.
Steudel, K., 1990, The Work and Energetic Cost of Locomotion. I. The Effects of Limb Mass Distribution in Quadrupeds. *Journal of Experimental Biology* 154:273–285.
Strang, K.T., and Steudel, K., 1990, Explaining the Scaling Frequency of Transport Costs: the Role of Stride Frequency and Stride Length. *Journal of Zoology, London* 221:343–358.
Suzuki, R., 1985, Human Adult Walking. In *Primate Morphophysiology, Locomotor Analysis and Human Bipedalism*, edited by S. Kondo, pp. 3–24. University of Tokyo Press, Tokyo.
Taylor, C.R., Shkolnik, A., Dmi'el, R., Baharav, D., and Borut, A., 1974, Running in Cheetahs, Gazelles, and Goats: Energy Cost and Limb Configuration. *American Journal of Physiology* 227:848–850.
Trinkaus, E., 1981, Neanderthal Limb Proportions and Cold Adaptation. In *Aspects of Human Evolution*, edited by C.B. Stringer, pp. 187–224. Taylor & Francis, London.
Trinkaus, E., and Rhoads, M.L., 1999, Neandertal Knees: Power Lifters in the Pleistocene? *Journal of Human Evolution* 37:833–859.
Trinkaus, E., and Ruff, C.B., 1999, Diaphyseal Cross-Sectional Geometry of Near Eastern Middle Palaeolithic Humans: the Tibia. *Journal of Archaeological Science* 26:1289–1300.
Trinkaus, E., Ruff, C.B., and Churchill, S.E., 1998, Upper Limb vs. Lower Limb Loading Patterns Among Near Eastern Middle Paleolithic Hominids. In *Neandertals and Modern Humans in Western Asia*, edited by T. Akazawa, K. Aoki, and O. Bar-Yosef, pp. 391–404. Plenum Press, New York.
Walker, A., and Leakey, R.E.F., 1993, The Postcranial Bones. In *The Nariokotome Homo erectus Skeleton*, edited by A. Walker and R.E.F. Leakey, pp. 95–160. Harvard University Press, Cambridge.
Witte, H., Preuschoft, H., and Recknagel, S., 1991, Human Body Proportions Explained on the Basis of Biomechanical Principles. *Zeitschrift Für Morphologie und Anthropologie* 78:407–423.
Wolpoff, M., 1999, *Paleoanthropology*. McGraw-Hill, Boston.

Chapter 9

The Running-Fighting Dichotomy and the Evolution of Aggression in Hominids

DAVID CARRIER

Abstract. A suite of functional tradeoffs involving specialization for fighting versus locomotor economy is hypothesized to have influenced the evolutionary pathways of hominids. Characters such as pronounced forelimb strength, robust distal limbs, short stature, wide hips, robust head and neck, and fully bipedal stance and gait gave australopithecines a body configuration that is consistent with specialization for fighting with the forelimbs. This, plus a high level of sexual dimorphism in body and forelimb size suggest that australopithecines were specialized for male-male aggression. Nevertheless, traits that made australopithecines effective fighters would also have made them inefficient runners by modern human standards. Given the evidence of specialization for male–male aggression in australopithecines, the evolution of high locomotor stamina in *Homo* represents a dramatic change in evolutionary trajectory. I suggest that the evolution of improved locomotor stamina in *Homo* was made possible by the invention of new weapon technology that reduced functional conflicts between specialization for economical transport and specialization for male-male fighting. Invention of the first effective weapons would have shifted selection for male-male aggression from favoring greater physical strength and agility to favoring creative innovation, allowing independent evolution of cursorial specialization. In this scenario, the emergence of *Homo* 1) is dependent on the invention of new weapons, and 2) represents both greatly improved distance transport and increased lethality in a lineage already highly specialized for fighting.

INTRODUCTION

The demands of survival and reproduction often require compromises in the design of organisms (Gans, 1988; Lauder, 1991; Maynard Smith et al., 1985). One category of functional tradeoff that may be particularly important in the evolution of vertebrate diversity is the set of compromises entailed by the requirements of locomotion versus the requirements of fighting. Both locomotion and fighting are critical to survival and reproductive fitness in many species. Yet, characters that make an individual good at fighting may, in many cases, limit locomotor performance and vice versa (Hamilton, 1979; Pasi and Carrier, 2003). For example, among apes, sexual dimorphism in body size and male-male fighting are most

---

From Biped to Strider: The Emergence of Modern Human Walking, Running, and Resource Transport, edited by D. Jeffrey Meldrum and Charles E. Hilton. Kluwer Academic/Plenum Publishers, New York, 2004.

Figure 1. Individuals of the two breeds of domestic dogs, pit bulls and greyhounds, which are used to illustrate functional conflicts between specialization for running versus specialization for fighting. Both individuals in this photograph are adult males with a body mass of 33 kg. Modified from Chase et al. (2002).

dramatic in gorillas (Nowak and Paradiso, 1983). Large body size allows dominant male gorillas to defend multi-female groups against lone males intent on killing the infants and attracting the females (Fossey, 1983, 1984; Watts, 1989). Large size, however, severely limits the ability of male gorillas to climb trees (Schaller, 1963). In contrast, both male and female gibbons brachiate with spectacular grace and agility. Gibbons exhibit little or no sexual dimorphism in body size and mate in monogamous pairs (MacKinnon and MacKinnon, 1984; Nowak and Paradiso, 1983). Both male and female gibbons participate in defense of territory and aggressive encounters rarely involve physical contact (Mitani, 1987; Preuschoft et al., 1984). This comparison illustrates what may be a general trend; locomotor performance is constrained in fighting specialists whereas fighting ability is limited in locomotor specialists.

The ideas presented in this chapter emerged from an investigation of functional trade-offs in the musculoskeletal system of two breeds of domestic dogs selected as examples of specialization for fighting versus specialization for running performance: pit bulls and greyhounds (Fig. 1; Pasi and Carrier, 2003). The more we learned about the anatomy of these two breeds, the more I began to see them as a possible analogy for many of the structural differences between australopithecines and *Homo*. For example, the short stature and broad stance of pit bulls is similar to that of australopithecines, whereas the longer limbed and more gracile body of greyhounds is comparable to the build of *Homo*. In the case of the two dog breeds, we can be confident of the evolutionary basis of the anatomical differences; intense artificial selection for fighting ability in pit bulls versus intense artificial selection for economical, high speed running in greyhounds. Could functional tradeoffs associated with locomotor and fighting ability also have played a role in the evolution of hominids?

The origin of *Homo* has been argued to be associated with locomotor adaptations that improved economical transport and increased locomotor stamina (Bramble, 1990, 2000; Carrier, 1984; Jungers, 1991; McHenry, 1994; Porter, 1993; Ruff, 1991; Wolpoff, 1999). Much of the anatomy of australopithecines, in contrast, appears to be related to specialization for arboreal climbing (Stern and Susman, 1983; Susman, et al., 1984). It is clear, however, that australopithecines were highly specialized for terrestrial bipedality (Dart, 1925; Heglund and Schepens, 2002; Kramer, 1999; Kramer and Eck, 2000; Latimer et al., 1987; Latimer and Lovejoy 1989, 1990a, b; Lovejoy, 1981; 1988; Lovejoy, et al., 1973; Wolpoff, 1999). Although australopithecines were likely to have been much better climbers than any lineage of *Homo*, their specialization for terrestrial locomotion clearly compromised their climbing performance relative to that of their more fully arboreal ancestors (Latimer et al., 1987; Latimer and Lovejoy 1989, 1990a, b; Lovejoy, 1978). Thus, as is the case with *Homo*, the evolutionary origin of *Australopithecus* is associated with adaptations for terrestrial bipedalism.

Here I propose that australopithecines can also be distinguished from their arboreal climbing ancestors and from the genus *Homo* by characters that are related to specialization for fighting. Specifically, I suggest that australopithecines were anatomically specialized for male-male aggression, and that the evolution of cursorial specialization in *Homo* was made possible by the invention of new weapon technology that greatly reduced functional conflicts between specialization for endurance locomotion and specialization for fighting.

WERE AUSTRALOPITHECINES SPECIALIZED FOR FIGHTING?

There are a number of musculoskeletal differences between australopithecines (including *Paranthropus*) and *Homo* that are consistent with fighting specialization in the australopithecines and adaptation for locomotor economy in *Homo* (Table 1). Much of what gave australopithecines a fighting physique, however, were characters that clearly represent a legacy from arboreal ancestors. Indeed, a component of the fighting hypothesis, presented here, is that adaptations for suspension climbing, forelimb hanging, and brachiation preadapt a lineage for fighting with the forelimbs in the manner of chimps, gorillas, and humans. Thus, the aspect of the fighting hypothesis that is hardest to evaluate is not whether physical combat among male australopithecines was common and evolutionarily important, or if australopithecines had more of a fighting physique than any group of *Homo*, but whether or not the genus *Australopithecus* was specialized for fighting relative to their more arboreal ancestors. In the discussion that follows, I attempt to distinguish plesiomorphic, arboreal characters of *Australopithecus* that would have conveyed high fighting ability, from those characters that would have enhanced fighting ability but would likely have interfered with climbing performance (Table 1). This latter set of characters represents the strongest argument for fighting specialization in australopithecines.

Proximal to Distal Distribution of Limb Mass

The mechanical work of locomotion can be divided into two components, the work required to accelerate and decelerate the mass of the animal during each step and the work

Table 1. Comparison of expected musculoskeletal differences in primates specialized for terrestrial bipedal locomotion, fighting with forelimbs, and arboreal suspension climbing and brachiation.

	Bipedal locomotion	Forelimb fighting	Suspension climbing
Distal limb segments	Gracile	Robust	Robust
Forelimb/hindlimb robustness*	Low	High	Moderate
Elastic storage	High	Low	Low
Hip width*	Narrow	Wide	Narrow
Hindlimb length*	Short or long	Short	Long
Forelimb length	Short	Long	Long
Cranio-scapular musculature*	Limited	Robust	Moderate
Body size sexual dimorphism*	Low	High	Low
Forelimb sexual dimorphism*	Low	High	Low
Skull buttressing*	Low	High	Low
Habitual bipedal posture*	Well developed	Well developed	Poorly developed

*Characters that have potential to distinguish fighting specialization from climbing specialization.

required to swing the limbs relative to the center of mass of the animal. This latter component is referred to as the internal work of locomotion. During high speed running or in animals with heavy limbs, such as hominids, the internal work can constitute a significant portion of the total mechanical work of running (Cavagna and Kaneko, 1977; Fedak et al., 1982; Willems et al., 1995). This leads to the expectation that animals specialized for economical running will have relatively less muscle mass in their distal limbs to reduce the rotational inertia of the oscillating limbs (Hildebrand and Hurley, 1985; Steudel, 1991). In contrast, the distal limbs of animals specialized for fighting can be expected to be well muscled to allow the production of large forces and high power for opponent manipulation and agile movement (Table 1). This distinction appears to hold among many species comparisons. Consider the comparison of apes given above, or the comparison of lions and cheetahs. The distinction also exists between pit bulls and greyhounds (Pasi and Carrier, 2003). Pit bulls have a greater percentage of their limb muscle mass distal to the elbow and knee joints than do greyhounds (Fig. 2A). Hence, the relative robustness of distal limbs may help distinguish fighting from locomotor specialists.

The distal segments of both the fore- and hindlimbs of australopithecines appear to have been robust and massive compared to those of *Homo* (Fig. 3). Mid-shaft diameters of the radius and ulna suggest that the forearms of australopithecines were often exceptionally robust (Hartwig-Scherer, 1993; McHenry, 1986). Evidence of relatively massive forearms, compared to *Homo*, also emerges from analysis of the articular surfaces of the elbow joint (McHenry, 1978, 1992; McHenry and Berger, 1998). Forearms of australopithecines also appear to have been very long relative to humeral length (reviewed by McHenery and Coffing, 2000). Similarly, australopithecines are known to have had relatively larger hands and feet (Stern and Susman, 1983) and larger peroneal muscles (Stern and Susman, 1983; Tuttle, 1981) than *Homo*. Additionally, the structure of the metacarpals and wrist bones of australopithecines suggest that the muscles of the forearm associated with wrist and finger flexion were relatively large (Stern and Susman, 1983).

Certainly, a component of the relatively enlarged distal forelimbs of australopithecines may be explained as a legacy from their arboreal ancestors as well as specialization for forelimb dominated climbing given that they no longer possessed a significant capacity for grasping with their feet (Latimer and Lovejoy, 1990a, b; Wolpoff, 1999; Table 1).

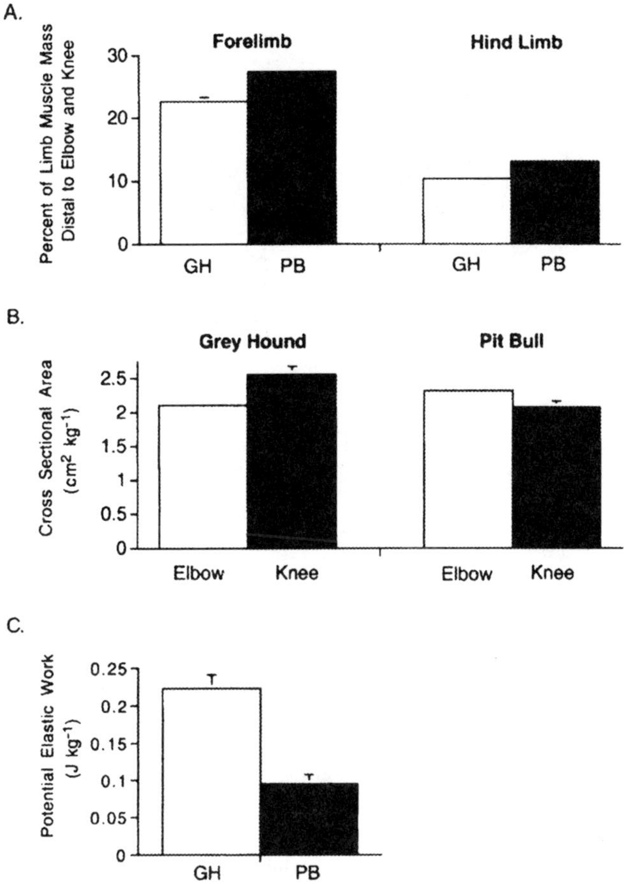

Figure 2. A) Percent of total limb muscle mass distal to the elbow and knee joints in four greyhounds and five pit bulls. In each comparison, the percent of muscle mass in the distal limb is significantly greater in the pit bulls. B) Comparison of the average cross-sectional area of the extensor muscles of the elbow and knee joints in four greyhounds and five pit bulls. The greyhounds have stronger muscles in their hindlimbs, whereas the pit bulls have stronger muscles in their forelimbs. C) Comparison of the average potential for storage and recovery of elastic strain energy in the tendons of the extensor muscles of the ankle joint in four greyhounds and five pit bulls. (Pasi and Carrier, in press).

Nevertheless, the exceptional sexual dimorphism in distal limb size and robustness observed among australopithecines (McHenry, 1991a, 1996, discussed below) is more consistent with specialization for male-male aggression than with an arboreal lifestyle. In contrast, the relative reduction in distal limb mass that characterizes *Homo* is consistent with improved locomotor economy (Hildebrand and Hurley, 1985; Steudel, 1991).

Relative Muscular Strength of Fore- and Hindlimbs

The contractile physiology of muscle leads to the expectation that the relative strength of fore and hindlimb muscles will be different in quadrupedal runners and fighters. Because active skeletal muscle generates much greater force when it is stretched than when it shortens (Katz, 1939) we expect quadrupedal animals specialized for running to have less muscle

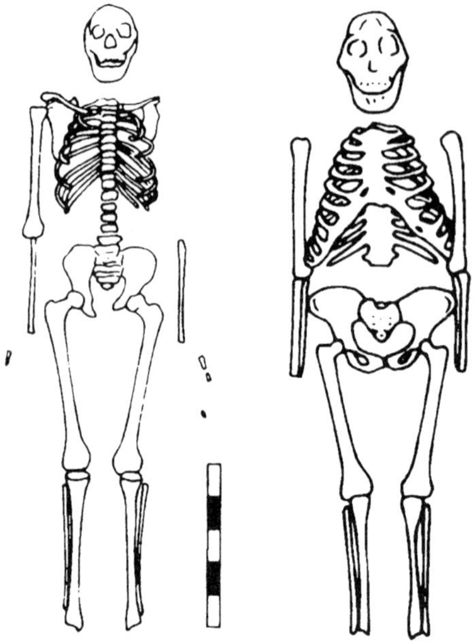

Figure 3. Illustrations of the skeletal anatomy of *Homo ergaster* and *Australopithecus afarensis*. The drawing of *A. afarensis* has been scales up to the same height as *H. ergaster* to better illustrate the differences in proportions. Modified from Klein (1999).

strength in their forelimbs than in their hindlimbs (Pasi and Carrier, 2003). This hypothesis emerges from the observed division of labor in the limbs in which the forelimbs of running animals play a greater role in deceleration and hindlimbs play a greater role in acceleration (Blickhan and Full, 1987; Cavagna et al., 1977; Cruse, 1976; Full et al., 1991; Heglund et al., 1982; Jayes and Alexander, 1978). Because the extensor muscles of the limbs must actively stretch to absorb energy during deceleration, but must actively shorten to produce acceleration, less muscle strength will be required in the forelimbs than the hind limbs. In contrast, the extensor muscles of the forelimbs of animals specialized for fighting can be expected to be as large or larger than those of the hindlimbs because forelimbs are often used to initiate rapid turns as well as to strike and manipulate an opponent during fighting.

Here also comparison of closely related species seems to support this expectation. The hindlimbs of cheetahs appear to have more muscle mass than the forelimbs, whereas the muscle mass of the forelimbs of lions appears to be greater than that of their hindlimbs. Similarly, our comparison of the relative strength (i.e., cross sectional area) of the extensor muscles in the forelimb and hindlimb of dogs shows that greyhounds have stronger extensor muscles in their hindlimbs than in their forelimbs, whereas the pattern is reversed in pit bulls, who have greater or equal strength in the extensor muscles in their forelimbs than in their hindlimbs (Fig. 2B; Pasi and Carrier, 2003).

Although hominids walk and run bipedally the analogy for relative limb strength holds and is useful when comparing *Australopithecus* and *Homo*. Hominids specialized for fighting are expected to have relatively strong forelimbs for opponent manipulation (Table 1).

In contrast, those specialized for running are expected to have relatively weak forelimbs, to reduce body mass, and strong hindlimbs to generate the needed locomotor forces and power. Compared to *Homo*, australopithecines had relatively less robust hindlimbs and significantly more massive forelimbs (McHenry, 1992; Fig. 3). Relative to their hindlimbs and sacral vertebrae both *A. afarensis* and *A. africanus* have robust forelimbs (McHenry and Berger, 1998). In this case, as in the previous comparison of distal limb robustness, the difference between *Australopithecus* and *Homo* is likely due, at least in part, to specialization for forelimb dominated climbing in *Australopithecus* and to specialization for economical terrestrial transport in *Homo*. Nevertheless, the dramatic sexual dimorphism in the size and robustness of the forelimbs of australopithecines (discussed below) suggests that specialization for fighting maybe more important than specialization for climbing in explaining the robustness of the forelimbs.

Potential for Storage and Recovery of Elastic Strain Energy

Storage and recovery of elastic strain energy in the tendons of distal limb muscles is thought to greatly enhance the economy of transport during both low and high speed running (Alexander, 1984; Biewener, 1998; Cavagna et al., 1964; Dawson and Taylor, 1973; Roberts et al., 1997; Taylor, 1994). For this reason we expect animals specialized for distance and/or high speed running to have a high capacity for elastic storage, i.e., long stretchy tendons. In contrast, long stretchy tendons in series with extensor muscles would likely pose a handicap for any animal attempting to over-power an opponent during physical combat. Hence, we expect animals specialized for fighting to have relatively limited abilities to use elastic storage during running (Table 1). Breeds of domestic dogs selected for running and fighting appear to fit these expectations. Greyhounds have more than a two-fold greater capacity to store and recover elastic strain energy in the extensor muscles of their ankle than do pit bulls (Fig. 2C; Pasi and Carrier, 2003).

Analysis of modern humans suggests that storage and recovery of elastic strain energy is an effective and important mechanism of energy conservation in human runners (Alexander, 1988; Cavagna et al., 1964; Ker et al., 1987). The potential of australopithecines to use elastic storage is difficult to access, but it can be assumed to have been substantially less than that observed in modern humans (Bramble pers. com.). It is known that the major sites of elastic strain are the in-series tendons, likely including apponeroses, rather than the muscle tissue itself (Alexander, 1988). Furthermore, the amount of energy that can be stored in a tendon is a direct function of its length. Because tendon length is related to leg length, animals with relatively long legs tend to have longer tendons and a greater capacity for elastic storage than animals with relatively short legs. In our comparison of greyhounds and pit bulls, tendon length played the greatest role in determining the estimated two-fold difference in potential for elastic storage (Pasi and Carrier, 2003). Hence, the relatively short hindlimbs of australopithecines relative to humans (Jungers, 1982; Ruff, 1988; Wolpoff, 1983, 1999; Fig. 3) lead to the suggestion that the potential for elastic storage was probably much lower in australopithecines than in modern humans.

Stature and Girdle Breadth

Although the two dogs shown in Figure 1 both weighed 330 N at the time the photograph was taken, they differ dramatically in shape. The greyhound is long limbed and narrow in the thoracic girdle. In contrast, the pit bull is shorter in stature, much broader in the thoracic girdle, and has a circular thorax. Narrow girdles and deep chests are common in animals specialized for running (Hildebrand and Goslow, 2001). Narrow girdles allow the limbs to swing in a parasagittal plane and reduce lateral moments at the limb joints and on the trunk. Hence, narrow girdles are thought to reduce the cost of transport. Long limbs also decrease the cost of transport by increasing stride length, thereby reducing the number of times the locomotor muscles must turn on and off to cover a given distance (Heglund and Taylor, 1988; Kram and Taylor, 1990). The broad girdles exhibited by pit bulls are thought to increase postural stability both by increasing the capacity of the limbs to apply laterally directed forces to the substrate and by increasing limb mobility in abduction and adduction. The circular shape of the pit bull thorax also likely allows greater adduction and abduction of the forelimb (see Ward, 1998). The combination of broad girdles and short limbs lowers the center of gravity of pit bulls, further increasing lateral stability. Thus, the contrasts in stature and girdle breadth illustrated in Figure 1 are thought to reflect selection for economical rapid locomotion in the case of the greyhound and postural stability, lateral strength, and agility in the case of the pit bull.

Compared to *Homo*, australopithecines were characterized by extraordinarily broad pelvic girdles (Fig. 3). The broad hips of australopithecines were due to a relatively wide interacetabular distance as well as a pronounced lateral flare of the iliac blades (Berge and Kazmierczak, 1986; Lovejoy, 1988; Lovejoy et al., 1973; Rak, 1991). Broad hips appear to be a derived condition of *Australopithecus*. Those of nonhuman apes and fossil hominoids are much narrower (Berge and Kazmierczak, 1986; Rak, 1991; Ward, 1997).

The broad hips of australopithecines were unlikely to have conveyed an energetic or mechanical advantage to bipedal walking and running. First, a broad distance between the hip joints increases the gravitational moment arm at the hip joint and would increase the muscular force required of the hip abductor muscles. This may have been partially or fully offset by the long muscle moment arm of the abductor muscles (Berge and Kazmierczak, 1986; Jungers, 1991), but broad hips certainly do not facilitate support of body weight against gravity at the hip joint. Second, broad hips would also increase moments at the hip due to fore-aft ground forces (Carrier, 1990; Fife et al., 2001). In the case of an upright biped these forces would tend to cause body rotation about a vertical axis through the hip joint (i.e., yaw) first in one direction as the body decelerated during the first half of the step and then in the opposite direction as the body accelerated during the second half of the step. To stabilize the body against these moments, relatively larger forces would have been required from the muscles that produce lateral and medial rotation of the limb at the hip joint in australopithecines than in the narrower hipped *Homo*. Finally, if as Rak (1991) suggests broad hips decrease the vertical oscillation of the center of mass during a walking step, the economy of walking would be diminished due to reduced energy cycling through the transfer of potential and kinetic energy (Cavagna et al., 1977).

The impact that the broad hips of australopithecines may have on obstetric function is controversial. A number of people have argued that the pelvis of australopithecines was shaped more to meet the demands of erect posture and bipedal locomotion than to allow an increase in fetal size, which occurred much later in the evolution of hominids (Abitbol, 1987; Rosenberg and Trevathan, 1995; Tague and Lovejoy, 1986). Alternatively, the mediolaterally broad birth canal of early hominids may have been needed to allow passage of the broad shouldered infants (Rosenberg and Trevathan, 1995; Trevathan and Rosenberg, 2000; Wolpoff, 1999). Whether or not the broad hips of early hominids were a function of upright posture and/or obstetric function, the energetics and mechanics of terrestrial locomotion would have been negatively impacted as described above.

Broad hips, however, must have increased the postural stability of australopithecines in two ways. First, broad hips provide a large area of support for the abdominal viscera. This in combination with the cone shaped thorax of australopithecines resulted in a low center of gravity by positioning the viscera low in the trunk (Hunt, 1998). Second, broad hips increase the lateral forces the limbs can apply to the substrate. Although the mechanical advantage of the hip abductors to support body weight in australopithecines appears to have been lower than (Jungers, 1991) or roughly equivalent to (Berge and Kazmierczak, 1986) that in modern humans, the mechanical advantage of the hip abductors for the production of laterally directed forces on the ground, as would be needed when applying a lateral force to an opponent, was much higher in australopithecines than in *Homo*. The muscle moment arms of the *Gluteus medius* and *Gluteus minimus* muscles are determined by the distance from the greater trochanter to the center of rotation of the hip joint, in other words, by the length of the femoral neck. For the production of laterally directed ground forces, the ground force moment arm of the hip abductors is a function of the length of the leg. Hence, the long femoral neck and shorter legs of australopithecines (Jungers, 1982; Ruff, 1988; Wolpoff, 1983, 1999) gave their hip abductors a much greater mechanical advantage to apply lateral forces to the substrate than was present in early *Homo*.

In summary, the exceptionally broad hips of australopithecines are a derived trait of early hominids that does not appear to be related to selection for economical or rapid terrestrial locomotion. Broad hips combined with short limbs do, however, increase postural stability, particularly against laterally directed perturbations. These are characteristics that would likely benefit individuals engaged in physical combat (Table 1). Alternatively, broad hips could have been selected because they increased stability during climbing. However, if this were true why are broad hips a derived character of australopithecines, not found in chimpanzees, gorillas, or the arboreal ancestors of *Australopithecus*? Thus, hip width is a character that should help to distinguish fighting from climbing specialization (Table 1). In contrast, the relatively narrower hips and greater limb length of *Homo* are both consistent with improved locomotor economy.

Hindlimb Length

The short legs of australopithecines have often been equated with inefficient terrestrial locomotion. The argument is that short legs result in short stride lengths and this necessitates a higher stride frequency for a given walking or running speed. As observed by Kramer (1999), however, energetic expenditure is not solely a matter of the number of strides taken, but also of the energy consumed in a single stride. Recent modeling of the gait of australopithecines (Kramer, 1999; Kramer and Eck, 2000) and comparisons to modern children of the similar body proportions to australopithecines (Heglund and Schepens, 2002) suggest that the terrestrial locomotion of australopithecines was not inefficient at low to moderate speeds. The difference between australopithecines and *Homo* suggested by both the modeling and the comparisons to human children is that the energetically optimal walking speed would have been lower in australopithecines, and australopithecines would not have been able to walk and run efficiently at high speeds.

Based on these results, Kramer and Eck (2000) suggest that australopithecines may have been "optimized for a particular ecological niche—slow speed foraging". Although this is certainly a possibility, I view it as unlikely. Specialization for economical slow locomotion requires low-mass limbs (e.g., short limbs of low muscle mass) and slow contracting muscle fibers. These are traits found in low energy specialists such as tortoises (Woledge,

1968). The downside of specialization for this type of economical transport is that low-mass limbs have muscles of small cross-sectional area that cannot produce the large forces needed for rapid acceleration, and slow-contracting muscle fibers that limit maximum running speeds. Thus, if australopithecines were specialized for slow speed foraging, their capacity for acceleration and sprint velocity would have been significantly constrained, as in tortoises. Such specialization seems unlikely in a relatively defenseless terrestrial forager that was confronted by larger and swift terrestrial predators. Alternatively, if their limbs were equipped with large, fast-contracting muscles, australopithecines would have been inefficient at high speeds, but they would have accelerated rapidly and they may have run at relatively high speeds over short distances. This arrangement of short hindlimbs equipped with large, fast-contracting muscles is entirely consistent with specialization for fighting (Table 1).

The argument that short hindlimbs are an adaptation for climbing (Cant, 1987; Cartmill, 1974; Hunt, 1991a; Jungers, 1978) is inconsistent with the relatively long hindlimbs of exceptional climbers such as gibbons and spider monkeys. Indeed, long hindlimbs can be expected to be advantageous in an environment in which the locomotor substrate is limited and unpredictable in its location, allowing more support options than would be available with shorter hindlimbs (Table 1). Thus, the relatively short hindlimbs of chimpanzees and gorillas (Jungers, 1984), as well as those of early hominids, may reflect selection for fighting rather than climbing.

Head and Neck

Another obvious difference between greyhounds and pit bulls is the size and robustness of the head and neck (Fig. 1). Relative to the basal "wolf-like" morphology of domestic dogs the greyhound lineage has lost head mass whereas the reverse is true in the pit bull lineage. In our sample, the mass specific size of the jaw adductors was almost three-fold greater in the pit bulls (11.9 g kg^{-1}, $N = 5$) than in the greyhounds (4.1 g kg^{-1}, $N = 4$; $P < 0.0001$, unpaired t-test; Carrier, unpublished data). In terms of adaptation, we can speculate that the light, streamlined head of greyhounds improves locomotor economy, whereas the robust head and neck of pit bulls provides large biting force and enhances the ability to apply torques to an opponent.

Information that would allow a comparison of the head and neck of australopithecines to that of their arboreal hominoid ancestors does not exist in the current fossil record. Thus, for the sake of argument, and because no other comparison is currently possible, I will assume that the head and neck of the arboreal ancestors of hominids resembled that of modern chimpanzees. Relative to chimpanzees, australopithecines had large heads, due primarily to their exceptionally large faces, and a large area for the attachment of nuchal muscles (Adams and Moore, 1975; Wolpoff, 1999; Fig. 3). In *A. afarensis*, the nuchal plane of the occipital bones was much broader than in chimpanzees, suggesting that the nuchal muscles were well-developed, and reflecting an "emphysis on muscles important in lifting and pulling at the shoulder" (Wolpoff, 1999). Additionally, the mastoid processes of *A. afarensis* are also enlarged relative to that in modern chimpanzees and gorillas (Olson, 1985). Specimens from Hadar (A.L. 333–106 and A.L. 333w-14) show that the spinous processes of the cervical vertebrae were long and robust (Lovejoy et al., 1982), suggesting that the cervical epaxial muscles were also relatively large in *A. afarensis*.

In contrast to most tetrapods, including many primates, the hominid trend towards reduction in canine size (Wolpoff, 1999) suggests that biting and opponent manipulation were not important functions of the head and neck during fighting in early hominids. Nevertheless, during fighting, the head and neck of hominids serves 1) as the attachment site for external appendicular muscles that raise the shoulder and forelimb, allowing upward (i.e., craniad) swings, strikes, and pushes from the forelimb, and 2) as a relatively vulnerable and fragile target. Powerful, craniad directed forelimb punches or pushes require large force production from the muscles that raise the pectoral girdle on the trunk: sternocleidomastoid, trapezius, levator scapula, rhomboideus minor. When these muscles contract they exert moments on the axial skeleton, specifically on the cervical vertebrae and skull. To stabilize the axial skeleton against these appendicular moments the epaxial and hypaxial muscles must be active (Fife et al., 2001; Gray, 1968). Thus, if we assume that the head and neck of the arboreal ancestors of hominids resembled chimpanzees in build and proportions, then the relatively large cervical and nuchal muscles of australopithecines would suggest fighting specialization (Table 1).

Given that the head and neck are important targets in fighting (see Goodall, 1986; LaBlanc, 1999; Weidenreich, 1943, 1951), buttressing of the face and cranium could be expected if fighting was an important behavior. Although buttressing of the cranium does not appear to have been present in australopithecines (Wolpoff, 1999), buttressing of the face remains a possibility. The large prognathic face that characterizes australopithecines is both a legacy from their hominoid ancestors and a function of their expanded cheek teeth/chewing complex (Wolpoff, 1999). Although it may be very difficult to distinguish masticatory function from protective buttressing, future analyses of structures such as the anterior pillars, zygomatic prominences, and supraorbital ridges should also include functional analysis as possible defensive shielding.

Sexual Dimorphism in Early Hominids

Australopithecines appear to have had a higher level of sexual dimorphism in body size than is observed in *Homo* (Table 2). (Given the lack of consensus on the identity of the limited number of postcranial specimens, I view the level of sexual dimorphism in *A. habilis* and *A. rudolfensis* to be unresolved.) Sexual dimorphism in australopithecines

Table 2. Sexual dimorphism in body size in hominids and extant hominoids (McHenry, 1996; McHenry and Coffing, 2000).

Taxon	Dates (mya)	Male Mass (kg)	Female Mass (kg)	Male/Female
Australopithecus anamensis	4.2–3.9	51	33	1.54
Australopithecus afarensis	3.9–3.0	45	29	1.55
Australopithecus africanus	3.0–2.4	41	30	1.37
Paranthropus boisei	2.3–1.4	49	34	1.44
Paranthropus robustus	1.9–1.4	40	32	1.25
Australopithecus habilis	1.9–1.6	?	?	?
Australopithecus rudolfensis	2.4–1.6	?	?	?
Homo ergaster	1.9–1.7	66	56	1.18
Homo sapiens	extant	64.9	53.2	1.18
Hylobates lar	extant	6.3	5.6	1.1
Pongo pygmaeus	extant	77.5	37.0	2.1
Gorilla gorilla	extant	169.5	71.5	2.4
Pan paniscus	extant	39.0	31.0	1.3
Pan troglodytes	extant	49	41	1.20

is apparent in their dentition (Kimbel and White, 1988; Wolpoff, 1976), limb bone lengths (McHenry, 1991b), limb bone robustness (Hartwig-Scherer, 1993; Lockwood et al., 1996), and limb joint size (McHenry 1991a, 1992). Based on an analysis of the mandibular corpus, proximal femur and humerus Lockwood and collaborators (1996) suggest that the level of body size sexual dimorphism in *A. afarensis* approached, but did not exceed that of the most dimorphic modern hominoids: gorillas and orangutans. An analysis of the robustness of both fore and hindlimb bones by Hartwig-Scherer (1993) suggests that the ratio of male to female body weight of *A. afarensis* was greater than that of any modern hominoid. Using metrics of the size of hindlimb joints, McHenry (1991a) found that *A. afarensis* had a moderate level of body size sexual dimorphism that was well above that seen in modern humans, somewhat greater than that of the two extant species of chimpanzee, and below the sexual dimorphism of gorilla and orang-utan. Furthermore, the extensive analysis of body weight dimorphism in early hominids by McHenry (1978, 1986, 1991a, b, 1992, 1994, 1996) leads to the conclusion that dimorphism declines through time from *A. afarensis* to *A. africanus* to *P. robustus* with a dramatic reduction in *H. erectus* (McHenry, 1996).

A high level of body size sexual dimorphism may be a basal characteristic of the great ape/hominid clade (Begun et al., 1997). *Oreopithecus bambolii* is a large bodied, late Miocene (8.5–9.0 Ma.) member of the great ape/hominid clade (Harrison, 1987; Sarmineto, 1987), and is represented by a relatively extensive fossil record (Harrision, 1991). Based on the dentition, as well as the cranial and postcranial material, Harrison (1991) estimates that the body mass of males ranged from 30–40 kg, whereas females were ranged from 15–20 kg. Thus, a level of body size sexual dimorphism that exceeds that of *Homo* may be basal for Hominoidea.

The forelimbs of australopithecines appear to have been more dimorphic than the jaws, teeth, or hindlimbs (McHenry, 1996). In *A. afarensis*, for example, the difference between large and small ulnae, radii, and capitates is as great or greater than that between male and female means of the most dimorphic extant apes (McHenry, 1986; 1991a; 1996). Forelimbs also appear to have been relatively dimorphic in both *A. africanus and P. boisei* as well (McHenry, 1996). The fact that McHenry's comparisons are based primarily on metrics of articular surfaces at joints is significant because articular surface area is independent of mechanical loading history (Lieberman et al., 2001; Trinkaus et al., 1994). Thus, the relatively high level of sexual dimorphism in the forelimbs of australopithecines must be related to evolution at the species level rather than to gender differences in activity level. Most importantly, the higher level of sexual dimorphism in the forelimb joints than the hindlimb joints suggests that the forelimb dimorphism was due to more than just allometry associated with males being physically larger in body size than females.

Is the relatively high level of sexual dimorphism in body and forelimb size consistent with the hypothesis of high levels of male–male aggression in australopithecines? The functional and behavioral implications of sexual dimorphism in hominids have been discussed by number of authors (Darwin, 1871; Foley and Lee, 1989; Frayer and Wolpoff, 1985; Lovejoy, 1981; McHenry, 1994; 1996; Plavcan and van Schaik, 1997; Wolpoff, 1976). The discussion that follows is similar in theme to that presented by Wolpoff (1976) and McHenry (1994, 1996).

Darwin (1871) proposed two mechanisms of sexual selection that can explain sexual dimorphism. One in which males compete with each other for access to females through direct fighting [or more subtle mechanisms such as sperm competition (Ridely, 1993), and the other in which females choose particular males based on male characters that convey some mating advantage. In cases of female choice, females may prefer handicapped males because the male's survival indicates his high quality, or females may choose healthy, unparasitized males (Ridley, 1993). Darwin believed that sexual selection would operate more powerfully in polygynous than in monogamous species because the level of selection for adaptations that enable males to gain access to females would be proportional to the number of females a male could possibly monopolize. Darwin's original analysis and subsequent work by others (Andersson, 1994; reviewed in Ridley, 1993) has shown this is largely true; sexual dimorphism is mainly found in polygynous species, and body size sexual dimorphism is generally found is polygynous species in which males compete through fighting or the threat of fighting. Within the Order Primates, monogamous species are always monomorphic and polygynous species are usually dimorphic with males often weighing twice as much as females (Clutton-Brock et al., 1977). Thus, we can be relatively confident that australopithecines were polygynous. But is the relatively high level of sexual dimorphism in body and forelimb size observed in australopithecines consistent with selection due to male–male aggression or with selection due to female choice?

Analysis of extant taxa suggests that the sexual dimorphism in body and forelimb size in australopithecines was a product of selection for male–male aggression. Although the belief that competing males rarely injure each other remains strongly held by many biologists (Barnett, 1967; Lorenz, 1963; Tinbergen, 1969), male–male aggression that results in serious injury or death is common among both vertebrates and invertebrates (Andersson, 1994; Berzin, 1972; Clutton-Brock, 1982; Daly and Wilson, 1988; de Waal, 1986; Enquist and Leimar, 1990; Geist, 1971; Hamilton 1979; Huntingford and Turner, 1987; Silverman and Dunbar, 1980; Wrangham and Peterson, 1996). When males compete physically, the potential for serious injury needs to be real for a male to achieve dominance (Andersson, 1994; Darwin, 1871; Geist, 1971; Parker, 1983). Hence, it should not come as a surprise that the relative size of male weapons is strongly correlated with both the extent to which males are larger in body size than females and with the level of polygyny (Andersson, 1994; Clutton-Brock et al, 1977, 1980; Hamilton, 1979; Jarman, 1983; Parker, 1983). Kangaroos and gorillas represent particularly relevant examples. Kangaroos use their forelimbs as weapons in fighting. Among species of kangaroos, dimorphism in the mass of the forelimb musculature and length of the forelimbs is most pronounced in polygynous species (Jarman, 1983). Similarly, gorillas fight with their forelimbs (Wrangham and Peterson, 1996). Zihlman and McFarland (2000) found that the greatest sexual dimorphism in lowland gorillas is in the weight of the forelimbs, weight of the forelimb trunk binding muscles, and the epaxial muscles. Finally, given that extant apes most closely related to australopithecines, *Homo, Pan*, and *Gorilla*, all use their forelimbs as weapons during male–male aggression (Wrangham and Peterson, 1996), we can be relatively confident that australopithecines did as well. These observations suggest that the high levels of sexual dimorphism in body and arm size in australopithecines were associated with selection for improved performance in male–male aggression.

Robustness of forelimb joint size in both genders might be argued to be consistent with selection for forelimb dominated arboreal behavior in australopithecines. Selection for climbing performance, however, cannot explain the greater sexual dimorphism in forelimb joint size than hindlimb joint size and spine robustness. Thus, the dramatic sexual dimorphism in forelimb robustness of australopithecines is more consistent with specialization for male–male aggression than with specialization for suspension climbing (Table 1). Similarly, given that male gorillas climb less than female gorillas, climbing cannot account for the sexual dimorphism in forelimb robustness in gorillas (Zihlman and McFarland, 2000; see also results in McHenry, 1992).

Relationship between Vertical Climbing, Bipedal Posture, and Fighting with Forelimbs

Specialization for vertical climbing and forelimb suspension preadapts a lineage for fighting with the forelimbs, both because it predisposes an organism to bipedal posture and because it results in long and powerful forelimbs with high mobility. The ancestor of modern hominoids is thought to have been an arboreal catarrhine adapted for vertical climbing and forelimb suspension (Begun et al., 1997; Harrison, 1991; Rose, 1997; Ward, 1997). The out group to extant hominoids for which post-cranial anatomy is best known, *Oreopithecus*, has a suite of characters that suggest it was an adept vertical climber that relied on forelimb suspension (Harrison, 1991). These include features that: 1) strongly differentiated usage of the hind and forelimbs (i.e., long and robust forelimbs relative to hindlimbs), 2) increased ability to raise the forelimbs above the head, 3) greater potential for circumduction at the shoulder and pronation-supination at the elbow and wrist, 4) increased potential for powerful grasping of large diameter vertical supports with the hands, 5) adoption of a more orthograde trunk posture, 6) an increased potential for full extension of the hip and knee joints, 7) greater ranges of rotation at the hip and knee joints and inversion-eversion at the ankle joint, and 8) an increased potential for body weight to be supported by a single hindlimb. Although these eight abilities, present in extant great apes (Gregory, 1916, 1928; Hunt, 1991b; Morton, 1926) and inferred from the postcranial anatomy of Miocene hominoids (Begun et al., 1997; Harrison, 1991; Rose, 1997; Ward, 1997), are associated with selection for climbing and forearm suspension, each ability can be argued to enhance fighting performance. Highly mobile, long and robust forelimbs and hands would increase the potential for striking, grasping, and manipulating opponents with the forelimbs. An increase in orthograde trunk posture (5) and an increased ability to support the body bipedally on the hindlimbs (6, 7, and 8) would free the forelimbs to be used as weapons. Finally, associated with the arboreal behaviors of vertical climbing and forelimb suspension are powerful shoulder and brachial muscles that produce hoisting and lateral transfer of the body (Tuttle et al., 1983; Tuttle and Basmajian, 1974, 1978). These are the same muscles that can be inferred to be employed in arching, over-the-head and roundhouse punches that chimpanzees and gorillas are observed to use when fighting (Goodall, 1986; Wrangham and Peterson, 1996). Thus, selection for improved vertical climbing and forelimb suspension can be argued to preadapt a species for fighting with its forelimbs.

Many quadrupedal mammals adopt a bipedal posture when fighting, such as rats and mice, squirrels, dogs, bears, tigers, and horses. Among apes, both chimpanzees and gorillas adopt a bipedal posture for threat displays (Jablonski and Chaplin, 1993), beating an opponent with the forelimbs (Goodall, 1986; Livingstone, 1962; Wrangham and Peterson, 1996), jumping on an opponent that is lying on the ground (Goodall, 1986), and grabbing an opponent to deliver bites (Goodall, 1986; Wrangham and Peterson, 1996). Chimpanzees also stand bipedally to use weapons such as rocks and clubs (de Waal, 1989; Goodall, 1986; Kortlandt, 1980; Wrangham and Peterson, 1996). Thus, bipedal posture allows the forelimbs to be used as weapons to beat, restrain, and manipulate opponents. For a species that is normally quadrupedal, adoption of an upright bipedal posture also significantly reduces rotational inertia of the body (Carrier et al., 2001; Lee et al., 2001), increasing the speed at which the individual can rotate to face a new direction, as well as increasing the speed, and likely the power with which a round-house blow can be applied to an opponent. Hence, characters that improve balance and strength in a bipedal stance would likely enhance fighting ability in a lineage of quadrupeds that uses forelimbs as weapons. These arguments lead one to the suggestion that the bipedal posture and locomotion of early australopithecines may have improved their fighting performance relative to that of their more fully arboreal ancestors (Table 1).

The idea that selection for aggressive behavior may have played a role in the evolution of hominid bipedalism is not new. Wescott (1967) suggested the evolution of bipedalism allowed agonistic exhibitionism in which bipedal behavior made the individual appear larger and more threatening. Jablonski and Chaplin (1993) have argued that the bipedal activities that were critical to the differentiation of the Hominidae were those involved in the control of intragroup aggression, the successful resolution of intergroup conflicts, and the acquisition of mates. The most important behaviors in this context, they suggest, were bipedal displays, bipedal charges, and bipedal mock fights and deference to these behaviors. Livingstone (1962) used observations of aggressive bipedal chest-thumping, charging, and fighting in gorillas to argue that aggression was fundamental to the evolution of hominid bipedalism. Both Washburn (1967, cited in Guthrie (1970)) and Guthrie, (1970) have suggested that as forelimbs became more important in aggression in the lineage from which hominids evolved, a bipedal posture was adopted not only in fighting but also to present a more formidable appearance in height and breadth. Kortlandt (1980) suggested that early hominids adopted an upright posture and gait to be able to wield thorn branches as a defense against their predators.

In addition to using bipedal posture when fighting, male chimpanzees and bonobos appear to prefer to fight while on the ground rather than from an arboreal perch. The most violent and serious fights in both chimpanzees and bonobos occur on the ground (de Waal, 1986, 1989; Goodall, 1986; Kano, 1992; Wrangham and Peterson, 1996). In contrast, when attempting to avoid physical aggression from other males, both chimpanzees and bonobos retreat by climbing into trees (de Waal, 1989; Goodall, 1986; Kano, 1992). These observations highlight the importance of the terrestrial substrate during fighting in our closest living relatives and are consistent with the hypothesis that specialization for balance and strength in bipedal posture may have improved fighting ability in early hominids.

Thus, selection for increased male–male aggression may have helped draw the early hominids to the ground. Alternatively, the necessity of greater terrestrial specialization, due to the opening of the habitat, may have opened the door for greater fighting specialization in early australopithecines by freeing the forelimbs from their primary role in climbing. In either case, it is likely that as a consequence of terrestrial specializations early australopithecines had greater fighting abilities and were more dangerous opponents than were their arboreal ancestors.

Summary

Several lines of evidence lead to the suggestion that australopithecines were specialized for male–male aggression. First, relative to *Homo*, many aspects of the australopithecine musculoskeletal system appear to have conferred high potential for fighting with the forelimbs: long powerful forelimbs, short hindlimbs, broad pelvic girdles, limited potential for storage and recovery of elastic strain energy in muscle-tendon systems, robust muscles that raise the pectoral girdle on the trunk, and proficiency in bipedal stance and gait. Although some of these features were characters inherited from arboreal ancestors, they would have conferred high fighting performance in the first hominids. Most importantly, the high ratio of forelimb to hindlimb joint robustness, the very wide hips, relatively short hindlimbs, and the robust axial and appendicular muscles of the head and neck are more consistent with specialization for fighting with the forelimbs than with specialization for suspension climbing or terrestrial locomotion. Second, the high level of sexual dimorphism in body and forelimb size in early australopithocines is also more consistent with high levels of male–male aggression than with specialization for climbing. Third, it can be argued that specialization for terrestrial locomotion and bipedal posture likely improved the fighting performance of australopithecines relative to their arboreal ancestors. Thus, australopithecines appear to have been apes that were specialized for both a terrestrial habitat and for male–male aggression.

IMPLICATIONS OF THE RUNNING-FIGHTING DICHOTOMY TO THE EVOLUTION OF *HOMO*

Homo sapiens is a violent species, and the pattern of modern violence indicates that homicide and warfare are often causally tied to male–male competition for access to reproductive females (Daly and Wilson, 1988; Keeley, 1996; Wilkinson, 1997; Willey, 1990; Wrangham and Peterson, 1996). The analyses by Daly and Wilson (1988) have shown that the predominant form of homicide by humans involves males killing other males. Across cultures, same-sex homicides are on average 27 times more likely to be among males than among females. Daly and Wilson (1988) argue that sexual selection theory (Bateman, 1948; Trivers, 1972) predicts the cross-cultural universality of this enormous sexual difference in the incidence of same-sex homicide. They also relate a young man's social standing and life prospects to the probability that he will turn to violence in his pursuit of status, resources, and women. Hence, evidence from modern humans indicates that we are a violent species and the evolutionary basis of much human violence appears to be the result of sexual selection for male–male competition.

Yet a comparison of *Homo* and *Australopithecus*, indicates there is less anatomical specialization for fighting in *Homo* than was the case in early hominids. Some of the most conspicuous changes that occurred in the evolution of *Homo* from an australopithecine ancestor include an increase in body size, a reduction in relative forelimb length and robustness, a lengthening of the hindlimb, a reduction in the relative size of the feet, a relative narrowing of the pelvis, development of a more barrel-shaped chest, reduction in relative head mass, reduction of relative gut size, larger surface to volume ratio, and a drop in sexual dimorphism. Each of these changes can be argued to represent anatomical changes that would have facilitated economical transport and locomotor stamina (see above discussion; Bramble, 1990, 2000; Carrier, 1984; Jungers, 1991; Kramer and Eck, 2000; McHenry, 1994; Ruff, 1991, 1994; Wolpoff, 1999). Within the context of the locomotor-fighting dichotomy, however, they are also changes in body configuration (excluding the increase in body mass) that would reduce male fighting ability. Nevertheless, if modern *Homo sapiens* are representative of the genus, male–male aggression and propensity for male violence are unlikely to have abated with the emergence of *Homo* (see for example Daly and Wilson, 1988; Frayer, 1997; Keeley, 1996; LeBlanc, 1999; Wrangham and Peterson, 1996). Thus, our species is specialized both for locomotor stamina and male–male aggression, and this appears to be in conflict with the suggestion that there are functional tradeoffs between specialization for fighting performance and specialization for locomotor economy. Are the functional tradeoffs between running and fighting a false dichotomy?

The distinctions that I wish to raise between *Australopithecus* and *Homo* are based on body shape and locomotor anatomy. Hence, in this discussion I will follow the division proposed by Wood and Collard (1999) and Wolpoff (1999) in which *Paranthropus* and *H. habilis* and *H. rudolfensis* are grouped with the australopithecines and the genus *Homo* is represented by *H. ergaster*, *H. erectus*, *H. neaderthalensis*, and *H. sapiens* [i.e., *H. sapiens* (Wolpoff, 1999)].

Cervids, Bovids, Antilocapra, and Macropods

Humans are not the only species that are both locomotor and fighting specialists. In terms of running speed and stamina three families of artiodactyls appear to be the top performers. Although cheetahs are often recognized as being the fastest runners on the planet, species in the artiodactyls families Cervidae, Bovidae, and Antilocapridae run almost as fast, and are known to have much greater stamina than the fastest cat (Lindstedt et al., 1991; Nowak and Paradiso, 1983). Yet these three families of artiodactyls are also recognized for their polygynous mating systems with intense male competition and male–male aggression (Andersson, 1994; Geist, 1971; Jarman, 1989). At first glance, these artiodactyl families appear to falsify the hypothesis of a running-fighting dichotomy.

Alternatively, these artiodactyl lineages may have circumvented the limits imposed by the dichotomy. Each of these groups, apparently independently (Janus, 1982), evolved a new weapon system, horns or antlers, which is largely independent of the locomotor system. The basal weapons of mammals are teeth, claws, and limbs. But cervids, bovids, and antilocaprids evolved teeth that are grinding machines for tough, fibrous plant material, and evolved limbs that function primarily as pogo sticks for economical transport. In other words, their basal weapons have become so specialized for feeding and locomotion that they no longer function as effective weapons. The evolution of new weapons, horns or antlers, in each of these lineages very likely made extreme specialization of teeth and limbs possible.

If the evolution of new weapons systems that are independent of locomotor function can circumvent the functional conflicts between running and fighting, then two predictions can be made. First, artiodactyls with horns or antlers should be better runners than artiodactyls without these weapons. Second, artiodactyls with horns or antlers should be better runners than other cursorial mammals. Figure 4 shows that these predictions are largely true in regards to running speed. Thus, the evolution of horns or antlers in three families of artiodactyls may have circumvented the locomotor-fighting dichotomy, allowing greater specialization for running speed and stamina than is possible in other mammalian lineages.

In this context, kangaroos also warrant consideration. Although not as fast as many species of artiodactyls, kangaroos are among the most economical of runners. The hindlimbs of kangaroos function so well in the storage and recovery of elastic strain energy (Biewener et al., 1998) that the energetic cost to run a given distance actually decreases as running speed increases in the species *Macropus rufus* (Dawson and Taylor, 1973). This high capacity for elastic storage is expected to diminish fighting ability (see discussion above), yet larger species of kangaroo such as *M. rufus* are noted for fierce fighting among males over access to females. Kangaroos fight with their forelimbs. They use their forelimbs to punch, and to hold their opponent around the head, neck and shoulders while they bite (Jarman, 1989). Other species hold the opponent at arm's length, clawing at the face and positioning him for a kick with the hind feet. Consistent with forelimbs being weapons, sexual dimorphism in muscular strength and length of forelimbs is highly correlated with the level of sexual dimorphism in body mass (Jarman, 1989). Hence, adoption of a bipedal hopping gait has allowed the forelimbs to become specialized for fighting. Specialization of the forelimbs

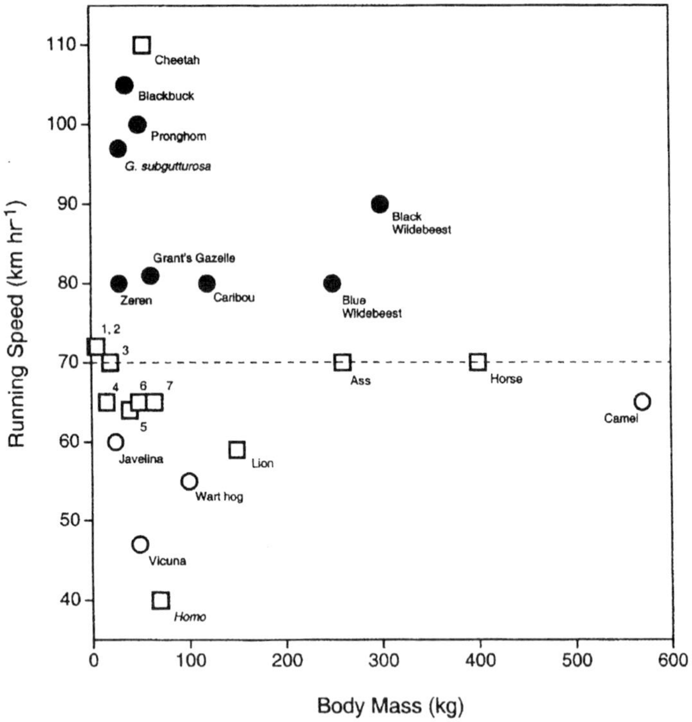

Figure 4. Maximum running speeds for representative cursorial mammals plotted against the average body mass of the species. Solid circles are species with horns or antlers from three artiodactyl families: Bovidae, Antilocapridae, and Cervidae. Open circles are species of artiodactyls that lack horns or antlers from two families: Suidae and Camelidae. The open squares are species of other cursorial mammals that also lack horns or antlers: 1) antelope jackrabbit, 2) fox (*Vulpes fulva*), 3) African hunting dog, 4) coyote, 5) gray wolf, 6) red kangaroo, 7) spotted hyena. Thus, the fastest artiodactyls with horns or antlers are faster runners than artiodactyls without these weapons. The fastest artiodactyls with horns or antlers are also faster runners than other cursorial mammals, with the exception of cheetahs. All speed records come from a compilation by Garland (1983), except for those of the Suidae and Camelidae that are from Nowak and Paradiso (1983). The speed of javelinas is an estimate based on their ability to out maneuver and escape coyotes (Nowak and Paradiso, 1983).

for fighting may, in turn, have allowed the hindlimbs to become more highly specialized for storage and recovery of elastic strain energy during running.

Invention of New Weapon Technology and the Evolution of *Homo*

Analogous to the evolution of new weapons in bovids, cervids, and antilocapra, the invention of weapon technology by early hominids may have reduced the conflict between selection for locomotor stamina versus selection for male–male aggression, and allowed a greater independence in the evolution of characters associated with running and fighting. In other words, the invention of new weapons may have allowed the evolution of locomotor specialization in the human lineage.

Among cursorial mammals, modern humans are one of the best distance runners (Carrier, 1984). Humans are not fast runners, but we do have an ability to cover great distances more rapidly than most species. This ability is remarkable given our relatively recent arboreal ancestors and the relatively high energetic cost of human running. The physiological mechanisms that allow humans to be elite distance runners, in spite of their high cost of transport, are not well understood. Humans have been suggested to have 1) superior abilities to dissipate metabolic heat (Carrier, 1984; Porter, 1993; Wheeler, 1984, 1991), and 2) an uncoupling of locomotor, ventilatory, and thermoregulatory functions that facilitates sustained vigorous running (Carrier, 1984). Recent work in sheep and dogs indicates that, in species that pant to thermoregulate and couple their ventilation to the locomotor cycle, regulation of body temperature during running does compete with both the minimization of ventilatory work and pH balance (Entin et al., 1998,1999; Wagner et al., 1997). Thus, independence of locomotor and ventilatory cycles, as well as cutaneous evaporative cooling, may indeed give humans an advantage in endurance running relative to many other species. In any case, humans are elite endurance runners and hominids evolved this capacity in a relatively short period of time from ancestors that were not specialized for running.

Anatomical features indicating cursorial specialization of *Homo* appeared in the fossil record approximately 2.0 to 1.8 my ago (Klein, 1999; Wolpoff, 1999). The earliest stone tools appeared somewhat earlier, just over 2.5 my ago (Harris and Capaldo, 1993). Foley (1987) has argued on the basis of tool use by chimpanzees (Boesch and Boesch, 1983; Goodall, 1970), that simple tool use is an ancestral hominid trait that may have had its origin in Miocene hominoids. Use of weapons, a category of tool, also likely predate the origin of hominids. Although the primary weapons of our closest living relatives are forelimbs and teeth, chimpanzees also occasionally attack by throwing rocks (de Waal, 1989; Goodall, 1986; Kortlandt, 1980) and wielding branches as clubs (Kortlandt, 1980). Given that the two extant hominoids most closely related to basal hominids, chimps and humans, use weapons, we can be relatively confident that early australopithecines used simple weapons as well.

The first weapons used by hominids in intraspecific aggression would have been relatively simple, such as the stones and tree branches used rather ineffectively by extant chimpanzees. Nevertheless, at the inception of weapons that concentrate force in order to pierce tissue, such as pointed wooden spears, or weapons that optimize the energy imparted in a blow to an opponent, such as clubs of the correct mass and rotational inertia, success in physical combat would have begun to shift from the opponent with the advantage in physical strength and quickness to the opponent with the best weapon design. In this context, it is worth noting that the history of human warfare, and therefore the course of human history, has been substantially influenced by weapon innovation and dissemination (LeBlanc, 1999; O'Connell, 1989; van Creveld, 1991). Thus, the invention of new weapon technology may have shifted the effect of selection for aggressive behavior from favoring physical performance to favoring creativity and technical innovation in the lineage from which *Homo* evolved.

If the invention of weapon technology did allow greater specialization for locomotor stamina in hominids, two predictions can be made. First, one would predict a reduction in sexual dimorphism in body and forelimb size with the appearance of specialization for locomotor stamina. Second, evidence of injuries from weapon use could be expected in the fossilized bone of individuals that have specializations for locomotor stamina.

The changes in sexual dimorphism in body size during hominid evolution are consistent with weapon innovation achieving greater importance in male–male aggression at the time of emergence of *Homo*. Sexual dimorphism in body size declined from the early australopithecines to *Homo ergaster* (McHenry, 1996). This decline has been interpreted to represent a decline in male–male aggression (McHenry, 1996). Alternatively, the invention of new weapon technology may have diminished the selective value of relatively large brawny males, and favored males and females of more similar size and shape for the sake of economical distance transport.

The incidence of healed cranial injuries also appears to be consistent with weapon innovation being associated with the evolution of *Homo*. Although there appears to be a high incidence of healed cranial fractures in early *Homo*, such injuries are not known in the fossil record of *Australopithecus* (Wolpoff, 1999). The clearest examples of healed cranial fractures come from the Zhou-Kou-Dian and Ngandong Middle Pleistocene *Homo erectus* sample (Weidenreich, 1943, 1951) and from the Neandertal sample (Berger and Trinkaus, 1995; Zollikofer et al., 2002). Of the eleven crania represented in the Ngandong sample, Weindenreich (1951) discusses healed cranial injuries in three skulls (I, IV, and VI). He attributes the injuries in skulls IV and VI to interpersonal violence, and "to have resulted from the effect of cutting blows (Skull IV), and the wielding of blunt weapons (Skull VI) . . . ". In an analysis of the anatomical distribution of traumatic lesions, Berger and Trinkaus (1995) found a similar incidence of head and neck injuries in Neandertals. Thirty percent of the specimens exhibited evidence of head or neck trauma. They found a tight correspondence between the pattern of injuries in the Neandertal sample and that of modern Rodeo atheletes and, therefore, speculated that the Neandertal pattern of injury might have been the result of trauma sustained during predation on large game animals. Berger and Trinkaus, however, do acknowledge the possibility that the Neandertal injuries were caused by interpersonal violence and there does not appear to be evidence to exclude this possibility. The recently described healed fracture in the St. Cesaire Neandertal is suggested to have resulted from interpersonal violence (Zollikofer et al., 2002). Finally, the Dmanisi 2280 skull provides evidence of large cranial injuries in *Homo* dating back to 1.7 million years ago (Wolpoff, per. com.). Given that neither modern humans nor nonhuman apes have the ability to crush the skulls of adult conspecifics with their bare hands, these examples of cranial injury in early *Homo* may be evidence of head-crushing weapons.

Structural buttressing and thickening of the cranium distinguishes early *Homo* from *Australopithecus* and may be related to the increased incidence of cranial injury. Cranial buttressing that separate *H. ergaster* and *H. erectus* from *Australopithecus* (including *A. habilis*) include the formation of a pronounced supraorbital torus (i.e., browridge) with a supratoral sulcus; an occipital torus (Klein, 1999); a torus along the temporal lines; one along the top of the cranium, and tori above the mastoids and ears (Weidenreich, 1943; Wolpoff, 1999). This buttressing, in particular the supraorbital torus, does not appear to be a function of mechanical loading due to mastication (Hylander and Ravosa, 1992; Lieberman, 2000; Ross and Hylander, 1996). Lieberman (2000) provides a review of the developmental basis of the supraorbital torus and relates differences in projection of the torus among groups of *Homo* to differences in facial projection and the cranial base. Formation of large occipital tori in the crania of early *Homo* has been suggested to be associated with enlarged occipital muscles necessary for head stabilization during running (Bramble, 1991, 2000). Independent of the developmental basis of cranial buttressing and possible functional implications to mastication and locomotion, cranial buttressing in early *Homo* must have resulted in a general strengthening of the cranial vault (Weidenreich, 1943; Wolpoff, 1999). Additionally, the cranial bones of early *Homo* were often much thicker than those in either *A. africanus* or *A. hablis* (Weidenreich, 1943; Wolpoff, 1999). This thickening of the cranium may be a developmental response to increased intensity of locomotor activity, as suggested by

Lieberman (1996). Nevertheless, evidence from the limb bones of domestic dogs indicates that bone thickness can be highly heritable (Chase et al., 2002; Lark pers. com.). These observations raise the possibility that the buttressing and increased thickness of the crania observed in *H. ergaster* and *H. erectus* may have been an adaptive response to an increased likelihood of cranial injury.

Summary

The observation that modern humans are both elite distance runners and a physically violent species appears to defy the hypothesis of functional tradeoffs between specialization for locomotion and specialization for fighting. Humans, as well as bovids, cervids, and antilocapra, however, may be the exceptions that prove the running-fighting dichotomy. All four groups appear to have evolved exceptional locomotor performance in association with the evolution or invention of new weapon systems that are largely independent of locomotion. In the case of hominids, the evolution of *Homo* may have been dependent on the invention of weapons that allowed specialization for locomotion and fighting to proceed independently.

CONCLUSIONS

Many aspects of their musculoskeletal system, as well as a high level of sexual dimorphism in body and forelimb size, suggest that australopithecines had a high level of male–male aggression. Furthermore, the adoption of bipedal posture and a terrestrial habitat are likely to have improved fighting performance in the early australopithecines relative to their arboreal ancestors. Indeed, selection for fighting performance may be one of the factors that led the ancestors of hominids to spend more time on the ground and adopt a bipedal striding gait. Thus, australopithecines appear to have been fighting specialists.

The characters that made australopithecines adept fighters, however, must also have limited their terrestrial locomotion. Their pronounced forelimb strength, short stature, limited potential for storage and recovery of elastic strain energy in the muscle-tendon systems of the limbs, wide hips, and robust head and neck would likely have made australopithecines inefficient runners by modern human standards. Nevertheless, in a relatively short period of a few million years, hominids evolved from arboreal climbers into a species that is one of the best distance runners on the planet. If the functional tradeoffs entailed in the running-fighting dichotomy are real, the evolution of elite endurance capacity in a lineage that was specialized for fighting represents a fundamental change in evolutionary trajectory. Gaining an understanding of the causes of this change in evolutionary pathway will likely improve our understanding of human nature and help explain why humans evolved.

Rather than trading fighting performance for locomotor stamina, I suggest that selection for male–male aggression resulted in the invention of weapon technology in the ancestors of *Homo*. By reducing inherent functional conflicts, the new weapons allowed the evolution of increased locomotor speed and stamina. With the invention of the first effective weapons selection for aggression in early hominids would have shifted from favoring greater physical strength and agility to favoring creative innovation. In this scenario, the emergence of *Homo* is: 1) dependent on the invention of new weapons, and 2) represents both greatly improved distance transport and increased lethality in a lineage that was already specialized for fighting.

ACKNOWLEDGEMENTS I wish to thank Jeff Meldrum and Charles Hilton for inviting me to participate in this symposium. Milford Wolpoff provided much appreciated early encouragement and advice, as well as many helpful comments on the manuscript. Charles Hilton and Dan Lieberman also generously responded to an endless stream of questions and provided very helpful feedback on early versions of the manuscript. I wish to also thank Dr. Norman Heglund, Dr. Patrick Willems, and the Université catholique de Louvain for their support during the completion of this chapter. This study was partially supported by The National Science Foundation: IBN-0212141.

REFERENCES

Abitbol, M. M., 1987, Obstetrics and Posture in Pelvic Anatomy. *Journal of Human Evolution* 16:243–255.

Adams, L. M,. and Moore, W. J., 1975, Biomechanical Appraisal of Some Skeletal Features Associated with Head Balance and Posture in the Hominoidea. *Acta Anatomica* 92:580–594.

Alexander, R. Mcn., 1984, Elastic Energy Stores in Running Vertebrates. *American Zoologist* 24:85–94.

Alexander, R. Mcn., 1988, *Elastic Mechanisms in Animal Movement*. Cambridge University Press, Cambridge.

Andersson, M., 1994, *Sexual Selection*. Princeton University Press, Princeton.

Barnett, S. A., 1967, *Instinct and Intelligence*. Macgibbon and Kee, London.

Bateman, A. J., 1948, Intra-sexual Selection in Drosophila. *Heredity* 2:349–368.

Begun, D. R., Ward, C. V., and Rose, M. D., 1997, Events in Hominoid Evolution. In *Function, Phylogeny and Fossils, Miocene Hominoid Evolution and Adaptations*, edited by D. R. Begun, C. V. Ward, and M. D. Rose, pp. 389–415. Plenum Press, New York.

Berge, C., and Kazminerczak, J. B., 1986, Effects of Size and Locomotor Adaptations on the Hominid Pelvis: Evaluation of Australopithecine Bipedality with a New Multivariate Method. *Folia Primatologia* 46:185–204.

Berger, T. D., and Trinkaus, E., 1995, Patterns of Trauma among Neandertals. *Journal of Archeological Science* 22:841–852.

Berzin, A. A., 1972, *The Sperm Whale*. Israel Program For Scientific Translations, Jerusalem.

Biewener, A. A., 1998, Muscle-Tendon Stresses and Elastic Energy Storage During Locomotion in the Horse. *Comparative Biochemical Physiology B* 120:73–87.

Biewener, A. A., Konieczynski, D. D., and Baudinette, R. V., 1998, In-Vivo Muscle Force-Length Behavior During Steady-Speed Hopping in Tammar Wallabies. *Journal of Experimental Biology* 201:1681–1694.

Blickhan, R. and Full, R. J., 1987, Locomotion Energetics of the Ghost Crab. II. Mechanics of the Center of Mass During Walking and Running. *Journal of Experimental Biology* 130:155–174.

Boesch, C., and Boesch, H., 1983, Optimization of Nut-Cracking with Natural Hammers by Wild Chimpanzees. *Behavior* 83:265–286.

Bramble, D.M., 1990, Head Stabilization and Locomotor Behavior in the Hominidae (Abstract). *American Journal of Physical Anthropology* 8:197–198.

Bramble, D. M., 2000, Head Stabilization in Human Running: Implications for Human Evolution. *American Journal of Physical Anthropology* Sup. 30:111.

Cant, J. G. H., 1987, Positional Behavior of Female Bornean Orangutans (*Pongo pygmaeus*). *American Journal of Primatology* 12:71–90.

Carrier, D. R., 1984, The Energetic Paradox of Human Running and Hominid Evolution. *Current Anthropology* 25:483–495.

Carrier, D. R., 1990, Activity of the Hypaxial Muscles During Walking in the Lizard *Iguana iguana*. *Journal of Experimental Biology* 152:453–470.

Carrier, D. R., Lee, D. V., and Walter, R. M., 2001, Influence of Rotational Inertia on the Turning Performance of Theropod Dinosaurs: Clues from Humans with Increased Rotational Inertia. *Journal of Experimental Biology* 204:3917–3926.

Cartmill, M., 1974, Pads and Claws in Arboreal Locomotion. In *Primate Locomotion*, edited by F. A. Jenkins, Jr, pp. 45–83. Academic Press, New York.

Cavagna, G. A., Heglund, N. C., and Taylor. C. R., 1977, Mechanical Work in Terrestrial Locomotion: Two Basic Mechanisms for Minimizing Energy Expenditure. *American Journal of Physiology* 233:R243–261.
Cavagna, G. A., and Kaneko, M., 1977, Mechanical Work and Efficiency in Level Walking and Running. *Journal of Physiology* 268:467–481.
Cavagna, G. A., Saibene, F. P., and Margaria, R., 1964, Mechanical Work in Running. *Journal of Applied Physiology* 19:249–256.
Chase, K., Carrier, D. R., Adler, F. R., Jarvik, T., Ostrander, E. A., Lorentzen, T., and Lark, K. G., 2002, Genetic Basis for Systems of Skeletal Quantitative Traits: Principle Component Analysis of the Canid Skeleton. *Proceedings of the National Academy of Sciences* 99:9930–9935.
Clutton-Brock, T. H., 1982, The Function of Antlers. *Behavior* 70:108–125.
Clutton-Brock, T. H., Harvey, P. H., and Rudder, B., 1977, Sexual Dimorphism, Socionomic Sex Ratio and Body Weight in Primates. *Nature* 269:797–800.
Clutton-Brock, T. H., Albon, S. D., and Harvey, P. H., 1980, Antlers, Body Size and Breeding Group Size in the Cervidae. *Nature* 285:565–566.
Cruse, H., 1976, The Function of Legs in the Free Walking Stick Insect, *Carausius morosus*. *Journal of Comparative Physiology* 112:135–162.
Daly, M., and Wilson, M., 1988, *Homicide*. Aldine De Gruyter, New York.
Dart, R. A., 1925, *Australopithecus africanus*: the Man-ape of South Africa. *Nature* 115:195–199.
Darwin, C., 1871, *The Descent of Man, and Selection in Relation to Sex*. John Murray, London.
Dawson, T. J., and Taylor, C. R., 1973, Energetic Cost of Locomotion in Kangaroos. *Nature* 240:313–314.
De Waal, F. B. M., 1986, The Brutal Elimination of a Rival Among Captive Male Chimpanzees. *Ethology and Sociobiology* 7:237–251.
De Waal, F. B. M., 1982, *Chimpanzee Politics: Power and Sex among the Apes*. Harper and Row, New York.
Dickinson, M. H., Farley, C. T., Full, R. J., Koehl, M. A. R., Kram R., and Lehman, S., 2000, How Animals Move: An Integrative View. *Science* 288:100–106.
Enquist, M., and Leimar, O., 1990, The Evolution of Fatal Fighting. *Animal Behavior.* 39:1–9.
Entin, P. L., Robertshaw, D., and Rawson, R. E., 1998, Thermal Drive Contributes to Hyperventilation During Exercise in Sheep. *Journal of Applied Physiology* 85:318–325.
Entin, P. L., Robertshaw, D., and Rawson, R. E., 1999, Effect of Locomotor Respiratory Coupling on Respiratory Evaporative Heat Loss in the Sheep. *Journal of Applied Physiology* 87:1887–1893.
Evans, H. E., 1993, *Miller's Anatomy of the Dog*. W. B. Saunders Company, Philadelphia.
Fedak, M. A., Heglund, N. C., and Taylor, C. R., 1982, Energetics and Mechanics of Terrestrial Locomotion: II. Kinetic Energy Changes of the Limbs and Body as a Function of Speed and Body Size in Birds and Mammals. *Journal of Experimental Biology* 79:23–40.
Fife, M. M., Bailey, C. L., Lee, D. V., and Carrier D. R., 2001, Function of the Oblique Hypaxial Muscles in Trotting Dogs. *Journal of Experimental Biology* 204:2371–2381.
Foley, R. A., 1987, *Another Unique Species*. Longman, Harlow.
Foley, R. A., 1989, Finite Social Space, Evolutionary Pathways, and Reconstructing Hominid Behavior. *Science* 243:901–906.
Fossey, D., 1983, *Gorillas in the Mist*. Houghton Mifflin, New York.
Fossey, D., 1984, Infanticide in Mountain Gorillas (*Gorilla gorilla beringei*) with Comparative Notes on Chimpanzees. In *Infanticide: Comparative and Evolutionary Perspectives*, edited by G. Hausfater and S. B. Hrdy, pp. 217–236. Hawthorne, New York.
Frayer, D. W., 1997, Ofntet: Evidence for a Mesolithic Masacre. In *Troubled Times: Violence and Warfare in the Past*, edited by D. L. Martin and D. W. Frayer, pp. 181–216. Gordon and Breach Publishers, Amsterdam.
Frayer, D. W., and Wolpoff, M. H., 1985, Sexual Dimorphism. *Annual Review of Anthropology* 14:429–474.
Full, R. J., Blickhan, R., and Ting, L. H., 1991, Leg Design in Hexapedal Runners. *Journal of Experimental Biology* 158:369–390.
Gans, C., 1988, Adaptation and the Form–Function Relation. *American Zoologist* 28:681–97.
Garland, T. Jr., 1983, The Relation Between Maximum Running Speed and Body Mass in Terrestrial Mammals. *Journal of Zoology, London* 199:157–170.
Garland, T. Jr., 2002, Selection Experiments: An Underutilized Tool in Biomechanics and Organismal Biology. In *Biomechanics and Evolution*, edited by V. L. Bels, J. P. Gasc, and A. Casinos. Bios Scientific Publishers, Oxford, U.K.
Garland, T. Jr., and Adolph, S. C., 1994, Why Not to Do Two-Species Comparative Studies: Limitations on Inferring Adaptation. *Physiological Zoology* 67:797–828.
Geist, V., 1971, *Mountain Sheep*. University of Chicago Press, Chicago.

Goodall, J., 1986, *The Chimpanzees of Gombe, Patterns of Behavior.* Belknap Press of Harvard University Press, Cambridge.

Gray, J., 1968, *Animal Locomotion.* Norton, New York.

Gregory, W. K., 1916, Studies on the Evolution of the Primates. *Bulletin of the American Museum of Natural History* 35:239–355.

Gregory, W. K., 1928, Were the Ancestors of Man Primative Brachiators? *Proceedings of the American Philosophical Society* 67:129–150.

Guthrie, R. D., 1970, Evolution of Human Threat Display Organs. In *Evolutionary Biology Vol. 4*, edited by Dobzhansky, T., Hecht, M. K., and Steere. W. C. Meredith Corp, New York.

Hamilton, W. D., 1979, Wingless and Fighting Males in Fig Wasps and Other Insects. In *Sexual Selection and Reproductive Competition in Insects*, edited by M. S. Blum and N. A. Blum, pp. 167–220. Academic Press, New York.

Harris, J. W. K., and Capaldo, S. D., 1993, The Earliest Stone Tools: Their Implications for an Understanding of the Activities and Behavior of Late Pliocene Hominids. In *The Use of Tools by Human and Non-Human Primates*, edited by A. Berthelet, and J. Chavaillon, pp. 196–220. Clarendon Press, Oxford.

Harrison, T., 1987, A Reassessment of the Phylogenetic Relationships of *Oreopithecus bambolii* Gervais. *Journal of Human Evolution* 15:541–583.

Harrison, T., 1991, The Implications of *Oreopithecus bambolii* for the Origins of Bipedalism. In *Origine(S) De La Bipedie Chez Les Hominides*, edited by Y. Coppen and B. Senut, pp. 235–244. Cahiers De Paleoanthropologie, Centre National De La Recherche Scientifique, Paris.

Hartwig-Scherer, S., 1993, Body Weight Prediction in Early Fossil Homiinds: Towards a Taxon-"Independent" Approach. *American Journal of Physical Anthropology* 92:17–36.

Heglund, N. C., Cavagna, G. A., and Taylor, C. R., 1982, Energetics and Mechanics of Terrestrial Locomotion. III. Energy Changes in The Center of Mass as a Function of Speed and Body Size in Birds and Mammals. *Journal of Experimental Biology* 79:41–56.

Heglund, N. C., and Schepens, B., 2002, Ontogeny Recapitulates Phylogeny? Locomotion in Children and Other Primative Hominids. In *Vertebrate Biomechanics and Evolution*, edited by Bels, Gase, and Casinos. BIOS Scientific Publishers, Oxford.

Heglund, N. C., and Taylor, C. R., 1988, Speed, Stride Frequency and Energy Cost Per Stride: How Do They Change with Body Size and Gait? *Journal of Experimental Biology* 138:301–318.

Hildebrand, M., and Goslow, G., 2001, *Analysis of Vertebrate Structure.* John Wilely and Sons, New York.

Hildebrand, M., and Hurley, J. P., 1985, Energy of the Oscillating Legs of a Fast-Moving Cheetah (*Acinonyx jubatus*), Pronghorn (*Antilocapra americana*), Jackrabbit (*Lepus californicus*) and Elephant (*Elephas maximus*). *Journal of Morphology* 184:23–32.

Hunt, K. D., 1991a, Positional Behavior in the Hominoidea. *International Journal of Primatology* 12:95–118.

Hunt, K. D., 1991b, Mechanical Implications of Chimpanzee Positional Behavior. *American Journal of Physical Anthropology* 86:521–536.

Hunt, K. D., 1998, Ecological Morphology of *Australopithecus afarensis*: Traveling Terrestrially, Eating Arboreally. In *Primate Locomotion*, edited by E. Strasser, J. Fleagle, A. Rosenberger, and H. McHenry, pp. 397–418. Plenum Press, New York.

Huntingford, F. A., and Turner, A. K., 1987, *Animal Conflict.* Chapman and Hall, London.

Hylander, W. L., and Ravosa, M. J., 1992, An Analysis of the Supraorbital Region of Primates: a Morphometric and Experimental Aproach. In *Structure, Function, and Evolution of Teeth*, edited by P. Smith and E. Tchernov, pp. 223–255. Freund, London.

Jablonski, N. G., and Chaplin, G., 1993, Origin of Habitual Terrestrial Bipedalism in the Ancestor of the Hominidae. *Journal of Human Evolution* 24:259–280.

Janis, C., 1982, Evolution of Horns in Ungulates: Ecology and Paleoecology. *Biological Review* 57:261–318.

Jarman, P. J., 1983, Mating System and Sexual Dimorphism in Large, Terrestrial, Mammalian Herbivores. *Biological Review* 58:485–520.

Jarman, P. J., 1989, Sexual Dimorphism in Macropodoidea. In *Kangaroos, Wallabies, and Rat-Kangaroos*, edited by Grigg, G., Jarman, P., and Hume, I., pp. 433–447. Surrwy Beatty & Sons, New South Wales.

Jayes, A. S., and Alexander, R. Mcn., 1978, Mechanics of Locomotion of Dogs (*Canis familiaris*) and Sheep (*Ovis aries*). *Journal of Zoology, London* 185:289–308.

Jungers, W. L., 1978, The Functional Significance of Skeletal Allometry in *Megaladapis* in Comparison to Living prosimians. *Journal of Physical Anthropology* 49:303–314.

Jungers, W. L., 1982, Lucy's Limbs: Skeletal Allometry and Locomotion in *Australopithecus afarensis. Nature* 297:676–678.

Jungers, W. L., 1984, Scaling of the Hominoid Locomotor Skeleton with Special Reference to the Lesser Apes. In *The Lesser Apes: Evolutionary and Behavioral Biology*, edited by H. Preuschoft, D. J. CHivers, W. Y. Brockelman, and N. Creel, pp. 146–169. Edinburgh University Press, Edinburgh.

Jungers, W. L., 1991, A Pygmy Perspective On Body Size and Shape in *Australopithecus afarensis* (AL 288–1, "Lucy"). In *Origine(S) De La Bipedie Chez Les Hominides*, edited by Y. Coppen and B. Senut, pp. 215–224. Cahiers de Paleoanthropologie, Centre National de la Recherche Scientifique, Paris.

Kano, T., 1992, *The Last Ape: Pygmy Chimpanzee Behavior and Ecology*. Translated by Evelyn Ono Vineberg. Stanford University Press, Stanford.

Katz, B., 1939, The Relation Between Force and Speed in Muscular Contraction. *Journal of Physiology* 96:45–64.

Keeley, L., 1996, *War Before Civilization: The Myth of the Peaceful Savage*. Oxford University Press, New York.

Ker, R. F., Bennett, M. B., Bibby, S. R., Kester, R. C., and Alexander, R. Mcn., 1987, The Spring in the Arch of the Human Foot. *Nature* 325:147–149.

Kimbel, W. H., and White, T. D., 1988, Variation, Sexual Dimorphism and the Taxonomy of *Australopithecus*. In *Evolutionary History of the Robust Australopithecines*, edited by F. E. Grine, pp. 175–192. Aldine de Gruyter, New York.

Klein, R. G., 1999, *The Human Career, Human Biological and Cultural Origins*. University of Chicago Press, Chicago.

Kortlandt, A., 1980, How Might Early Hominids Have Defended Themselves Against Large Predators and Food Competitors? *Journal of Human Evolution* 9:79–112.

Kram, R., and Taylor, C. R., 1990, Energetics of Running: A New Perspective. *Nature* 346:265–267.

Kramer, P. A., 1999, Modeling the Locomotor Energetics of Extinct Hominids. *Journal of Experimental Biology* 202:2807–2818.

Kramer, P. A., and Eck, G. G., 2000, Locomotor Energetics and Leg Length in Hominid Bipedality. *Journal of Human Evolution* 38:651–666.

LaBlanc, S. A., 1999, *Prehistoric Warfare in the American Southwest*. University of Utah Press, Salt Lake City.

Latimer, B., and Lovejoy, C. O., 1989, The Calcaneous of *Australopithecus afarensis* and its Implications for the Evolution of Bipedality. *American Journal of Physical Anthropology* 78:369–386.

Latimer, B., and Lovejoy, C. O., 1990a, Hallucal Tarsometatarsal Joint in *Australopithecus afaresis. American Journal of Physical Anthropology* 82:125–133.

Latimer, B., and Lovejoy, C. O., 1990b, Metatarsophalangeal joints of *Australopithecus afarensis. American Journal of Physical Anthropology* 83:13–23.

Latimer, B., Ohman, J. C., and Lovejoy, C. O., 1987, Talocural join in African Hominids: Implications for *Australopithecus afarensis. American Journal of Physical Anthropology* 74:155–175.

Lauder, G. V., 1991, An Evolutionary Perspective on the Concept of Efficiency: How Does Function Evolve? In *Efficiency and Economy in Animal Physiology*, edited by R. W. Blake, pp. 169–184. Cambridge University Press, Cambridge.

Lee, D. V., Walter, R. M., Deban, S. M., and Carrier, D. R., 2001, Influence of Increased Rotational Inertia on the Turning Performance of Humans. *Journal of Experimental Biology* 204:3927–3934.

Lieberman, D. E., 1996, How and Why Humans Grow Thin Skulls: Experimental Evidence for Systemic Cotrical Robusticity. *American Journal of Physical Anthropology* 101:217–236.

Lieberman, D. E., 2000, Ontogeny, Homology, and Phylogeny in the Hominid Craniofacial Skeleton: the Problem of the Browridge. In *Development, Growth, and Evolution: Implications for the Study of Hominid Skeletal Evolution*, edited by P. O'Higgins and M. Cohn, pp. 85–122. Academic Press, London.

Lieberman, D. E., Devlin, M. J., and Pearson, O. M., 2001, Articular Area Responses to Mechanical Loading: Effects of Exercise, Age, and Skeletal Location. *American Journal of Physical Anthropology* 116:266–277.

Lindstedt, S. L., Hokanson, J. F., Wells, D. J., Swain, S. D., Hoppeler, H., and Navarro, V., 1991, Running Energetics in the Pronghorn Antelope. *Nature* 353:748–750.

Livingstone, F. B., 1962, Reconstructing Man's Pliocene Pongid Ancestor. *American Anthropologist* 64:301–305.

Lockwood, C. A., Richmond, B. G., Jungers, W. L., and Kimbel, W. H., 1996, Randomization Procedures and Sexual Dimorphism in *Australopithecus afarensis. Journal of Human Evolution* 31:537–548.

Lorenz, K., 1963, *Das Sogenannte Bose*. Dr. G. Borotha-Schoeler-Verlag, Vienna.

Lovejoy, C. O., 1978, A Biomechanical Review of the Locomotor Diversity of Early Hominids. In *Early Hominids of Africa*, edited by C. J. Jolly, pp. 403–429. St. Marin's Press, New York.

Lovejoy, C. O., 1981, The Origin of Man. *Science* 211:341–350.
Lovejoy, C. O., 1988, Evolution of Human Walking. *Scientific American* 259:118–125.
Lovejoy, C. O., Heiple, K. G., and Burstein, A. H., 1973, The Gait of *Australopithecus*. *American Journal of Physical Anthropology* 38:757–780.
Lovejoy, C. O., Johanson, D. C., and Coppens, Y., 1982, Elements of the Axial Skeleton Recovered from the Hadar Formation: 1974–1977 Collections. *American Journal of Physical Anthropology* 57:631–635.
Mackinnon, J. R., and Mackinnon, K. S., 1984, Territorality, Monogamy and Song in Gibbons and Tarsiers. In *The Lesser Apes: Evolutionary and Behavioral Biology*, edited by H. Preuschoft, D. J. Chivers, W. Y. Brockelman, and N. Creel, pp. 291–297. Edinburgh University Press, Edinburgh.
Maynard Smith, J., Burian, R., Kauffman, S., Alberch, P., Campbell, J., Goodwin, B., Lande, R., Raup, D., and Wolpert, L., 1985, Developmental Constraints and Evolution. *Quarterly Review of Biology* 60:265–287.
McHenry, H. M., 1978, Fore- and Hindlimb Proportions in Plio-Pleistocene Hominids. *American Journal of Physical Anthropology* 49:15–22.
McHenry, H. M., 1986, Size Variation in the Postcranium of *Australopithecus afarensis* and Extant Species of Hominoidea. *Human Evolution* 1:149–156.
McHenry, H. M., 1991, Sexual Dimorphism in *Australopithecus afarensis*. *Journal of Human Evolution* 20:21–32.
McHenry, H. M., 1991, Femoral Lengths and Stature in Plio-Pleistocene Hominids. *American Journal of Physical Anthropology* 85:149–158.
McHenry, H. M., 1992, Body Size and Proportions in Early Hominids. *American Journal of Physical Anthropology* 87:407–431.
McHenry, H. M., 1994, Behavioral Ecological Implications of Early Hominid Body Size. *Journal of Human Evolution* 27:77–87.
McHenry, H. M., 1996, Sexual Dimorphism in Fossil Hominids and Its Socioecological Implications. In *The Archaeology of Human Ancestry: Power, Sex and Tradition*, edited by J. Steele and S. Shennan, pp. 91–109. Routledge, New York.
McHenry, H. M., and Berger, L. R., 1998, Body Proportions in *Australopithecus afarensis* and *A. africanus* and the Origin of the Genus *Homo*. *Journal of Human Evolution* 35:1–22.
McHenry, H. M., and Coffing, K., 2000, *Australopithecus* To *Homo*: Transformations in Body and Mind. *Annual Review of Anthropology* 29:125–146.
Mitani, J. C., 1987, Territoriality and Monogamy Among Agile Gibbons (*Hylobates agilis*). *Behav. Ecol. Sociobiol.* 20:265–269.
Morton, D. J., 1926, Evolution of Man's Erect Posture (Preliminary Report). *Journal of Morphology and Physiology* 43:147–179.
Nowak, R. M., and Paradiso, J. L., 1983, *Walker's Mammals of the World. Vol. 1, 4th Edition*. Johns Hopkins University Press, Baltimore.
O'Connell, R. L., 1989, *Of Arms and Men: A History of War, Weapons, and Aggression*. Oxford University Press, New York.
Olsen, T. R.. 1985, Cranial Morphology and Systematics of the Hadar Formation Hominids and "*Australopithecus*" *africanus*. In *Ancestors: The Hard Evidence*, edited by E. Delson, pp. 102–119. Alan R. Liss, Inc, New York.
Pasi, B. M., and Carrier, D. R., 2003, Functional Tradeoffs in the Limb Muscles of Dogs Bred For Running Versus Fighting. *Journal of Evolutionary Biology* 162:339–347.
Parker, G. A., 1983, Arms Races in Evolution-An ESS to the Opponent-Independent Costs Game. *Journal of Theoretical Biology* 101:619–648.
Plavcan, J. M., and Van Schaik, C. P., 1997, Interpreting Hominid Behavior on the Basis of Sexual Dimorphism. *Journal of Human Evolution* 32:345–374.
Porter, A. M. W., 1993, Sweat and Thermoregulation in Hominids. Comments Prompted by the Publications of P. E. Wheeler 1984–1993. *Journal of Human Evolution* 25:417–423.
Preuschoft, H., Chivers, D. J., Brockelman, W. Y., and Creel, N., 1984, *The Lesser Apes: Evolutionary and Behavioral Biology*. Edinburgh University Press, Edinburgh.
Rak, Y., 1991, Lucy's Pelvic Anatomy: Its Role in Bipedal Gait. *Journal of Human Evolution* 20:283–290.
Ridley, M., 1993, *Evolution*. Blackwell Scientific Publications, Boston.
Roberts, T. J., Marsh, R. L., Weyand, P. G., and Taylor, C. R., 1997, Muscular Force in Running Turkeys: The Economy of Minimizing Work. *Science* 275:1113–1115.
Rose, M. D., 1997, Functional and Phylogenetic Features of the Forelimbs in Miocene Hominoids. In *Function, Phylogeny and Fossils, Miocene Hominoid Evolution and Adaptations*, edited by D. R. Begun, C. V. Ward, and M. D. Rose, pp. 79–99. Plenum Press, New York.

Rosenberg, K. R., and Trevathan, W., 1995, Bipedalism and Human Birth: The Obstetrical Dilemma Revisited. *Evolutionary Anthropology* 4:161–168.

Ross, C. F., and Hylander, W. L., 1996, *In vivo* and *in vitro* Bone Strain Analysis of the Circumorbital Region of *Aotus* and the Function of the Postorbital Septum. *American Journal of Physical Anthropology* 101:183–215.

Ruff, C. B., 1988, Hindlimb Articular Surface Allometry in Hominoidea and *Macaca*, with Comparisons to Diaphyseal Scaling. *Journal of Human Evolution* 17:687–714.

Ruff, C. B., 1991, Climate and Body Shape in Hominid Evolution. *Journal of Human Evolution* 21:81–105.

Ruff, C. B., 1994, Morphological Adaptation To Climate in Modern and Fossil Hominids. *Yearbook of Physical Anthropology* 37:65–107.

Sarmiento, E. E., 1987, The Phylogenetic Position of *Oreopithecus* and Its Significance in the Origin of the Hominoidea. *American Museum Noveate* 2881:1–44.

Schaller, G. B., 1963, *The Mountain Gorilla.* University of Chicago Press, Chicago.

Silverman, H. B., and Dunbar, M. J., 1980, Aggressive Tusk Use by the Narwhal (*Monodon monoceros l.*). *Nature* 284:57–58.

Stern, J. T., and Susman, R. L., 1983, The Locomotor Anatomy of *Australopithecus afarensis. American Journal of Physical Anthropology* 60:279–317.

Steudel, K., 1991, The Work and Energetic Cost of Locomotion: I. The Effects of Limb Mass Distribution in Quadrupeds. *Journal of Experimental Biology* 154:273–286.

Susman, R. L., Stern, J. T., and Jungers, W. L., 1984, Arboreality and Bipedality in the Hadar Hominids. *Folia Primatologia* 43:113–156.

Tague, R. G., and Lovejoy, C. O., 1986, The Obstetric Pelvis of A.L. 288–1 (Lucy). *Journal of Human Evolution* 15:237–255.

Taylor, C. R., 1994, Relating Mechanics and Energetics During Exercise. *Adv. Vet. Sci. Comp. Med.* 38A:181–215.

Tinbergen, N., 1969, On War and Peace in Animals and Man. *Science* 160:1411–1418.

Trevathan, W., and Rosenberg, K., 2000, The Shoulders Follow the Head: Postcranial Constraints on Human Childbirth. *Journal of Human Evolution* 39:583–586.

Trinkaus, E., Churchill, S. E., and Ruff, C. B., 1994, Postcranial Robusticity in *Homo.* II: Humeral Bilateral Asymmetry and Bone Plasticity. *American Journal of Physical Anthropology* 93:1–34.

Trivers, R. L., 1972, Parental Investment and Sexual Selection. In *Sexual Selection and the Descent of Man 1871–1971*, edited by B. Campbell. Aldine, Chicago.

Tuttle, R. H., 1981, Evolution of Hominid Bipedalism and Prehensile Capabilities. *Philosophical Transactions of the Royal Society* 292:89–94.

Tuttle, R. H., and Basmajian, J. V., 1974, Electromyography of Brachial Muscles in *Pan Gorilla* and Hominoid Evolution. *American Journal of Physical Anthropology* 41:71–90.

Tuttle, R. H., and Basmajian, J. V., 1978, Electromyography of Pongid Shoulder Muscles, II. Deltoid, Rhomboid and "Rotator Cuff". *American Journal of Physical Anthropology* 49:47–56.

Tuttle, R. H., Velte, M. J., and Basmajian, J. V., 1983, Electromyography of Brachial Muscles in *Pan troglodytes* and *Pongo pygmaeus. American Journal of Physical Anthropology* 61:75–83.

Van Creveld, M., 1991, *Technology and War From 2000 B. C. to the Present.* The Free Press, New York.

Wagner, J. A., Horvath, S. M., and Dahms, T. E., 1997, Cardiovascular, Respiratory, and Metabolic Adjustments to Exercise in Dogs. *Journal of Applied Physiology* 42:403–407.

Ward, C. V., 1997, Functional Anatomy and Phyletic Implications of the Hominoid Trunk and Hindlimb. In *Function, Phylogeny and Fossils, Miocene Hominoid Evolution and Adaptations*, edited by D. R. Begun, C. V. Ward, and M. D. Rose, pp. 101–129. Plenum Press, New York.

Ward, C. V., 1998, *Afropithecus*, *Proconsul*, and the Primitive Hominoid Skeleton. In *Primate Locomotion*, edited by E. Strasser, J. Fleagle, A. Rosenberger, and H. McHenry, pp. 337–353. Plenum Press, New York.

Watts, D. P., 1989, Infanticide in Mountain Gorillas: New Cases a Reconsideration of the Evidence. *Ethology* 81:1–18.

Weidenreich, F., 1943, The Skull of *Sinanthropus pekinensis*: A Comparative Study of a Primitive Hominid Skull. *Palaeontologia Sinica*, 10 (Whole Series 127).

Weidenreich, F., 1951, Morphology of Solo Man. *Anthropological Papers of the American Museum of Natural History* 43:201–290.

Wescott, R. W., 1967, The Exhibitionistic Origin of Human Bipedalism. *Man* 2:630.

Wheeler, P. E., 1984, The Evolution of Bipedality and Loss of Functional Body Hair in Hominids. *Journal of Human Evolution* 13:91–98.

Wheeler, P. E., 1991, The Influence of Bipedalism On the Energy and Water Budgets of Early Hominids. *Journal of Human Evolution* 21:117–136.

Wilkinson, R. G., 1997, Violence Against Women: Raiding and Abduction in Prehistoric Michigan. In *Troubled Times: Violence and Warfare in the Past*, edited by D. L. Martin and D. W. Frayer, pp. 21–43. Gordon and Breach Publishers, Amsterdam.

Willems, P. A., Cavagna, G. A., and Heglund, N. C., 1995, External, Internal and Total Work in Human Locomotion. *Journal of Experimental Biology*, 198, 379–393.

Willey, P., 1990, *Prehistoric Warfare on the Great Plains: Skeletal Analysis of the Crow Creek Massacr Victims.* Garland Publishing, Inc., New York.

Woledge, R. C., 1968, The Energetics of Tortoise Muscle. *Journal of Physiology* 197, 685–707.

Wolpoff, M. H., 1976, Some Aspects of the Evolution of Early Hominid Sexual Dimorphism. *Current Anthropology* 17:579–606.

Wood, B., and Collard, M., 1999, The Human Genus. *Science* 284:65–71.

Wrangham, R., and Peterson, D., 1996, *Demonic Males: Apes and the Origins of Human Violence*. Houghton Mifflin Co, Boston.

Zihlman, A. L., and McFarland, R. K., 2000, Body Mass in Lowland Gorillas: A Quantitative Analysis. *American Journal of Physical Anthropology* 113:61–78.

Zollikofer, C. P. E., de Leon, M. S. P., Vandermeersch, B., and Leveque, F., 2002, Evidence for Interpersonal Violence in the St. Cesaire Neanderthal. *Proceedings of the National Academy of Sciences* 99:6444–6448.

Chapter **10**

Age, Sex, and Resource Transport in Venezuelan Foragers

Charles E. Hilton and Russell D. Greaves

Abstract. Enhancing our understanding of the skeletal biology of modern hunter-gatherers and developing more sophisticated models of fossil and prehistoric hominin locomotor behavior and subsistence activities requires information on male and female forager mobility patterns. Unlike other primates, modern human foragers expend considerable energy in activities involving the transport of resources across the landscape. Although male foragers are often associated with high mobility in comparison to their female counterparts, female foragers are seen to engage in subsistence tasks incorporating a high frequency of burden carrying. This paper examines the influence of age and sex on mobility and resource transport in a group of Pumé foragers located in the savanna-wetlands of southwestern Venezuela.

INTRODUCTION

The evolution of bipedalism as the predominant form of locomotion within the hominin lineage relaxed some of the selective pressures on the upper limbs and allowed for additional manipulative capabilities during locomotion. One benefit of bipedalism is that it enhanced the ability of hominins to transport resources across the landscape. In fact, in comparison to other primates, extensive resource transport is an important feature of hominin behavioral ecology (Binford, 1983; Foley, 1987; Potts, 1988, 1991). Although the earliest hominins most certainly carried organic burdens, the first "evidence" of resource transportation occurs in the late Pliocene with the appearance of lithics removed from their source locations and carried short distances (Isaac, 1978; Potts, 1991). Such burden carrying is in marked contrast to contemporary hominin foragers who carry a range of resources including large and heavy burdens for extended distances (Lee, 1979; Hilton and Greaves, 1995).

Resource transportation influences several key factors in the economic organization of central place foragers. Transporting food resources back to a central place offers greater opportunities for delayed consumption at a location that minimizes contact with predators (Isaac, 1978; Potts, 1984, 1988). In contemporary foragers, burden carrying is impacted by the sexual division of labor as pronounced differences in resource transportation are

From Biped to Strider: The Emergence of Modern Human Walking, Running, and Resource Transport, edited by D. Jeffrey Meldrum and Charles E. Hilton. Kluwer Academic/Plenum Publishers, New York, 2004.

seen to be associated with sex (Brown, 1970; Kelly, 1983, 1992, 1995; Murdock and Provost, 1973). Female foragers are seen often to carry large and heavy burdens such as dependent offspring, firewood, food and tools (Hilton and Greaves, 1995; Lee, 1979). In either sex, burden carrying presents problems in maintaining efficient locomotion as this behavior alters the location of the center of mass regardless of load position (Carsloo, 1972; Heglund et al., 1995; Maloiy et al., 1996; Winter, 1990). Due to problems associated with effective weight distribution and increases in overall transport weight, such behavior places additional requirements on an individual's anatomical and physiological systems particularly if these types of physical activities are a routine part of foraging (Ghesquiere et al., 1992; Heglund et al., 1995; Jenike et al., 1992; Maloiy et al., 1996).

Given the importance of resource transportation in shaping the economic organization of hominin foragers, it is surprising that little information is known regarding the age and sex patterns of resource transport, the size of loads, or the distances carried. To what extent does extensive burden carrying affect decisions regarding resource targeting and mobility? Without the aid of transportation technology or beasts of burden, how do foragers resolve the dilemma of carrying large resource burdens back to a central place? What age and sex groups are the most impacted by burden carrying? How does burden carrying affect the patterns of biomechanical loading on hominin locomotor systems?

This paper addresses such behavioral ecological issues associated with hominin foragers by examining the range of variation by sex and age of resource transportation and mobility in a community of Pumé Indians in the savannas (*llanos*) of southwest Venezuela. This analysis provides initial baseline documentation for a bio-behavioral complex that clearly is an important component of the foraging patterns of behaviorally modern humans.

STUDY COMMUNITY

The Pumé (Yaruro) are a foraging people living in the *llanos* of southwestern Venezuela (see Figure 1; Besnerais, 1948; Gragson, 1989; Greaves, 1997; Leeds, 1961, 1962, 1964; Lizzardle, 1988; Mitrani, 1973, 1975, 1988; Petrullo, 1939). A population of 5,400 Pumé live in the state of Apure primarily along the Capanaparo, Riecito, and Cinaruco Rivers and in the savanna between these drainages (Mitrani, 1988). Pumé subsistence economy is based on hunting, fishing, wild root collection, and some manioc horticulture. Communities adjacent to major drainages are large permanent villages that are reliant on agriculture, wage labor, and animal husbandry in addition to wild food resources. Savanna Pumé communities are smaller, more dispersed, and most dependent on hunting, fishing, and wild root collection.

Ethnoarchaeological research was conducted during 1990 and 1992–93 in the savanna community of Doro Aná, located south of the Capanaparo River (see Figure 2; Gragson, 1989; Greaves, 1997) while locomotor research was conducted from April 1992-September 1993 (Hilton, 1997). In 1992–93 the community contained approximately 22 adult females, 23 adult males, and 22 children. Investigations focused on collecting time allocation data on mobile activities. Travel distances, resource encounters, tool use, and other data were recorded routinely as part of these behavior samples.

The Venezuelan llanos are a seasonally flooded grassland savanna that are part of the depression of the Apure Basin. The llanos range from 40–80 m above sea level. These grasslands are maintained by edaphic controls, fire, and hyperseasonal variation in rainfall (Gragson, 1989; Greaves, 1997). Mean daily temperatures vary little between the six months of wet and dry seasons. This hyperseasonal savanna is characterized by extreme variation in rainfall (Greaves, 1997). Rainfall data collected at the study site indicate that 77%

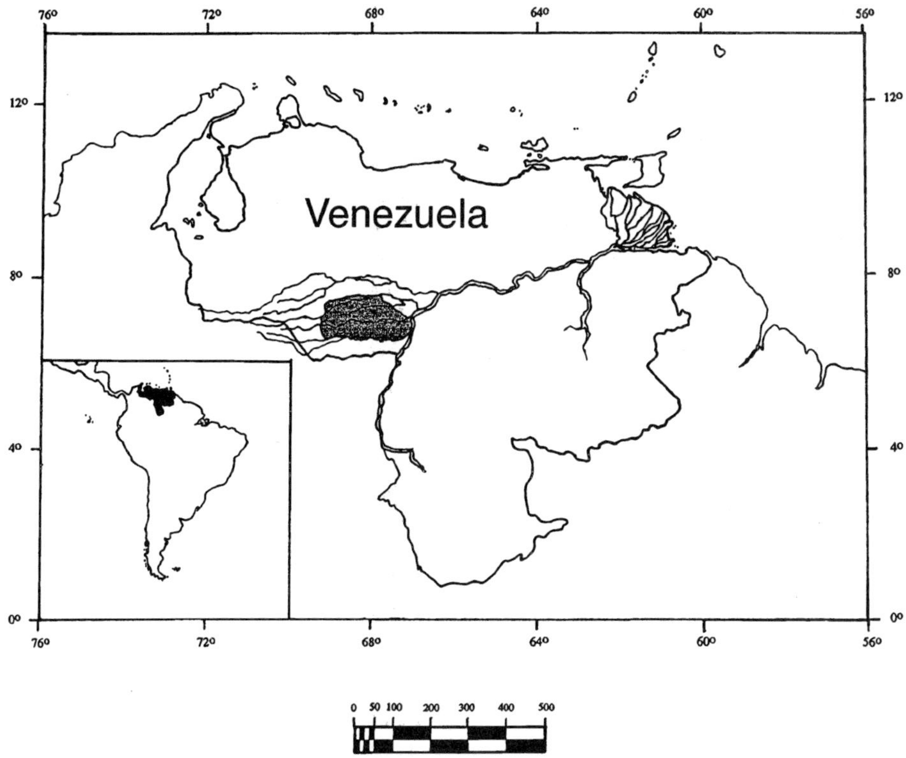

Figure 1. Approximate location of the river and savanna Pumé of southwestern Venezuela. Scale is in kilometers.

(1,404 mm) of the total annual precipitation (1,805 mm) occurred during the wet season months of May through October (Greaves, 1997). River levels change markedly in relation to the pronounced seasonality of rain.

Plant and animal responses to seasonal fluctuation in rainfall establish a biseasonal pattern of subsistence responses by the Pumé. Fishing, as in many parts of South America, is a dry season activity for the Pumé when fish are concentrated in restricted pools and small segments of streams (Gragson, 1989, 1992; Leeds, 1961; Mitrani, 1988). Women collect roots during both the dry and wet seasons. Wild roots are critical resources throughout the year (Gragson, 1997; Greaves, 1997). Some fruits are gathered during the dry season. Except for mangos, available primarily near the Capanaparo River, fruit does not constitute significant dietary bulk or energetic effort. Hunting of terresterial animals is performed during the wet season (Gragson, 1989; Greaves, 1997; Mitrani, 1988). The majority of agricultural labor and harvesting, performed by both sexes, is during the wet season. Manioc (*Manihot esculenta*) is the only important crop. The highest wild root returns also were collected during the agricultural season.

Pumé Residential Mobility and Settlement Pattern

The Pumé practice several kinds of residential mobility. The extreme seasonality differences in the llanos create marked differences in the plant and animal dynamics. This seasonality is associated with distinct shifts in Pumé subsistence activities and is responsible

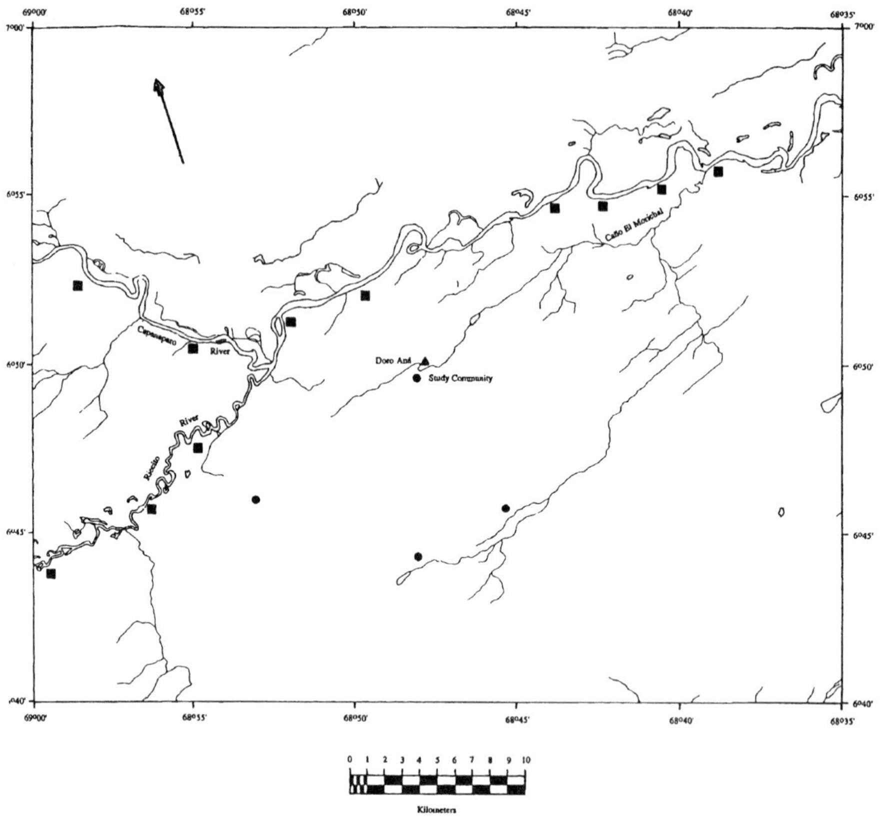

Figure 2. Location of the Doro Aná community (triangle) and the adjacent caño el Morichal. Pumé river communities along the Rio Capanaparo are indicated by squares. Pumé savanna communities are indicated by circles. Scale is in kilometers.

for a biseasonal settlement pattern (Gragson, 1989). Separate villages are occupied for approximately six months as central place bases during the dry and wet seasons. Dry season camps are located adjacent to streams to take advantage of water and proximity to fishing locations (Besnerais, 1948; Gragson, 1989; Mitrani, 1988; Petrullo, 1939). Architecture at dry season camps consists of brush sun shades and small circular or rectangular thatched structures that protect against the infrequent rain (Besnerais, 1948; Gragson, 1989; Mitrani, 1988; Petrullo, 1939). Dry season habitations usually contain only nuclear family units.

Dry season camps were situated within 20 m of this stream on its floodplain. While stream flow is intermittent in this caño during the dry season, isolated pools persist even during the periods of minimal rainfall (December–January). The proximity to fishing locations has been considered the most important reason for placing dry season camps near streams. However, camps near streams take advantage of water that is unavailable away from these ephemeral sources. The timing of moves to dry season camps is dictated by women's decisions regarding the decreasing availability of water at wet season camps and when it is procurable near potential dry season camp locations. The locations of dry season base camps change every year.

Wet season camps are situated away from these streams and are located on higher ground above the maximum levels of savanna flooding. Wet season camps were located approximately 750–1000 m from the caño el Morichal. Wet season houses are large, sturdy, rectangular, thatched habitations often containing extended families (Besnerais, 1950; Gragson, 1989; Mitrani, 1988). At the end of the wet season, the Doro Aná Pumé move their camps from areas that are above annual flooding of the savanna to locations adjacent to the ephemeral stream, caño el Morichal, as well water is not available near camp. Wet season camps are reoccupied for two or more years, probably because of the greater architectural investment. These campsites also were reoccupied after short abandonments (1–2 years) with rebuilding and reorientation of at least one, and usually several, of the houses already existing at that camp.

Subsistence Mobility

Mobility associated with subsistence activities can be described as primarily central place foraging. Few trips were performed across more than one day. During the dry season, residential mobility is associated with periodic intensive fishing bouts focused on one lagoon and palm raw material extraction.

Hunting

Pumé men hunt terrestricl game during the wet season (Gragson, 1989; Greaves, 1997; Mitrani, 1988). The switch away from aquatic resources appears to be due to dispersal of fish in the flooded savanna. Terrestrial game in the llanos is not abundant. When available, fish provide a more reliable food source. Terrestrial hunting involves much greater search efforts but similar risk compared with fishing. High search time during hunting is related to a very dispersed terrestrial biomass. Pumé hunting targets small body size animals, approximately 0.5–1.5 kilograms. Pumé terrestrial hunting emphasizes relatively high search mobility because of the low encounter rates with game. Hunting use areas are larger and exploit more different kinds of habitats than any other subsistence trips.

Fishing

Fishing is the main source of protein during the dry season (Gragson, 1989, 1992; Mitrani, 1988). The characterization of the Pumé as a fishing culture (Kirchoff, 1948), although a simplification of their biseasonal subsistence, is essentially accurate. Fishing is the most productive male protein acquisition activity, and it is not practiced during the wet season only because of the impracticability of pursuing the dispersed fish. Pumé take a wide variety of fish and other aquatic fauna using bow and arrow, hook and line, pescicides, and dams (Gragson, 1992; Greaves, 1997; Leeds, 1962). Pumé men practice bow and arrow intercept fishing from trees and platform facilities.

Wild Root Collecting

Female collection of wild roots provides major portions of Pumé diet. Wild roots were a very important food even during the agricultural season. Wild root collection involved generally lower total round trip distances than male food resource acquisition trips. However, the intensive physical activities involved in excavating and carrying wild root resources represents a considerable energetic effort even under conditions of lower mobility. Roots were most frequently collected during the dry season. Large roots were gathered from patches found in open flooded savanna during the wet season. Root trips concentrated on a small spectrum of food species, using one exploitation technique to access those resources.

Plant resources acquired by women are carried back to camp with the use of a lightweight burden basket. Women suspend the basket from their forehead using a tump line, therby allowing the basket to rest against the back. Baskets are woven from *Moriche* palm fronds. Larger burden baskets are employed when large resource loads are anticipated but the basket itself still remains lightweight.

METHODS

Analysis of resource transportation and mobility for the Doro Aná Pumé during foraging trips required combining several different types of anthropological data. Daily foraging trips were sampled by focal follows of both men and women (see Greaves, 1997 for details). Scan sampling was conducted for foraging trips involving multiple individuals such as female root gathering and fruit collecting trips. The key variables analyzed, described in further detail below, include distances traveled, weights of resources obtained and transported, relative resource weight, and an indicator of "Effort".

Travel distances during male and female foraging trips were measured using pedometers. In a number of cases, these measurements were verified using a meter wheel and/or topographic maps of the Doro Aná area.

Weights of all resources obtained during foraging were recorded using Pesola® and Homs® hanging spring scales. Relative resource weight was calculated by taking the absolute weight of each carried resource and dividing it by the body weight of the carrier. Forager body weights were measured using a standard bathroom spring scale that was calibrated prior to each anthropometric data collection session.

An "Effort" indicator provides a proxy measure of the number and intensity of lower limb biomechanical loadings for each individual and accounts for variation in three factors. The first factor is the resource weight that an individual chooses to carry. This implies that individual foragers make conscious decisions regarding the resource weight he/she may carry. This is likely the case when considering female foraging trips as root gathering and fruit collecting are accumulation processes of many individual food packages and each additional food package adds to the resource weight. For hunting, the weight of the prey obtained may exhibit a wide range of variation. If the prey weight is large enough, a hunter may return home after obtaining a single prey item. Such a decision may be influenced by an individual's own body weight. The "Effort" indicator also accounts for changes in locomotor behavior, particularly stride length, when walking with a heavy resource burden. Walking variables within the gait cycle are often modified by individuals who are transporting large burdens (Hilton, 1994, 1997). Thus, "Effort" accounts for the fact that individuals cover the same foraging distance differently due to differences in stride and the size of a resource burden.

The "Effort" indicator is calculated in the following procedure. Each foraging trip was evaluated to determine whether an individual returned with food resources. For any given foraging trip, it is expected that individuals will walk the first portion of the trip without any resource burden whereas the later portions of the trip should require carrying a resource burden. Male foragers returning from unsuccessful hunting trips will not have carried any resource burden. Thus, a percentage was calculated for every foraging trip using the body weight of each individual and the weight of the obtained resource to create burdened and unburdened weight percentage values. Individuals who returned with resources $\geq 15\%$ of their own body weight are considered burdened for half of the foraging trip. Individuals returning with resources $< 15\%$ of their body weight are considered unburdened for the entire foraging trip. The threshold of 15% for burdened versus unburdened walking is based on recent physiological work on burden carrying with traditional east African women who were shown to carry up to 20% of their body weight without incurring any additional energetic cost (Heglund et al., 1995; Maloiy et al., 1986). We assumed that the Pumé women might have a similar physiological response. However, since this type of physiological research has not been conducted with the Pumé, it was unclear whether a "burdened threshold" of 20% of body weight was appropriate and instead 15% of body weight was selected.

In controlled settings, measurements of stride length for each individual were collected during normal and burdened walking on a standardized trackway constructed at the field site (see Hilton, 1997 for details). The Pumé have been shown to modify their stride lengths when carrying large burdens (Hilton, 1994, 1997).

For those individuals who return with burdens, the foraging distances were divided in half in order to obtain unburdened and burdened distances. Foraging distances were then divided by the appropriate unburdened and burdened stride lengths for a given individual. This value represents the number of strides the individual was expected to use in order to walk that foraging distance with a burden.

Finally, the number of strides is multiplied against a "Burdened Weight" ratio to produce the "Effort" value. Thus, an unburdened forager has a "Burdened Weight" value of 1.0 whereas someone carrying 30% of their body weight has a ratio of 1.3. It is these values that are multiplied against the number of stride lengths.

All foraging trips sampled were divided into wet season trips and dry season trips. Each season was analyzed using two-way analysis of variance (ANOVA) with sex and age categories as the independent variables. The division between younger and older adults was set at 40 years of age. This age division is primarily meant to separate older females who are less likely to become pregnant from younger females who may be also involved in childcare and reproduction.

RESULTS

One hundred and one Pumé foraging trips were examined. Figure 3 presents the distance and resource return weights associated with 65 wet season and 36 dry season foraging observations. In numerous cases, data points exhibit the same distances (see Figure 3). Females often go on foraging trips together but may return with different resource loads. The presence of multiple females on a foraging trip may have some effect on how resource loads are distributed for the walk back to camp. In other instances, females may walk the same distance as a previous trip if they return to the same resource patch, e.g., the collection of mangos. A similar statement can be made regarding male hunting and fishing as males will hunt together or in cases of fishing return to the same spot. Hunting returns can often be low or entirely unsuccessful (see Figure 3).

Pumé foraging trips over 10 km are primarily male hunting trips or female mango gathering trips. Female root gathering trips generally involve walking less than 8 kilometers.

Comparative analyses of distance for 65 wet season foraging trips indicate a significant difference between the sexes but no significant differences between the age categories (see Table 1). As a group, males walk significantly greater distances during hunting in comparison to female root gathering (see Figure 4). Male hunting trips are seen to average 14.5 km whereas females averaged only 4.7 kilometers. This is not unexpected as males are often described as exhibiting greater mobility during hunting when compared to females.

Table 1. Pumé Wet Season Foraging Distance, Resource Return Weight, and Effort Values

	Women < 40	Women > 40	Men < 40	Men > 40	Two-Way ANOVA Sex and Age P-values
Foraging Distance (km):					
Sample Size	22	17	17	9	
Ave.	5.12	4.20	14.56	14.58	Sex (A): $p < .0001$
S.D.	4.4	2.6	4.9	5.9	Age (B): $p = .7006$
S.E.	.9	.6	1.2	1.9	AB: $p = .6867$
Min.-Max.	.53–13.97	1.18–10.5	3.85–21.94	3.57–21.94	
Resource Return Weights (kg):					
Sample Size	22	17	17	9	
Ave.	5.7	12.0	3.5	1.9	Sex (A): $p < .0001$
S.D.	4.3	7.0	4.6	2.0	Age (B): $p = .0773$
S.E.	.91	1.7	1.1	.7	AB: $p = .0039$
Min.-Max.	.6–18.7	1.9–28.7	0–12.3	0–4.5	
Relative Resource Weights:					
Sample Size	22	16	17	9	
Ave.	.12	.29	.06	.04	Sex (A): $p < .0001$
S.D.	.08	.18	.08	.04	Age (B): $p = .016$
S.E.	.02	.04	.02	.01	AB: $p = .0015$
Min.-Max.	.01–.35	.04–.68	0–.21	0–.09	
Effort Indicator:					
Sample Size	20	16	16	9	
Ave.	4720	3906	10792	11033	Sex (A): $p < .0001$
S.D.	4061	2992	3909	4617	Age (B): $p = .8523$
S.E.	908	748	977	1539	AB: $p = .6776$
Min.-Max.	469–12578	1001–10207	3122–18986	2541–17619	

Descriptive statistics for Doro Aná wet season foraging activities. Average values have been compared using two-way analysis of variance (ANOVA).

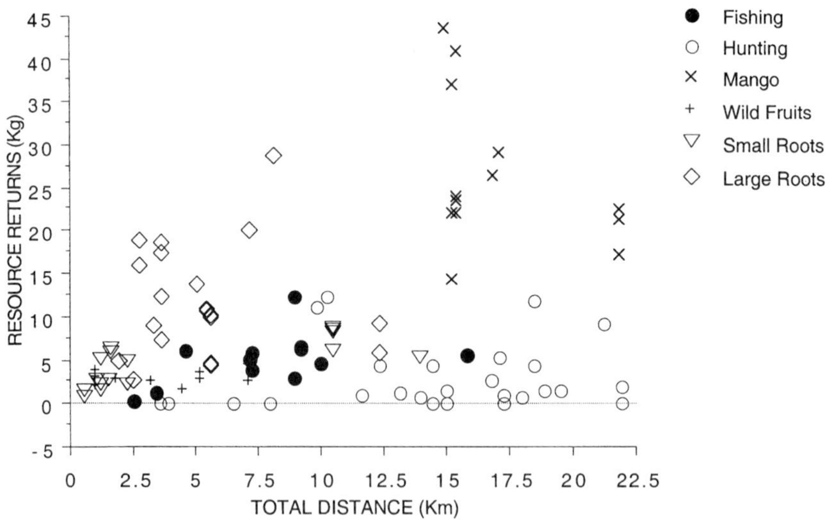

Figure 3. Bivariate plot of the 101 foraging observations of adult males and females of Doro Aná illustrating resource return weight against total trip distance. We season trips are indicated by open symbols and dry season trips by dark symbols.

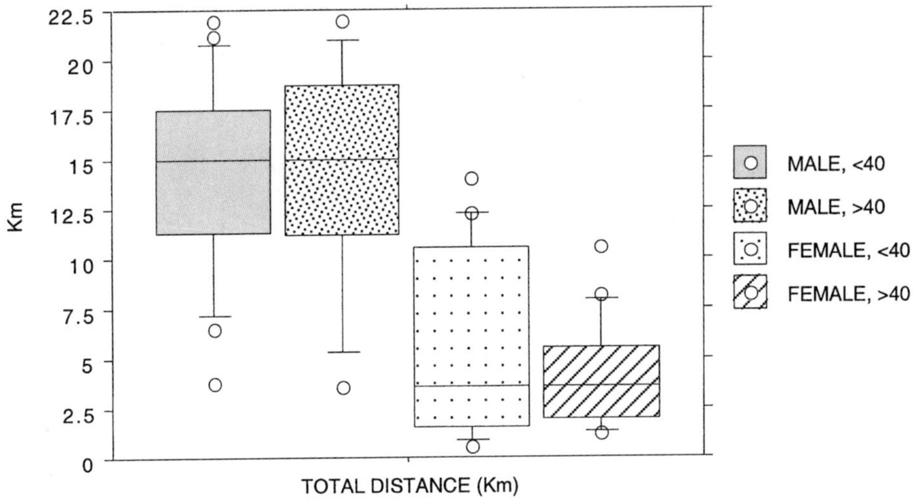

Figure 4. Box-plots of wet season foraging distances for the four subcategories.

High male mobility during the wet season is due primarily to the flooding of the llanos. During this time terrestrial animals are more dispersed and more difficult to locate. Hunting trips often last 6–8 hours of nearly constant search, primarily in the open savanna.

The targeting of both large and small roots which grow about ~5 km away occur in the wet season (see Table 1). These female activities often involve more direct line walking from the camp to the patch as opposed to the type of wandering associated with male hunting. Small roots grow in the low gallery forest near camp and in forested areas whereas large roots are ubiquitous in the open flooded savanna.

A significant interaction effect is seen in analysis of the absolute resource weights brought back to camp (see Table 1). Females >40 years are seen to return to camp with resource loads weighing 12 kilograms (see Table 1 and Figure 5). In contrast, younger females carry resources weighing almost 6 kg, followed by young males who return with

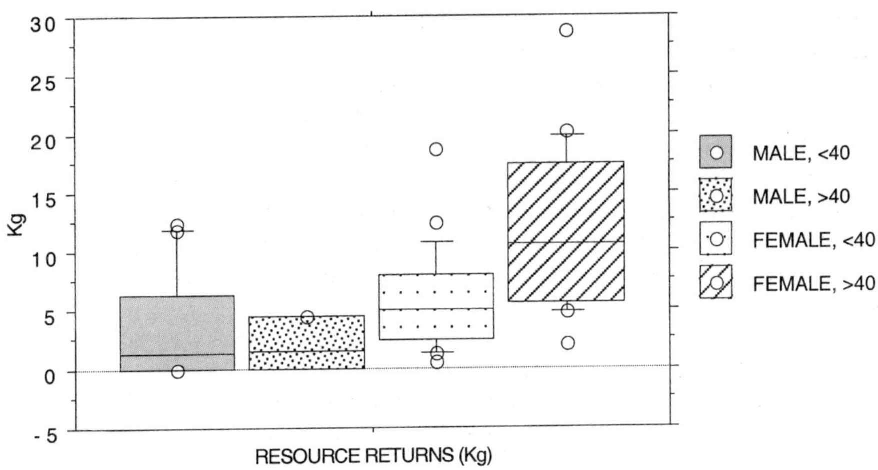

Figure 5. Box-plots of wet season resource return weights for the four subcategories.

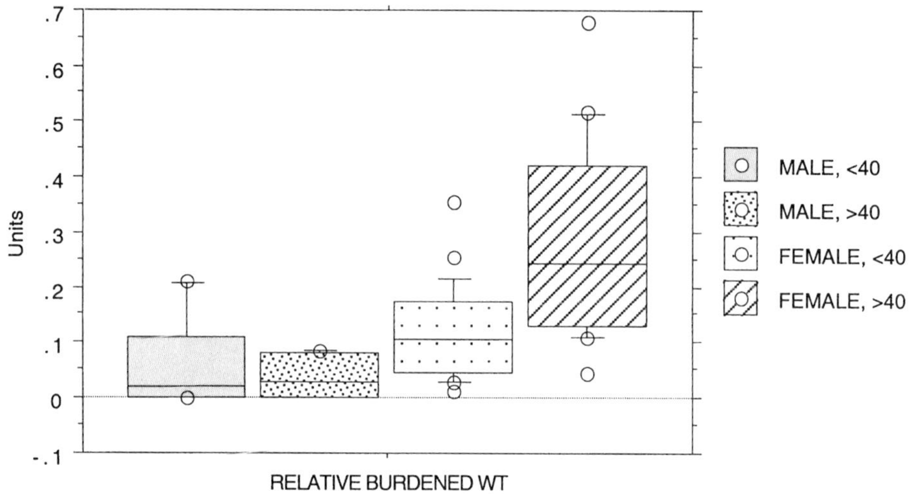

Figure 6. Box-plots of wet season relative resource weight values for the four subcategories.

resources weighing 3.5 kg, followed by older males who carry resources weighing only 2 kilograms. These values reflect the nature of the resources targeted by each sex. For example, out of 26 male hunting trips, 6 trips yielded 0 returns and 4 yielded returns weighing less than 1 kg whereas only 1 female foraging trip out of 39 yielded a resource return weight of less than 1 kilogram (see Figure 5).

As the absolute weight of a carried resource does not provide a true indication of whether individuals may have difficulty in transporting the resource burdens, relative resource weight for each individual was analyzed across the age and sex categories. This analysis indicates significant differences across the age and sex categories as well as a significant interaction (see Table 1). Older Pumé females during the wet season carry food resources that weigh 29% of their body weight (see Table 1 and Figure 6). The next highest values are those associated with the younger females who carry 12% of their body weight. Values for the younger versus older males are 6% and 4%, respectively. The relative resource weights for older women are more than 2x higher when compared to younger females and at least 5x higher in comparison to men.

Analysis of wet season "Effort" indicates that males exhibit significantly higher values (see Table 1). As noted in the Methods section, the "Effort" indicator takes into consideration that individuals may cover the same foraging distance rather differently, particularly due to differences in locomotor patterns and the relative size of the resources carried. While Pumé females carry heavy loads the low mobility seen on these foraging trips produces lower Effort values in comparison to males (see Figure 7). In males, the high Effort values are influenced more by the high mobility associated with hunting rather than weights of the resources brought back to camp.

Distance data for 36 dry season foraging trips shows a decrease in male mobility in comparison to the wet season whereas female mobility is seen to increase (see Table 2 and Figure 8). During this time of year, Pumé females target mangos located on ranches more than 8 km away from the community as well as targeting closer wild fruits. Pumé males target fish that become concentrated in drying streams. Additionally, males will often target the migratory birds that feed on these concentrated fish resources.

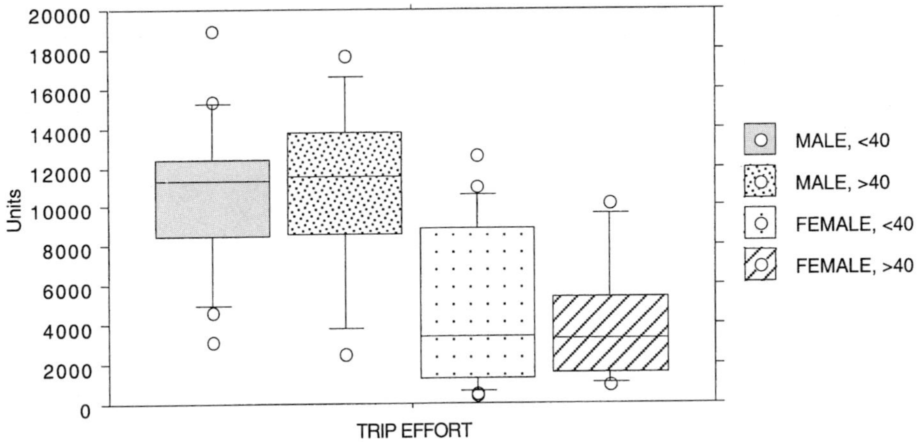

Figure 7. Box-plots of wet season "Effort" values for the four subcategories.

Statistical analysis indicates no significant differences across the age and sex categories (see Table 2). This result is surprising as it is generally assumed that at any given time males are more mobile in comparison to females.

Food resource return weights are seen to increase for both sexes during the dry season. Although older women are seen to carry the heaviest loads in comparison to younger females, 17 kg versus 12 kg respectively, no significant differences are seen between the age categories (see Table 2 and Figure 9). However, significant differences are seen across the sexes as women return to camp with significantly heavier resources in comparison to males. These food resource loads for females, primarily mango, average 15.5 kilograms. Fish and bird resources brought back to camp by males average approximately 5 kilograms.

Similar patterns are seen when food resource weights are recalculated as percentages of the carrier's body weight. Figure 10 shows the distribution of values for the relative weights of food resources for the age and sex categories (see Figure 10). On average, Pumé

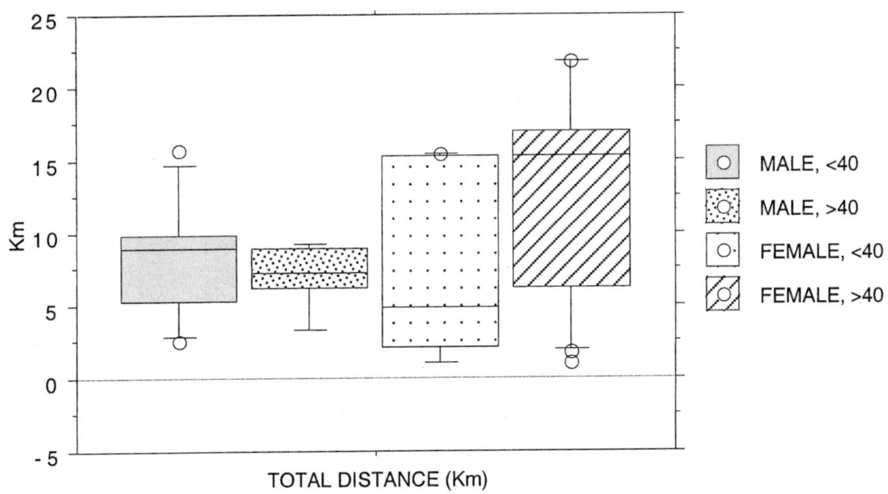

Figure 8. Box-plots of dry season foraging distances for the four subcategories.

Table 2. Pumé Dry Season Foraging Distance, Resource Return Weight, and Effort Values

	Women < 40	Women > 40	Men < 40	Men > 40	Two-Way ANOVA Sex and Age p-values
Foraging Distance (km):					
Sample Size	8	16	7	5	
Ave.	7.63	12.6	8.33	7.19	Sex (A): p = .3006
S.D.	6.5	7.21	4.25	2.3	Age (B): p = .4005
S.E.	2.3	1.8	1.6	1.04	AB: p = .1826
Min.-Max.	1.0–15.4	1.0–21.9	2.5–15.8	3.35–9.21	
Resource Return Weights (kg):					
Sample Size	8	16	7	5	
Ave.	12.41	17.25	5.69	4.29	Sex (A): p = .0255
S.D.	14.73	13.25	3.62	2.08	Age (B): p = .6846
S.E.	5.2	3.3	1.37	.93	AB: p = .4628
Min.-Max.	1.54–41.0	2.4–43.5	.36–12.4	1.36–6.27	
Relative Resource Weights:					
Sample Size	8	14	7	5	
Ave.	.291	.384	.098	.081	Sex (A): p = .0135
S.D.	.36	.28	.06	.04	Age (B): p = .69
S.E.	.13	.08	.02	.02	AB: p = .5673
Min.-Max.	.03–1.03	.06–.94	.006–.21	.02–.12	
Effort Indicator:					
Sample Size	7	10	6	5	
Ave.	8626	11508	5717	5178	Sex (A): p = .0966
S.D.	8360	8447	3202	1681	Age (B): p = .6649
S.E.	3160	2671	1307	752	AB: p = .5282
Min.-Max.	862–20171	875–24525	1973–11012	2386–6566	

females during the dry season carry food resources weighing 35% of their body weight and this value is significantly higher than the returns brought to camp by males which average only 9% of body weight (see Table 2). It is also seen that females on several occasions will carry food resources weighing over 100% of their body weight (see Figure 10). Although

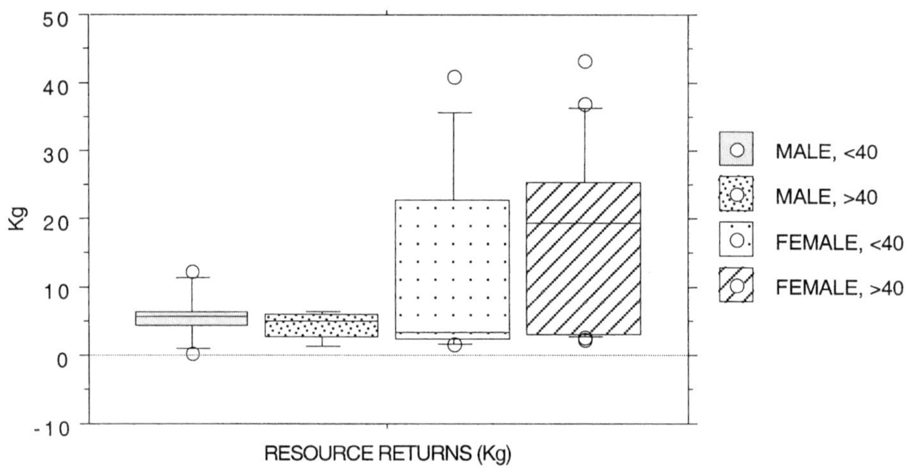

Figure 9. Box-plots of dry season resource return weights for the four subcategories.

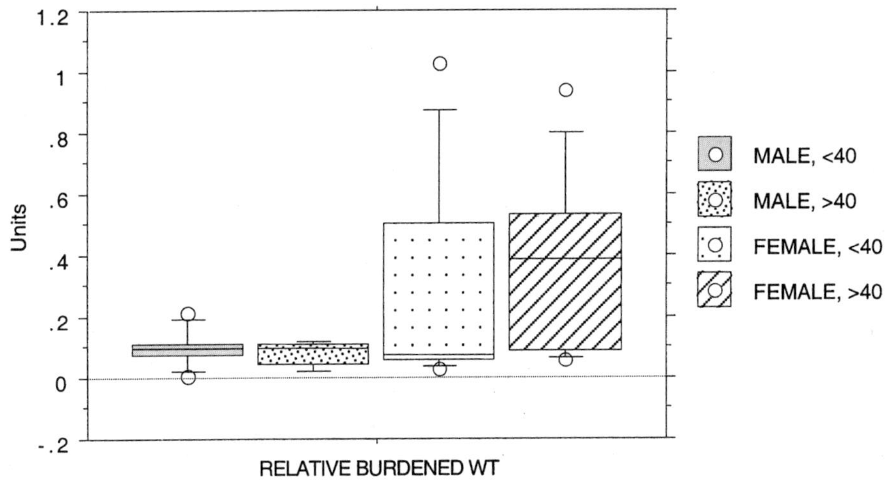

Figure 10. Box-plots of dry season relative resource weight values for the four subcategories.

Pumé females carry relatively heavier resources and are as mobile as their male counterparts during the dry season, analysis of these "Effort" values indicates no significant differences across the age and sex categories (see Table 2 and Figure 11).

DISCUSSION

A key factor towards understanding biomechanical loading patterns of the hominin locomotor system can be facilitated by determining the interaction between the range of mobility and subsistence behavior in contemporary foragers. Given that central place foraging and subsequent food sharing are important components of foraging strategies, transporting food resources is expected to have a biomechanical impact on the locomotor system. Male and female resource targeting, procurement, and transportation are clearly influenced by

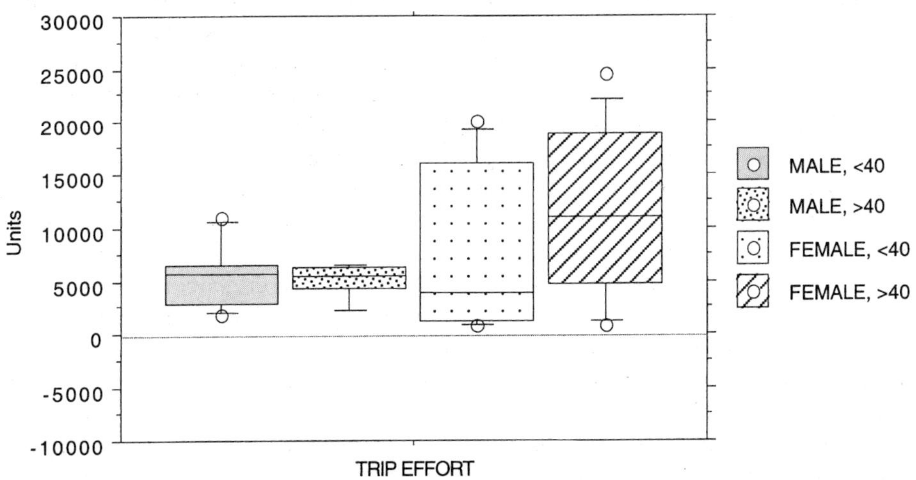

Figure 11. Box-plots of dry season "Effort" values for the four subcategories.

the sexual division of labor (Brown, 1970; Kelly, 1995; Murdock and Provost, 1973) leading to dramatic differences associated with physical activities as each sex carries out their respective subsistence tasks. Such physical activities are known to affect underlying anatomical and physiological systems throughout life, especially skeletal morphology (Bridges, 1991, 1992, 1994; Chamay and Tschantz, 1972; Currey, 1984; Lanyon, 1990; Lanyon et al., 1982; Lanyon and Rubin, 1984, 1985). While skeletal biologists have tried to assess biomechanical loading associated with sex-specific tasks in recent and prehistoric samples, limited research has been conducted by biological anthropologists that actually addresses the nature of biomechanical loading for living peoples in traditional groups (Bridges, 1991, 1992, 1994; Larsen et al., 1994; Lovejoy and Trinkaus, 1980; Ogilvie, 2000; Ruff, 1987; Ruff and Hayes, 1983a, b; Ruff and Larsen, 1990; Ruff et al., 1984; Ruff et al., 1993; Trinkaus, 1987, 1989, 1993; Trinkaus and Ruff, 1989a, 1989b; Wesson and Martin, 1995).

Pumé male and female mobility patterns during foraging are clearly affected by seasonal changes in resource availability. Pumé males increased their mobility during wet season foraging trips in comparison to the dry season. Wet season hunting is a difficult endeavor as game animals are more dispersed across the flooded savanna. Hunters can spend large amounts of time searching for prey and this requires long bouts of walking over a landscape that in some situations will be muddy and flooded. Yet, this type of male wet season walking can be viewed as relatively unencumbered locomotion since hunting trips often exhibit low or no food returns and hunters usually carry only a set of bows and arrows in their hands and a knife strapped to their waist. Biomechanical loading of the locomotor system during male wet season hunting trips is therefore influenced primarily by the long distances associated with a hunt and the number of strides necessary to cover those distances rather than a large weight associated with a food resource return. Thus, the male "Effort" values seen for the wet season are more a reflection of the number of strides associated with a hunting trip rather than the size of a food resource. In contrast, male dry season foraging distances are dramatically lower than wet season, reflecting the fact that in the dry season the targeted game animals (birds and fish) are concentrated in locations closer to camp. Male dry season food returns are also higher as the increased densities of bird and fish species make them easier to capture. Average dry season food returns are approximately 5 kg (a burden weighing slightly more than two liters of milk). Although the average weight associated with male dry season foraging returns approaches nearly 10% of male body weight (see Table 2), the overall level of biomechanical loading on the locomotor system can be viewed as less strenuous since "Effort" indicator values are lower.

One result of this analysis is that the patterns of female mobility and associated levels of biomechanical loading appear to be more complex than that of males. During the wet season, female foraging distances for both small and large root gathering trips are low but the food returns for these trips are consistently higher than those seen for males (see Table 1). The significantly heavier food resources associated with older females are notable, especially when these weights are transformed into relative resource weights. Older Pumé women carry food resources back to camp that weigh on average nearly 30% of their body weight. These relative resource weights are more than 2x higher than the values seen for younger females and at least 5x higher than those seen in males (see Table 2). These older women can transport food resources approaching nearly 70% of body weight. With these types of resource loads for older women, it is surprising that their "Effort" indicator values are the lowest for the four subgroups. One reason for their low "Effort" values is that their foraging distances are the lowest values. Older women in comparison to younger women appear to obtain larger amounts of the same resource by foraging in patches closer to camp. This probably reflects the greater experience that older women have in obtaining small and large root resources. Pumé women's knowledge of local conditions and seasonal variation in root condition allow them to select particular patches with a higher frequency of successful returns. Even if the productivity of those patches was below expectations, time spent in such a patch still yielded proportional returns on search and collection time.

With regard to the dry season, women are seen to significantly increase the walking associated with foraging (dry = 10.9 km; wet = 4.7 km) and the weight of the food resources that they bring back to camp (dry = 13.10 kg; wet = 8.07 kg). Younger women increase their food resource loads to nearly 30% of their body weight whereas the older women carry loads that are approximately 38% of their body weight. This increase is accomplished by targeting mangos, a fruit resource that has been introduced to the *llanos*. Gathering mangos involves the highest average mobility (17.5 km) and highest average weight return (23.9 kg) of any Pumé subsistence activity (see Figure 3). In fact, several younger and older women are willing to carry mango loads weighing nearly 100% of their body weight for a distance of almost 12 kilometers (see Figure 12). The other female dry season foraging activity, wild fruit collection, involves low mobility and low weight returns. Women appear to be balancing their long foraging trips with these shorter trips. Surprisingly, this increased female mobility and burden carrying does not produce "Effort" values significantly different from males.

It is noteworthy that throughout the year older females consistently carry food resources weighing over 30% of their body weight whereas the food resource burden size fluctuates for the younger females. This fluctuation may be related to the level of energy expenditure associated with foraging and locomotion. Wet season food resources are limited and during this season the Doro Aná Pumé appear to experience periodic bouts of food

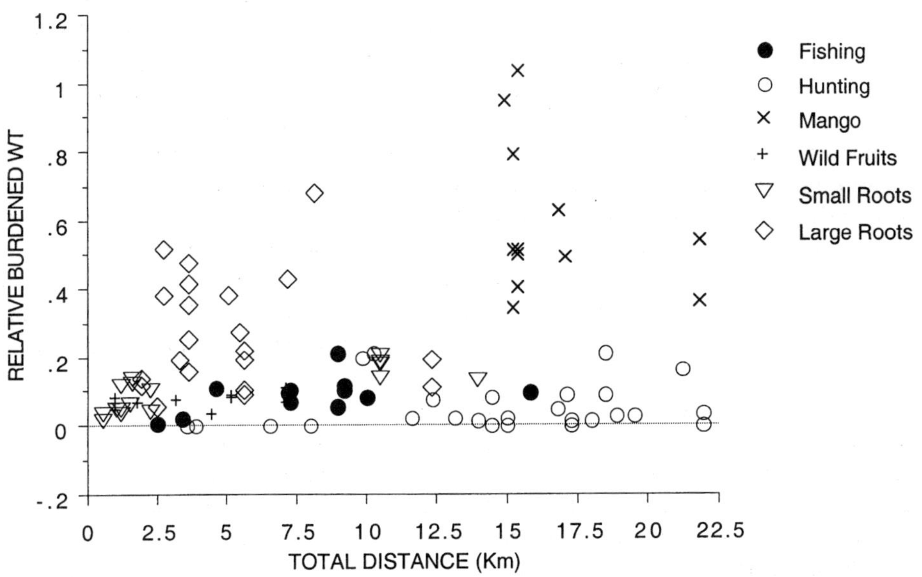

Figure 12. Bivariate plot of the 101 foraging observations of adult males and females of Doro Aná illustrating resource return weight relative to body weight against total trip distance. Wet season trips are indicated by open symbols and dry season trips by dark symbols.

stress. By reducing aspects of their foraging workload in the wet season, younger females may be conserving energy and redirecting it towards reproduction and lactation. However, investigation of this hypothesis requires more detailed data and analysis beyond the scope of this paper.

It is a common assumption in the anthropological literature that female foraging mobility and resource transportation is less than that of males (see Kelly, 1995 and references therein for a review). Female foraging mobility is generally viewed as being constrained by the demands of reproduction and child-care (Brown, 1970; Draper and Cashdan, 1988; Hurtado and Hill, 1990; Hurtado et al., 1992; Kelly, 1995; Murdock and Provost, 1973). The current analysis suggests that this may not always be the case as female foraging mobility can equal that seen in males and even exceed male mobility. Variation in female mobility appears to be influenced by seasonal resource availability whereas variation in female resource transport is influenced by the age of the forager. Older non-reproductive women are seen to transport the greater bulk of food resources during critical periods of food stress and their greater experience at foraging allows them to do this more effectively without having to walk longer distances.

These subtle differences in the sexual and age division of labor provide information for understanding the nature of biomechanical loading of the locomotor system in foragers. For some human female foragers, especially older women, lower limb skeletal loading is due to a combination of mobility and the increased body weight associated with burden carrying. In male foragers, lower limb skeletal loading is probably associated with higher levels of mobility and lower levels of burden carrying.

CONCLUSION

This study shows that for the Pumé foragers of Doro Aná age and sex have an affect on mobility and resource transportation. Older non-reproductive women consistently transport large resource burdens weighing 30% of their body weight throughout the year as a part of their regular foraging activities. Younger women of reproductive age show a greater range of variation for the size of resource loads. In contrast to females, males regardless of age usually have low resource returns relative to their body weight. In terms of annual mobility, males do walk greater distances than females, but when season is taken into consideration, female walking is seen to equal that of males during the dry season. For those who have worked with traditional peoples, these burden carrying and mobility patterns may not seem surprising. However, it is initially counter-intuitive that women carry their heaviest food resource burdens on their longest foraging trips. Women appear to be minimizing the number of mango collecting trips by maximizing the number of mangos brought back to camp for any one trip.

ACKNOWLEDGEMENTS We wish to thank the people of Doro Ana for responding to this intrusion into their lives with extreme patience. Their cooperation made this project a success. Thanks also go to P. Yu for collecting data on female foraging trips. M.D. Ogilvie, E. Trinkaus, and R. L. Nuger kindly provided numerous comments on this paper. This project was funded in part by the Graduate Student Association, the Latin American Institute, and the Department of Anthropology of the University of New Mexico, Sigma Xi, the Pumé Locomotor Research Fund, the L.S.B. Leakey Foundation, and the National Science Foundation (BNS 91-23875).

REFERENCES

Besnerais, H. L., 1948, Algunos Aspectos del rio Capanaparo y de Sus Indio Yaruros. *Memoria de la Sociedad de Ciencias Naturales La Salle* VIII:9–20.

Binford, L. R., 1983, *In Pursuit of the Past: Decoding the Archaeological Record.* Thames and Hudson, New York.

Brown, J. K., 1970, A Note on the Sexual Division of Labor. *American Anthropologist* 72:1073–78.

Bridges, P. S., 1991, Degenerative Joint Disease in Hunter-Gatherers and Agriculturalists from the Southeastern United States. *American Journal of Physical Anthropology* 14:379–391.

Bridges, P. S., 1992, Vertebral Arthritis in a Northwestern Alabama Skeletal Sample. *American Journal of Physical Anthropology*, Supplement 14:54.

Bridges, P. S., 1994, Vertebral Arthritis and Physical Activities in the Prehistoric Southeastern United States. *American Journal of Physical Anthropology* 93:83–93.

Carsloo, S., 1972, *How Man Moves: Kinesiological Studies and Methods.* Heinemann, London.

Chamay, A., and Tschantz, P., 1972, Mechanical Influences in Bone Remodeling. *Journal of Biomechanics* 5:173–180.

Currey, J., 1984, *The Mechanical Adaptations of Bones.* Princeton University Press, Princeton.

Draper, P., and Cashdan, E., 1988, Technological Change and Child Behavior among the !Kung. *Ethnology* 27:339–365.

Foley, R., 1987, *Another Unique Species: Patterns in Human Evolutionary Ecology.* Longman Scientific and Technical, New York.

Ghesquiere, J., Bisschop, A., Jenike, M., and Nkiama, E., 1992, Energy Expenditure of Daily Life Activities in Villagers of the Ituri Forest (NE Zaire). *American Journal of Physical Anthropology* Supplement 14:81.

Gragson, T. L., 1989, *Allocation of Time to Subsistence and Settlement in a Ciri Khonome Pumé Village of the Llanos of Apure, Venezuela.* University Microfilms International, Ann Arbor.

Gragson, T. L., 1992, Strategic Procurement of Fish by the Pumé: A South American "Fishing Culture". *Human Ecology* 20:109–130.

Greaves, R. D., 1997, *Ethnoarchaeological Investigations of Subsistence Mobility, Resource Targeting, and Technological Organization Among Pumé Foragers of Venezuela.* Ph.D. dissertation, University of New Mexico. University Microfilms, Ann Arbor.

Heglund, N. C., Willems, P. A., Penta, M., and Cavagna, G. A., 1995, Energy-Saving Gait Mechanics with Head-Supported Loads. *Nature* 375:52–54.

Hilton, C. E., 1994, Locomotor Kinesiology of Habitual Pedestrians of the Llanos of Venezuela (Abstract). *American Journal of Physical Anthropology* Supplement 18:108.

Hilton, C. E., 1997, *Comparative Locomotor Kinesiology in Two Contemporary Hominid Groups: Sedentary Americans and Mobile Venezuelan Foragers.* Ph.D. dissertation, University of New Mexico. University Microfilms, Ann Arbor.

Hilton, C. E., and Greaves, R. D., 1995, Mobility Patterns in Modern Human Foragers (Abstract). *American Journal of Physical Anthropology* Supplement 20:111.

Hurtado A. M., and Hill K. R., 1990, Seasonality in a Foraging Society: Variation in Diet, Work Effort, Fertility, and the Sexual Division of Labor Among the Hiwi of Venezuela. *Journal of Anthropological Research* 46(3):293–345.

Hurtado, A. M., Hill, K., Kaplan, H., and Hurtado, I., 1992, Trade-Offs Between Female Food Acquisition and Child Care among Hiwi and Ache Foragers. *Human Nature* 3(3):185–216.

Isaac, G. L., 1978, The Food Sharing Behavior of Proto-Human Hominids. *Scientific American* 238:90–108.

Jenike, M., Bisschop, A., Ghesquiere, J., and Nkiama, E., 1992, Energetic Cost of Activities and the Energetic Efficiency of Carrying Loads in the Ituri Forest, Zaire. Paper presented at the 4th Annual Meeting of the Human Behavior and Evolution Society, Albuquerque, New Mexico.

Kelly, R. L., 1983, Hunter-Gatherer Mobility Strategies. *Journal of Anthropological Research* 39:277–306.

Kelly, R. L., 1992, Mobility/Sedentism: Concepts, Archaeological Measures, and Effects. *Annual Review of Anthropology* 21:43–66.

Kelly, R. L., 1995, *The Forager Spectrum: Diversity in Hunter-gatherer Lifeways.* Smithsonian Institution Press, Washington, D.C.

Kirchoff P., 1948, The Yaruro. In *Handbook of South American Indians, Vol. 4, The Circum-Caribbean Tribes*, edited by J. H. Steward, pp. 456–463. Smithsonian Institution, Bureau of American Ethnology Bulletin 143. U.S. Government Printing Office, Washington, D.C.

Lanyon, L. E., 1990, The Relationship Between Functional Loading and Bone Architecture. In *Primate Life History and Evolution, Monographs in Primatology Volume 14*, edited by C.J. DeRousseau, pp. 269–284. Wiley-Liss, New York.

Lanyon, L. E., Goodship, A. E., Pye, C. J., and Macfie, J. H., 1982, Mechanically Adaptive Bone Remodelling. *Journal of Biomechanics* 15:141–154.

Lanyon, L. E., and Rubin, C. T., 1984, Static Vs Dynamic Loads as an Influence on Bone Remodelling. *Journal of Biomechanics* 17:897–905.

Lanyon, L. E., and Rubin, C. T., 1985, Functional Adaptation to Load-Bearing in Bone Tissue. In *Functional Vertebrate Morphology*, edited by Hildebrand, M., Bramble, D. M., Liem, K. F., Wake, D. B., pp. 1–25. Belknap/Harvard University Press, Cambridge.

Larsen, C. S., and Kelly, R. L., 1995, *Bioarchaeology of the Stillwater Marsh: Prehistoric Human Adaptation in the Western Great Basin. Anthropological Papers of the American Museum of Natural History, No. 77.* American Museum of Natural History, New York.

Larsen, C. S., Ruff, C. B., and Kelly, R. L., 1991, Structural Analysis of the Stillwater Postcranial Human Remains: Behavioral Implications of Articular Joint Pathology and Long Bone Diaphyseal Morphology. In *Bioarchaeology of the Stillwater Marsh: Prehistoric Human Adaptation in the Western Great Basin*, C. S. Larsen and R. L. Kelly, pp. 107–133. *Anthropological Papers of the American Museum of Natural History, Number 77.* American Museum of Natural History, New York.

Lee, R. B., 1979, *The !Kung San: Men, Women and Work in a Foraging Society*. Cambridge University Press, Cambridge.

Leeds, A., 1961, The Yaruro Incipient Tropical Forest Horticulture: Possibilities and Limits. In *The Evolution of Horticultural Systems in Native South America: Causes and Consequences*, edited by J. Wilbert, pp. 1–12. Anthropologica Supplement No. 2, Editorial Sucre, Caracas.

Leeds, A., 1962, Ecological Determinants of Chieftanship among the Yaruro Indians of Venezuela. *Atken des 34 Internationalen Amerikanistenkongresses*, pp. 597–608. Verlag Ferdinand Berger, Vienna.

Leeds, A., 1964, Some Problems of Yaruro Ethnohistory. *Actas y Memorias del 25 Congreso Internacional de Americanistas* 2:157–175.

Lizzarlde, R., 1988, Los Pumé. In *Los Aborîgenes de Venezuela, Vol. II, Ethnologia Contemporánea I*, edited by R. Lizarralde and H. Seijas, pp. 125–216. Fundacion La Salle de Ciencias Naturales, Instituto Caribe de Anthropologia y Sociologia, Caracas.

Lovejoy, C. O., and Trinkaus, E., 1980, Strength and Robusticity of the Neandertal Tibia. *American Journal of Physical Anthropology* 53:465–470.

Maloiy, M. O., Heglund, N. C., Prager, L. M., Cavagna, A., and Taylor, C. R., 1986, Energetic Cost of Carrying Loads: Have African Women Discovered an Economic Way? *Nature* 319:668–669.

Mitrani, P., 1973, Contribution a l'étude des Formes Reigieuses et Cultuelles chez les Yaruro de l'Apure. *Anthropologia* 35:25–67.

Mitrani, P., 1975, Remarques sur l'Organisation Sociale, la Parenté et l'Alliance des Yaruro de l'Apure. *Anthropologia* 40:2–24.

Mitrani, P., 1988, Los Pumé (Yaruro). In *Los Aborigenes de Venezuela, Vol. III, Ethnologia Contemporanea II.*, edited by J. Lizot, pp. 147–213. Fundacion La Salle de Ciencias Naturales, Caracas.

Murdock, G. P., and Provost, C., 1973, Factors in the Division of Labor by Sex: A Cross-Cultural Analysis. *Ethnology* 12:203–225.

Ogilvie, M. D., 2000, *A Biological Reconstruction of Mobility Patterns at the Foraging to Farming Transition in the American Southwest.* Ph.D. dissertation, University of New Mexico. University Microfilms, Ann Arbor.

Petrullo, V., 1939, The Yaruros of the Capanaparo River, Venezuela. *Anthropological Papers 11, Bureau of American Ethnology Bulletin* 123:161–290. Smithsonian Institution, Washington, D.C.

Potts, R., 1984, Home Bases and Early Hominids. *Scientific American* 72:338–347.

Potts, R., 1988, *Early Hominid Activities at Olduvai.* Aldine de Gruyter, New York.

Potts, R., 1991, Why the Oldowan? Plio-Pleistocene Toolmaking and the Transport of Resources. *Journal of Anthropological Research* 47:153–176.

Ruff, C. B., 1987, Sexual Dimorphism in Human Lower Limb Bone Structure: Relationship to Subsistence Strategy and Sexual Division of Labor. *Journal of Human Evolution* 16:391–416.

Ruff, C. B., and Hayes, W. C., 1983a, Cross-Sectional Geometry of Pecos Pueblo Femora and Tibiae-A Biomechanical Investigation: I. Method and General Patterns of Variation. *American Journal of Physical Anthropology* 60:359–381.

Ruff, C. B., and Hayes, W. C., 1983b, Cross-Sectional Geometry of Pecos Pueblo Femora and Tibiae, A Biomechanical Investigation: II. Sex, Age, and Side Differences. *American Journal of Physical Anthropology* 60:383–400.

Ruff, C. B., and Larsen, C. S., 1990, Postcranial Biomechanical Adaptations to Subsistence Strategy Changes on the Georgia Coast. In *The Archaeology of Mission Santa Catalina De Guale: 2. Biocultural Interpretations of a Population in Transition*, edited by C. S. Larsen, pp. 94–120. *Anthropological Papers of the American Museum of Natural History, Number 68*. American Museum of Natural History, New York.

Ruff, C. B., Larsen, C. S., and Hayes, W. C., 1984, Structural Changes in the Femur with the Transition to Agriculture on the Georgia Coast. *American Journal of Physical Anthropology* 64:125–136.

Ruff, C. B., Walker, A., Trinkaus, E., and Larsen, C. S., 1993, Robusticity in *Homo*, I: Temporal Trends and Mechanical Interpretation. *American Journal of Physical Anthropology*, 91:21–53.

Trinkaus, E., 1987, Bodies, Brawn, Brains and Noses: Human Ancestors and Human Predation. In *The Evolution of Human Hunting*, edited by M. H. Nitecki and D. V. Nitecki, pp. 107–145. Plenum, New York.

Trinkaus, E., 1989, The Upper Pleistocene Transition. In *The Emergence of Modern Humans*, edited by E. Trinkaus, pp. 42–66. Cambridge University Press, Cambridge.

Trinkaus, E., 1993, Femoral Neck-Shaft Angles of the Qafzeh-Skhul Early Modern Humans, and Activity Levels among Immature Near Eastern Middle Paleolithic Hominids. *Journal of Human Evolution* 25:393–416.

Trinkaus, E., and Ruff, C. B., 1989a, Cross-Sectional Geometry of Neandertal Femoral and Tibial Diaphyses: Implications for Locomotion. *American Journal of Physical Anthropology* (Abstract) 78:315–316.

Trinkaus, E., and Ruff, C. B., 1989b, Diaphyseal Cross-Sectional Morphology and Biomechanics Of The Fond-De-Foret and the Spy 2 Femur and Tibia. *Anthropologie et Prehistorie* 100:33–42.

Wesson, A. L., and Martin, D. L., 1995, Women Carried Heavy Loads while Men were Weaving: Precontact Sexual Division of Labor at Black Mesa (Abstract). *American Journal of Physical Anthropology*, Supplement 20:220.

Winter, D., 1990, *The Biomechanics and Motor Control of Human Movement*. Wiley-Liss, New York.

Chapter 11

Mobility and the Locomotor Skeleton at the Foraging to Farming Transition

MARSHA D. OGILVIE

Abstract. One of the most important economic, social and biological transitions in human history is the shift from food collecting to food producing systems. Competing models regarding the interpretation of the archaeological evidence bring the economic organization of incipient Southwestern farmers into question. Were they committed agriculturalists occupying settled villages, or seasonally mobile groups using agriculture as a buffer to wild resources? Reconstructing the changing economic strategies accompanying the first archaeologically detectable evidence of increased residential stability can be approached from a biological perspective. Bone biology research indicates a clear link between habitual physical activities and skeletal form. Cross-sectional data derived from computed tomographic scans of femoral diaphyses provide a means to test hypotheses regarding relative mobility patterns accompanying the foraging to farming transition.

This paper examines femoral cross-sectional property values for I_x/I_y, a proxy for in vivo locomotor behavior, in three samples. Comparative study groups consist of mobile foragers ($n = 42$) from the Trans- and Lower Pecos regions of southwest Texas, early agriculturalists ($n = 21$) from the Tucson Basin in southeastern Arizona, and sedentized late agriculturalists ($n = 76$) from the aggregated prehistoric pueblo of Pottery Mound, New Mexico. The Tucson Basin sample is currently the earliest skeletal series found in association with pre-200 B.C. maize.

Statistical analyses indicate that femoral robusticity significantly declined across subsistence categories with increasing dependence on maize. Partitioning by sex and subsistence revealed intriguing patterns. Early agricultural males maintained behaviors necessitating high mobility levels. Female mobility, however, dramatically declined at this time, suggesting that they were spending more time in proximity to residences incorporating the sedentary tasks associated with early domestication into the domain of women's work.

INTRODUCTION

There has been a dramatic change in the human locomotor skeleton during the Holocene with the trend towards increasing gracilization of the lower limb (Ruff et al., 1993; see Figure 1). It is generally recognized that this trend is tied to one of the most important economic, social and biological transitions in human history, the shift from food collecting to food producing systems (Larsen, 1997; Ruff, 2000a; Ruff et al., 1993). As agriculture

---

From Biped to Strider: The Emergence of Modern Human Walking, Running, and Resource Transport, edited by D. Jeffrey Meldrum and Charles E. Hilton. Kluwer Academic/Plenum Publishers, New York, 2004.

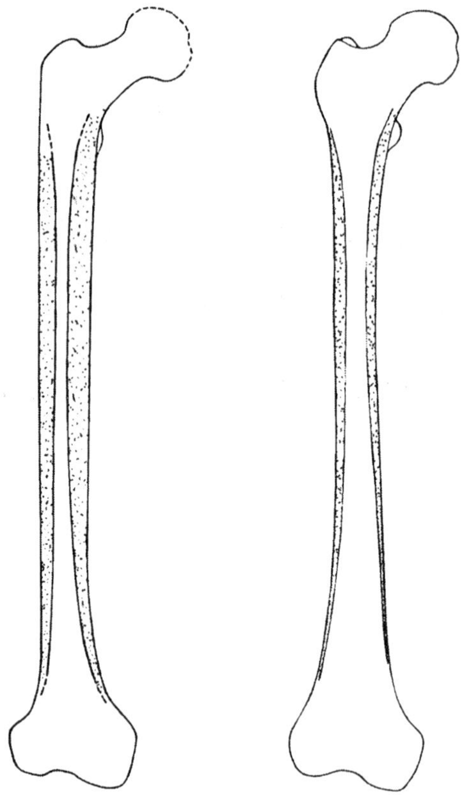

Figure 1. Longitudinal sections of femora illustrating differences in cortical bone thickness between an early Homo specimen (on left) and a contemporary specimen (on right). Modified from Ruff et al. (1993) and reproduced with permission of the author.

required some degree of sedentism, this transition had a major impact on the behavioral repertoire of highly mobile hunter-gatherer populations (Binford, 2001).

Reconstructing past mobility in groups making the foraging to farming transition requires a clear understanding of the sexual division of labor. The behavioral dichotomy in male and female locomotor roles can be elucidated through the link between cultural behavior and human biology (Ruff, 1987). Expectations from ethnographic documentation suggest that the constraints associated with female reproductive biology contribute to behavioral sex differences that shape the division of labor, resulting in limited spatial range for females (Draper, 1985; Draper and Cashden, 1988; Ember, 1983; Ember and Ember, 1976; Frayer, 1980; Hilton and Greaves, 1995; Hurtado and Hill, 1990; Hurtado et al., 1992; Monroe et al., 1984). Subsistence roles assigned to females are, therefore, those that are compatible with simultaneous child care and do not put children at risk (Brown, 1970). Conversely, males tend to perform subsistence tasks that, quite often, require long distance travel or absence from the home and increased risk to the individual (Murdock and Provost, 1973). In fact, activities involving risky long-range travel are performed almost exclusively by males (Frayer, 1980; see review in Gaulin and Hoffman, 1988).

Traditionally, attempts to reconstruct the economic organization of past human groups have relied on artifactual evidence in conjunction with ethnographic analogy. These approaches, generally, have not utilized an important archaeologically-derived resource, the human skeletal remains. Bone biology research suggests that behavioral changes resulting from shifts in subsistence strategies can equate with profound modification to skeletal anatomy (Ruff, 1992; Ruff and Larsen, 1990; Ruff et al., 1984; see review in Larsen, 1997).

The skeleton continually remodels throughout life in response to the biomechanical forces imposed during habitual physical activities, such as repetitive locomotor behavior (Chamay and Tschantz, 1972, 1983a, b; Carter et al., 1989; Curry, 1990; Ruben et al., 1990; Ruff and Hayes, 1982). The osseous loading history maintained in skeletal architecture can be elucidated through the use of computed tomography. Analysis of the structural geometry of the femoral diaphysis has proven to be a useful tool for addressing mobility-related questions (Nagurka and Hayes, 1980; Ruff and Hayes, 1983a, b; Ruff and Leo, 1986).

ARCHAEOLOGICAL CONTEXT

Recent archaeological investigations in southeastern Arizona unearthed the oldest skeletal series recovered in association with early maize agriculture. The economic context has been described as "transitional," based on lines of archaeologically derived evidence for mixed foraging and farming strategies (Huckell, 1995; Mabry, 1998).

Competing models concerning the interpretation of the archaeological evidence bring the economic organization of incipient farmers into question. Were they committed agriculturalists occupying settled villages, or seasonally mobile using agriculture as a buffer to wild resources? What were the initial changes in locomotor behavior that facilitated the incorporation of early domesticates into the hunting and gathering adaptive strategy?

The rare Tucson Basin skeletal series provide the opportunity to biologically examine the processes by which early agricultural systems developed. The presence of both males and females with intact femora can elucidate the concomitant impact of domestication on the sexual division of labor. As such, this paper proposes to elucidate the impact of early farming on male and female mobility roles in the prehistoric American Southwest.

Sample Selection

Three study groups were selected from the perspective of probable subsistence related locomotor behavior, based on lines of archaeological and ethnographic evidence in each study area. The study groups are analogs for three distinct economic stages seen in the foraging to farming transition in the American Southwest, foraging, seasonally occupied villages and permanent agricultural settlements. Bone structure provides an independent means to track changes in subsistence behavior in groups making the transition to food production. The foragers and late agriculturalists provide a behavioral baseline of comparison for the incipient agricultural sample. Figure 2 presents an overview of the study area.

The sample sizes involved in this study are a reality due to the paucity of preserved human remains from the Archaic period. For this reason, the chronological placement of all individuals sampled in each study group was not always contemporaneous, but are the "universe" for this temporal period. All samples were confined to similar geographic areas to minimize the confounding effects of terrain and genetic variation in local populations on bone biomechanics (Bridges, 1995b; Ruff, 2000a, c).

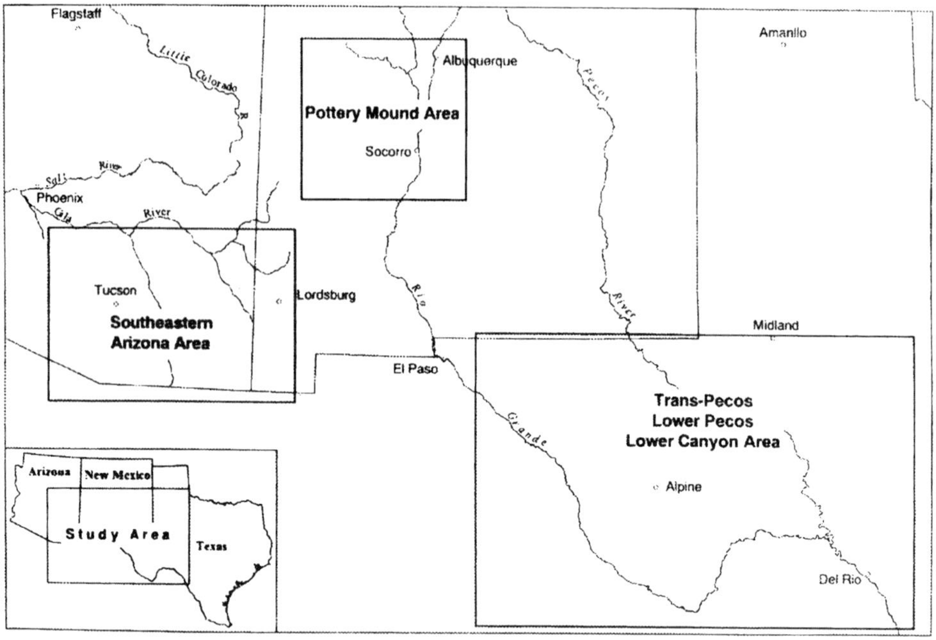

Figure 2. Overview map of the American Southwest with boxes highlighting the areas from which the skeletal samples are derived.

Skeletal Samples

The Preagriculturalists

The preagriculturalists (n = 42) are represented by 19 males and 23 females who traversed the arid landscape of the Trans- and Lower Pecos regions of southwest Texas where agriculture was never prehistorically incorporated into the economy. The majority of the burials were recovered from dry rock shelters dating to the Blue Hills Phase of the late Archaic, ca. 2300–1300 YBP (Collins and Labadie, 1999; Steele and Olive, 1989; Turpin, 1986, 1988, 1992, 1995). A drying trend at the end of the Pleistocene established arid grasslands mixed with desert succulent plant communities (Bryant, 1986a, b; Shafer, 1986a). Hundreds of caves and overhangs in this karstic system provided shelter for the small bands of foragers who first inhabited this area.

As in all arid regions, settlement and subsistence were tempered by the location and availability of water sources (Bayham and Morris, 1990; Carmichael, 1990; Hard, 1990; Speth, 1990; Taylor, 1964; Vierra, 1990). The Rio Grande, Devils' and Pecos rivers provided a reliable supply of water, critical in this otherwise arid setting. Foragers in such desert environments where there is no agriculture must move frequently to take advantage of seasonally available resources (Kelly, 1983, 1995). Repeated use sites near critical resource locations, rather than long-term occupations, attest to seasonally fluctuating hunting and gathering forays for small animals and desert plants (Binford, 1980; Marmaduke, 1978).

Dry conditions in the Pecos preserved perishable artifacts and a wealth of plant and animal remains in the cultural deposits of the dry rock shelters (Ward, 1992). Botanical and faunal studies revealed that the basic diet consisted of Carbon 4 (C_4) and Crassulacean Acid Metabolism (CAM) rich plants and the animals that consumed them, predominately rodents, lagomorphs and fish (Huebner, 1991, 1995a, b). Plants with a C_4 photosynthetic pathway include grasses and seeds, while CAM plants are primarily desert succulents. Palynological, paleofecal and dental microwear studies indicate that Sotol (*Dasylirion texanum*), agave (*Agave scabra*), prickly pear (*Opuntia phaecantha*) and lechuguilla (*Agave lechuguilla*) were heavily exploited CAM plants that were a mainstay in the Pecos diet for thousands of years (Bryant, 1974, 1986b; Bryant, 1986a, b; Danielson and Reinhard, 1998; Dering, 1979, 1999; Hartenady and Rose, 1991; Riskind, 1970; Stock, 1983; Story and Bryant, 1966; Williams-Dean, 1978). Sotol and lecheguilla were particularly valuable as they represent one of the few food resources that could be harvested year-round (Huebner, 1995).

The sexual division of labor in the Pecos region is projected from archaeological evidence, coupled with ethnographically documented cases of task specificity in male and female hunter-gatherers (Brown, 1970; Murdock and Provost, 1973; Shafer, 1986c). Observations of desert succulent processing by present-day groups living near the South Texas/Mexican border support the above documented cases (Kludt, n.d.). Procuring and processing fibrous desert succulents is a labor-intensive task requiring the participation of both males and females (Brown, 1991; Kludt, personal communication). High group mobility is implied by the necessity of travel to distant plant locations. As the roasting process took several days, all members of the group camped, foraged, and hunted near the earth oven locations (Dering, 1999; Shafer, 1986c). Burned rock middens and charred remains of lecheguilla and sotol have been found by the thousands in cultural deposits in southwest Texas (Vierra, 1998).

Strategic site locations in an ecotone provided hunting and foraging opportunities for males and females to exploit riverine, upland, and valley resources. This required frequent travel between these zones. The limited riparian and cienega plant communities, including seeds from a variety of grasses, were either clustered near the rivers or found in locations that necessitated strenuous travel to higher elevations (Dering, 1979; Stock, 1983).

Male mobility was likely tempered by logistically organized hunting forays (Binford, 1980). Excellent chert sources were abundant in the local environment. A portable tool kit geared for foraging and hunting small game was recovered during archaeological investigations at Horseshoe Cave. The kit contained multifunctional tools that could have been used to exploit any of the resources in the canyon (Hester, 1983). With the exception of stylistic changes in projectile points, the basic tool kit apparently remained adequate through time for use in a foraging subsistence economy (Dibble and Prewitt, 1967). Faunal remains suggest that many of the larger hunted animals, such as, white-tailed deer (*Odocoileus virginus macrourus*), grey fox (*Vulpes* sp.), raccoon (*Procyon lotor mexicanus*) and coyote (*Canis letrans texensis*), were taken during less arid climatic phases. Most of the protein in the diet, however, came from smaller animals, such as rodents and lagomorphs, likely trapped by women and children (Huebner, 1995a, b; Murdock and Provost, 1973).

Desert succulents, also widely available, provided the raw material for a flourishing fiber industry in the canyon (Shafer, 1986a). In addition to being dietary staples, fiber from succulent plant communities was used for weaving everything from technological items to clothing and children's' toys (Shafer, 1986c).

Resource abundance in this environment made it unnecessary to carry all burdens on routine excursions. It has been archaeologically noted that artifacts were "stashed and cached" near critical resource locations. The 9,000-year archaeological sequence, with no evidence of abandonment, clearly attests to the successful exploitation strategies employed by the inhabitants of this diverse biotic region (Lundelius, 1974; Reinhard et al., 1989; Van Devender and Spaulding, 1979; Van Devender et al., 1984).

The Early Agriculturalists

The early agriculturalists (n = 21) are represented by 14 males and 7 females from southeastern Arizona where agriculture was thought to be in the earliest stages (Huckell, 1988, 1995). They resided in the Tucson Basin during the Late Archaic, or Early Agricultural, period, approximately 3500–1850 Y.B.P. (Huckell, 1983, 1995). Their subsistence economy was "transitional," based on lines of paleobotanical and archaeological evidence for mixed foraging and farming strategies (L. Huckell 1995; Mabry, 1998).

The Tucson Basin is located on the eastern edge of the Sonoran Desert zone and is part of the Santa Cruz River drainage system. This low southern desert is surrounded by the Tucson Mountains on the west, the northern Serita mountains on the southwest and the southern end of the Tortolita and Santa Catalina Mountains on the north. On the east and southeast, the basin is bordered by the Rincon and Santa Rita Mountains.

Around the turn of the nineteenth century, an erosional episode along Cienega Creek and Pantano Wash in the southeastern part of the basin exposed Early Agricultural levels. Subsequently, during a highway expansion project in the vicinity of the Santa Cruz River, additional Early Agricultural sites were encountered. These discoveries caused us to rethink previous assumptions about socioeconomic systems during the late centuries B.C. (Mabry and Clark, 1994; Mabry, 1998). Prior to this time, the conventional wisdom was that domesticates had not yet been incorporated into the hunting and gathering adaptive strategy (Matson, 1991; Minnis, 1985, 1992; Wills, 1988a, b, 1992). However, archaeological investigations at these recently exposed sites revealed the presence of maize by at least 1000–500 B.C. (refer to references cited in Huckell, 1995; Mabry, 1998).

Cienega residential sites were strategically positioned along the perennial waterways of the Santa Cruz and its' tributaries, the most critical resource for the inhabitants of the arid basin. Small special-use sites were located in three surrounding ecological zones the riverine flood plain, the piedmont (the slopes leading from the major drainages to the mountain fronts) and the uplands bordering the basin (Huckell, 1995). Positioning themselves in this ecotone provided the inhabitants with, not only rich soil for agriculture, but a variety of microenvironments for hunting and gathering opportunities (L. Huckell, 1995, 1996).

Maize fields and agricultural products at the large residential sites provide solid evidence for early agriculture (Eddy, 1958; Eddy and Cooley, 1983; Hemmings et al., 1968; Huckell, 1977; Huckell and Haury, 1988; L. Huckell, 1996; Mabry et al., 1995; Mabry, 1998). Dentition recovered with the skeletal remains also suggests maize consumption by site inhabitants (Minturn and Babb, 1995; Minturn, 1998). Excavations revealed large pithouse villages with thick midden deposits, extramural storage, numerous groundstone artifacts and food processing features (Adams, 1996; Mabry and Clark, 1994; Mabry et al., 1995). Wild plant species recovered from bell-shaped storage pits representing multiple seasons of the year, demonstrated that the site inhabitants anticipated being in one location for extended periods of time (Binford, 1982; Eddy, 1958; Huckell and Haury, 1988; L. Huckell, 1995).

Because agriculture requires some degree of sedentism, it is thought that early agricultural populations were relatively more settled than highly mobile Middle Archaic groups (Huckell, 1995). The shift towards reduced residential mobility would have required a reliance on stored resources and use of logistic strategies to most efficiently exploit resources in different environmental zones (Binford, 1980; Kelly, 1983, 1992, 1995; Huckell, 1995). Mobility could decrease because stored resources were reliable and did not require daily long distance travel. More practical targeting of economically important resources could be accomplished by logistic task groups. This is archaeologically reflected at special use sites identified in the topographically diverse Tucson Basin. These short-term sites demonstrate that broad-spectrum hunting and gathering of wild plant and animal resources was still an indispensable component of the economy (Eddy, 1958; Huckell, 1984a, b; L. Huckell, 1995, 1996).

Paleoethnobotanical studies indicate that the plant species dominating assemblages were those available in late summer and fall—maize, amaranth (*Amaranthus* sp.), purslane (*Portulaca* sp.), agave (*Agave* sp.), yucca (*Yucca elata, baccata*), rushes (*Juncus* sp.), acorns (pericarps of *Quercus* sp.), wheelscale saltbush (*Atriplex elegans*) and Arizona walnut (*Juglans major*) (L. Huckell, 1996). The nutritionally important legume family (Chenopodiaceae and Amaranthaceae) thrived in the agriculturally disturbed soils along flood plains at the riverine sites (L. Huckell, 1995). The limited presence of spring plants, such as stickleaf (*Mentzelia* sp.), mesquite (*Prosopis* sp.), chenopods (Chenopodiaceae) and saguaro (*Carnegiea gigantea*), suggests partial site abandonment at this time, with people moving back and forth between multiple environmental zones during seasonal rounds.

Obviously, Early Agricultural groups retained some degree of mobility to procure resources at varying distances from home. The rich secondary biomass in the Sonoran biotic region has implications for reduced mobility in females. Some suggest a reduction in long-range mobility during foraging forays, because wild plant resources were widely distributed and easily accessible (Fish et al., 1992). Such clustered resources would equate with a reduction in travel and search time. With foraging costs reduced, the time formerly spent in long distance travel could be devoted to plant processing, making foraging more efficient. Moreover, this localized abundance of reliable and predictable plant communities was potentially storable, providing food sources during winter and spring months when most wild plants were not yet available (Fish et al., 1990; Fish and Fish, 1991; L. Huckell, 1995). In addition, many resources, such as acorns and agave were available within relatively short distances of residences and permanent water sources (Fish et al., 1992). Based on Binford's (1980) Economic Zonation model, many resources could have been efficiently targeted by simple foraging trips near settlements.

High male mobility was likely conditioned by hunting, a commonly logistically organized activity (Binford, 1978, 1980). Reaching the Sonoran desert zone and rugged uplands would have necessitated logistic mobility. Such forays are evidenced by small, extended hunting camps in the surrounding mountains and grasslands (Bayham et al., 1986; Huckell, 1995). Mule deer (*Odocoileus hemionus canus*), white-tailed deer (*Odocoileus virginianus macrourus*), antelope (*Antilocapra americana*), bighorn sheep (*Olgas kenadensus*), elk (*Cervus merriami*), bobcat (*Lynx rufus*), coyote (*Canis latrans*) and jackrabbit (*Lepus gaillardi*), comprised a significant portion of the diet (Eddy, 1958; Thiel, 1996). Procuring large game, more commonly targeted than smaller mammals in all biomes, required extensive travel to distant mountain ranges by male hunters, sometimes at distances of 70 km. or more (Huckell, 1995).

The topographic diversity and abundance of resources in the Sonoran Desert zone made for a rich environment for Early Agricultural populations. At the present time, our understanding of these newly investigated sites and related subsistence economies during this critical time period is still ongoing.

The Late Agriculturalists

The late agriculturalists (n = 76) are represented by 42 males and 34 females who resided at the aggregated community of Pottery Mound, New Mexico. This large 500 room pueblo was occupied during the Pueblo IV period (700–500 Y.B.P), and abandoned just prior to the Spanish entradas in A.D. 1540 (Hibben, 1955, 1961, 1996). The array of ceramic types littering the site landscape, predominately Glaze A, reinforces the PIV designation (Brody, 1964; Voll, 1967).

This massive three-tiered community is located in Valencia County in central New Mexico, at the confluence of the southern edge of the Rio Grande Valley and the northern periphery of the Chihuahuan Desert Zone (Cordell, n.d.; Hibben, n.d; Schorsch, 1962). The site locality is bordered on the north by Mt. Taylor, on the east by the Manzano Mountains, on the west by topographically broken high canyon country and on the south by the Magdalena and Ladron Mountains. The Rio Grande and its tributary, the Rio Puerco, course along the eastern periphery of the site.

Intensive maize agriculture with specialization of food processing was practiced at Pottery Mound (Cordell, 1984, 1997; Wills et al., 1996). Major investment in multi-storied adobe architecture and nonportable technology reflects the expectation of permanent residency (Wills, 1988b). Ceremonial structures, storage features, flora and fauna indicative of year-round occupation and a large burial population also demonstrate permanent ties to the land (Schorsch, 1962; Wills and Huckell, 1994).

This high southern desert is ecologically classified by the U.S. Department of Agriculture as the Upper Sonoran life zone (Bailie, n.d.). The predominant plant species on the plain included Russian thistle (*Salosa kalitinfolia*), salt brush (*Atriplex* sp.), sacaton grass (*Aristida* sp.) and juniper (*Juniperus* sp.) (Schorsch, 1962). The ecotone within which the site was positioned provided arable soil for agriculture and high quality wild plant and animal resources. Palynological reconstruction identified three biomes in the vicinity of Pottery Mound during PIV times, riparian, open grassland and montane (Emslie, 1981; Emslie and Hargrave, 1978). The headwaters of the Rio Puerco, of the east, approximately 100 km. to the north, provided a permanent source of water for the site residents.

Unique insight into the prehistoric lifeways at Pottery Mound comes from the over 800 spectacular kiva murals, the most abundant and well preserved in the Southwest (Brody, 1964, 1970; Crotty, 1979; Hibben, 1960, 1966, 1967, 1975, 1996; Leviness, 1959; Vivian, 1961). The importance of maize (*Zea* sp.) is reflected in mural depictions of corn plants and subsistence tasks associated with agricultural production. Hundreds of preserved charred cobs have been recovered from the cultural deposits (Hibben, 1996). It is likely that other cultigens, beans (*Phaseolus* sp.), squash (*Cucurbita* sp.) and, possibly cotton (*Gossypium* sp.) were also grown (Wills, personal communication). The grasslands adjacent to the site contained extensive agricultural fields. Additional fields were probably located more distant from the main site as insurance against crop disasters (Wills, personal communication).

The economic roles of males and females documented in the ethnographic literature are clearly depicted in the kiva paintings (Brown, 1970; Murdock and Provost, 1973). Males were more heavily engaged in agricultural pursuits at late prehistoric communities than in earlier time periods (Ember, 1983). Field preparation and crop tending likely reduced the total time males allotted for long distance mobility in favor of more sedentary agricultural duties.

Strictly in terms of farming tasks, the proportional contribution of females was declining relative to that of males (Ember, 1983). Women's total work effort, however, did not decrease. A wide variety of grinding implements at the site support ethnographic documentation that women were heavily involved in maize processing, often for six to eight hours per day (Adams, 1996; Hard, 1990; Hard et al., 1996; Lancaster, 1986; Murdock and Provost, 1973). A wide variety of grinding technologies, various metates, including open troughs, were used with one and two hand manos (Adams, 1996). Grinding specificity of such items as graters, grinders, pounders, mortars and pestles, suggested that women were grinding substances, other than maize, such as various seeds and pigments. Females in mural tabloids carried burden baskets for wild plant collecting trips, prepared foods, transported fuel and water and produced pottery (Brown, 1970; Hibben, 1976; Murdock and Provost, 1973).

Mural portrayals of male hunters and the over 10,000 recovered faunal remains brought in from surrounding grasslands and mountains indicated that hunting was engaged in at Pottery Mound. As hunting is usually logistically organized, long distance mobility of some portion of the male population is expected (Binford, 1980). Targeting large mammals would require travel to the Manzano Mountains 20 miles to the east and the broken high mesa country approximately 10 miles west, of the site. These locations were excellent for mule deer (*Odocoileus hemionus canus*), white-tailed deer (*Odocoileus virginianus macrourus*), Merriam's elk (*Cervus merriami*) and bighorn sheep (*Olgas kenadensus*). Pronghorn antelope (*Antilocapra americana*) were locally available, ranging very close to the site in the Puerco Valley. Antelope drives were historically documented in this same prime location.

Ethnobiological research identified over 50 species of birds in the avifauna assemblage, many of which were in mural depictions (Emslie and Hargrave, 1978). Pottery Mound was located on a "fly-way" for migrating birds, implying spring and fall bird hunting. The hundreds of birds and watercourses along the eastern edge of the site, in turn, attracted large numbers of mammals, providing additional hunting opportunities.

Surrounding montane zones with piñon pine (*Pinus edulis*) provided locales for piñon gathering trips in the fall by more mobile members of the community. Piñon nuts are a high return storable resource and facilitated "over-wintering" at a time when other wild resources were not available. Less mobile individuals, including older males and children, could perform tasks closer to home. This reduction in residential mobility with year-round site occupation made it possible for all members of the group to make labor contributions in this highly specialized prehistoric farming community.

The impressive richness of the site, i.e., the elaborate murals and thousands of decorative pot sherds from far-ranging locales, cause researchers to speculate that Pottery Mound was a center of trade and ceremonial significance in the American Southwest (Wills, personal communication). With only approximately 50% of the site excavated, the extent of our knowledge of Pottery Mound will be improved by further excavations (Hibben, 1996).

METHODS

Because long bone hypertrophy and atrophy reflect in vivo loadings imposed by habitual activities, examination of the amount and distribution of cortical bone in the femoral diaphysis makes it possible to discern its loading history. Morphological similarities among groups allow inferences to be made regarding the variability resulting from economic transitions and the concomitant impact on the sexual division of labor.

This functional approach uses an engineering beam model (Timoshinko and Gere 1970). The model can appropriately be applied to a structure whose length is long relative to its width, making it particularly suitable for structural analyses of long bone diaphyses (Lovejoy et al. 1976; Ruff 1992; Ruff and Hayes 1983a). When modeled as a beam, structural properties of the diaphysis can be calculated at biologically relevant section locations lying perpendicular to the long axis of the bone. Biomechanical analyses of cross-sections indicate the strength or rigidity of the beam in resistance to the forces generated during habitual physical activities and, thus, provide a means for interpreting past lifeways. Biomechanical models predict that the changes in physical activities associated with subsistence shifts will be reflected in behaviorally-conditioned skeletal morphology (Ruff 1992, 2000a).

Femora from 139 individuals were included in this study. The femora ($n = 199$) were selected on the basis of completeness, lack of pathological involvement, and availability of associated pelvic remains for secure sex assessment. Standard osteological techniques were employed to evaluate the sex of each individual (see Buikstra and Ubelaker 1994). Preference was always given to the primary sexual characteristics seen in the os coaxe. For the majority of the skeletal sample ($n = 129$), pelvic criteria predicted sex. Fragmentary pelves in 10 individuals necessitated sex assessment from intact cranial remains. Whenever possible, all skeletally preserved attributes of sex were collectively evaluated.

Since developmentally-complete bone length measurements are required for size standardization between the sexes, only skeletally mature adults were included in this study. An upper age limit of approximately 59 years was arbitrarily set to reduce the confounding effects of age-related bone changes (Ruff and Hayes 1983b). The majority of the individuals ($n = 109$) were aged using degenerative changes in auricular surface morphology (Lovejoy et al. 1985) and pubic symphysis morphology (Todd 1920).

Length measurements (mm) were taken of all femora. Bone lengths described here correspond to those previously described as biomechanical length' (Ruff and Hayes 1983a, b; Ruff et al. 1993). Length and planes of orientation in reasonably incomplete specimens were approximated by morphological landmarks in anatomical planes with bones from same-sex, same-sample individuals. High levels of symmetry in long bone lengths and articulations justify this procedure (Trinkaus et al. 1994).

Computed tomography was performed on all femora preserving the 50% section of bone length. After quantification of section properties by the software program SLICE (Eschman, 1992), the resultant values were used in the femoral shape index. The index, I_x/I_y, serves as a proxy for relative mobility levels. Larger resultant index values indicate greater bending strength in the A/P plane that translate to relatively higher levels of mobility (Ruff and Hayes, 1983a, b). Two-way Analysis of Variance (ANOVAs) was computed within and between sex and subsistence sub-sets. For all statistical comparisons, differences were considered statistically significant at the alpha level of $p < .05$.

RESULTS

Analyses of the femoral shape index (I_x/I_y) indicate significant sex differences ($p = .0011$) and highly significant differences for subsistence ($p < .0001$; see Table 1).

Table 1. I_x/I_y Values by Sex and Subsistence

	Foragers		Transitional		Agricultural		2-way ANOVA p-values	
Variable	Males	Females	Males	Females	Males	Females	Sex	Subsist.
I_x/I_y								
Mean	1.34	1.29	1.39	1.09	1.17	1.00	.0011	<.0001
S.D.	.29	.15	.37	.25	.22	.14		
S.E.	.07	.05	.12	.102	.04	.03		
C.V.	21.79	11.54	26.28	23.0	18.77	14.26		

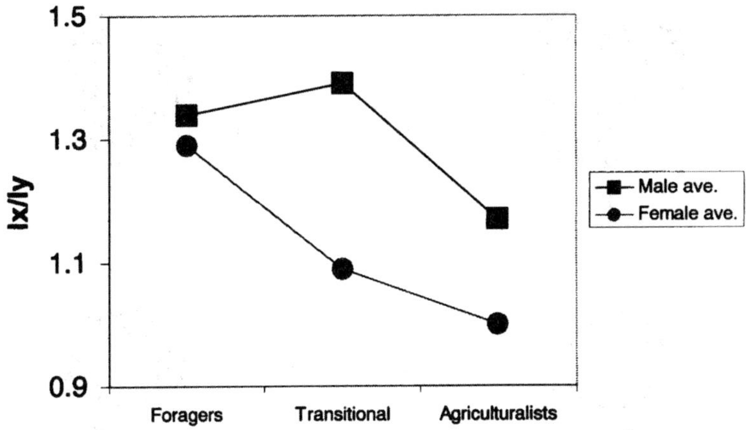

Figure 3. Femoral shape index (I_x/I_y) values reflecting decreasing mobility levels as dependence on maize increases. Note relative differences between the sexes in each subsistence category.

When grouped by subsistence, femoral shape index values reflect decreasing mobility levels through time as dependence on maize increases (see Figure 3). As would be expected, the preagricultural foragers, as a group, have the highest value, followed by the early agriculturalists, followed by the late agriculturalists. T-tests indicate a significant difference between the foragers and late agriculturalists, while the early agricultural value is not significantly different from either group.

The dichotomy in male and female index values provides insight into the processes by which incipient agricultural systems developed. Males, as a group, always have a larger value for the shape index than females, as a group. However, when partitioned by sex and subsistence, several interesting patterns emerge.

Although femoral index values for both males and females become increasingly smaller through time, they do so at very different rates (refer to Figure 3). This pattern is reflected in the amount of cortical bone in the cross-sections for each sub-sample (see Figure 4). During the foraging phase, values for males and females are not significantly different from each other, with those for males being only slightly larger.

Femoral index values diverge markedly between the sexes in the early agricultural phase. The male value basically remains at forager levels and the female value sharply declines.

With the transition to full-blown agriculture, index values for the sexes reconverge, driven by a rapidly decreasing male value from the early agricultural level. The late agricultural female value continues to decline from the early agricultural level.

RELEVANT RESEARCH

The long-term trend in decreasing femoral dimensions with increasingly sedentary lifeways is documented from early anatomically modern human fossils to Holocene human populations (Bridges, 1995b; Holt and Churchill, 2000; Lovejoy and Trinkaus, 1980; Ruff et al., 1993; Ruff et al., 1999; Trinkaus and Ruff, 1989; Trinkaus et al., 1994; refer to references cited in Larsen, 1997). This long-term trend is an appropriate context from which to consider the variability in long bone structure seen in more recent archaeological samples.

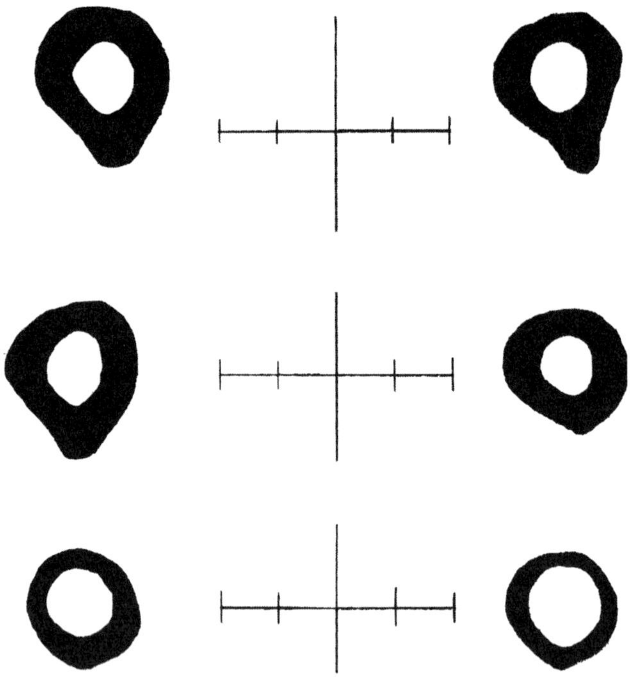

Figure 4. Cross-sections of right femora at 50% of length' showing size and shape changes in the femur from foraging (top) to early agriculture (center) to late agriculture (bottom). Note structural differences between males (on the left) and females (on the right). Cross-sections are oriented with anterior at top, posterior at bottom, medial at right, and lateral at left. Scale is in centimeters.

The reduction in locomotor behavior with increasing dependence on maize is biologically manifested as lower I_x/I_y values and more circular femoral diaphyses. Such short-term adaptation to subsistence change is consistently reported by biomechanical researchers. Although bone strength may vary by regional setting due to local ecology, terrain, and variability in subsistence practices. Reduction in femoral dimensions with farming is reported from the Georgia Coast (Ruff and Larsen, 1990), the American Southwest (Brock, 1985; Brock and Ruff, 1988; Ruff and Hayes, 1983a, b; Ogilvie, 2000), the Northern and Southern Great Plains (Ruff, 1994b), the Great Basin (Larsen and Kelly, 1995; Ruff, 2000c), the Tennessee and Ohio River Valleys (Bridges, 1989a, 1995a), and La Florida (Larsen et al., 1992; Ruff, 1997b).

DISCUSSION AND CONCLUSIONS

The biological evidence from the American Southwest demonstrated the trend in decreasing femoral dimensions with increased reliance on maize. It is of particular interest that application of these data to populations in economic transition proved useful for clarifying the processes by which early agricultural systems developed. The Trans- and Lower Pecos foragers were the most mobile of all the individuals examined. The males and females showed non-significant differences in lower limb loading, suggesting a residentially mobile strategy. Based on archaeological evidence, ethnohistoric documentation, and local ecology in the southwest Texas region, a subsistence strategy employing mixed sex resource procurement and processing would be expected (Binford, 2001).

The committed late agriculturalists from the sedentary setting of Pottery Mound, not surprisingly, were the least mobile. The late agricultural males, still more mobile than their female counterparts, likely engaged in small-scale hunting forays and agricultural pursuits some distance from the main community (Ember, 1983). The females displayed the lowest mobility levels in both sexes and all subsistence categories. Ethnographic projections suggest that they performed the sedentary domestic chores associated with the maintenance of agricultural production (Ember, 1983; Rafferty, 1985).

The occupants of the Tucson Basin during the late centuries B.C. were truly a population in economic transition. As a group, they demonstrated levels of locomotor behavior intermediate to those of the mobile foragers and sedentary late agriculturalists. Both the artifactual evidence and femoral index values suggest that Late Archaic groups in the Tucson Basin were still relying on wild resources to a great extent and were not yet fully committed to settled agriculture (Huckell, 1985).

It was during this economic transition that the sexual division of labor was noticeably impacted. During the incorporation of early maize into the economy, male femoral structure reflected high levels of mobility equivalent to those of the mobile male Texas foragers. The maintenance of behaviors necessitating high mobility by males is compatible with a strategy of logistical resource targeting. Lithic assemblages analyzed from hunting camps in distant mountain ranges surrounding the Tucson Basin also support implications for long range hunting forays by males (Vierra, 2001). The Arizona males provide the exception to the expected pattern of decreasing mobility with agriculture.

Interestingly, the Tucson Basin females showed a marked decline in mobility at this time. The femoral data demonstrate that their locomotor behavior was not significantly different from the sedentary female agriculturalists. Comparisons with the female foragers, however, revealed dramatic behavioral differences. The reduction in I_x/I_y values in early agricultural females is compatible with a reduction in residential mobility. The gathering and preparation of wild plant resources has consistently been assigned to the domain of women's work across the ethnographic record of foragers. A reduction in residential mobility in females would provide more extended periods of time in proximity to residences, presumably underwriting the costs of the initial incorporation of early domesticates into the subsistence economy. A division of labor incorporating increased logistical mobility in males and reduced residential mobility in females facilitated the addition of a new, perhaps more predictable, variant into the behavioral repertoire of late Archaic hunter-gatherer populations that would eventually lead to more sedentary lifeways. The female work effort, long overlooked in the anthropological literature, appears to have been the impetus for the most important economic, social and biological transition in human history, the shift from food collecting to food producing systems.

ACKNOWLEDGEMENTS I want to thank the affiliated Native tribes for their permission to conduct this research. Thanks are also extended to the University of New Mexico Department of Radiology, Albuquerque, the Maxwell Museum Laboratory of Human Osteology, Albuquerque, Desert Archaeology, Tucson, Arizona, Arizona State Museum, University Hospital, Tucson, Arizona, Texas Archaeological Research Laboratory Austin, Austin Radiological Association, and the Witte Museum, San Antonio, Texas. Additional support was provided by Arizona Archaeological and Historical Society, Tucson, the University of New Mexico Department of Anthropology, the Office of Graduate Studies and the Student

Research Allocations Committee at the University of New Mexico. Chris Ruff graciously permitted use of Figure 1. Production of map by R. Stauber. Other figure production was courtesy of C. E. Hilton.

REFERENCES

Adams, J. L., 1996, Refocusing the Role of Food Grinding Tools as Correlates for Subsistence Strategies of Gatherers and Early Agriculturalists in the American Southwest. Paper presented, Society For American Archaeology, New Orleans.

Bailie, B., n.d., Lifezones and Cropzones of New Mexico. *North American Fauna* 35:32–38.

Bayham, L. V., and Morris, D. H., 1990, Thermal Maxima and Episodic Occupation of the Picacho Reservoir Dune Field. In *Perspectives On Southwestern Prehistory*, edited by P. E. Minnis and C. L. Redman, pp. 26–37. Westview Press, Boulder.

Bayham, L. V., Morris, D. H., and Shackley, M. S., 1986, Prehistoric Hunter-Gatherers of South Central Arizona: The Picacho Reservoir Archaic Project. *Anthropological Field Studies 13*. Tempe: Office of Cultural Resource Management, Department of Anthropology, Arizona State.

Binford, L. R., 1978, *Nunamuit Ethnoarchaeology*. Academic Press, New York.

Binford, L. R., 1980, Willow Smoke and Dogs Tails: Hunter-Gatherer Settlement Systems in Archaeological Site Formation. *American Antiquity* 45:4–20.

Binford, L. R., 1982, Archaeology of Place. *Journal of Anthropological Archaeology* 1:5–31.

Binford, L. R., 2001, *Constructing Frames of Reference: An Analytical Method for Archaeological Use of Hunter-Gatherer and Environmental Data Sets*. University of California Press, Berkeley.

Bridges, P. S., 1989, Changes in Activities with the Shift to Agriculture in the Southeastern United States. *Current Anthropology* 30(3):385–394.

Bridges, P. S., 1995a, Biomechanical Changes in Long Bone Diaphyses with the Intensification of Agriculture in the Lower Illinois Valley. *American Journal of Physical Anthropology* Supplement 20:68.

Bridges, P. S., 1995b, Skeletal Biology and Behavior in Ancient Humans. *Evolutionary Anthropology* 4:112–120.

Brock, S. L., 1985, Biomechanical Adaptations of the Lower Limb Bones through Time in the Prehistoric Southwest. Ph.D. dissertation, University of New Mexico. University Microfilms, Ann Arbor.

Brock, S. L., and Ruff, C. B., 1988, Diachronic Patterns of Change in Structural Properties of the Femur in the Prehistoric American Southwest. *American Journal of Physical Anthropology* 75:113–127.

Brody, J. J., 1964, Design Analysis of the Rio Grande Glaze Pottery of Pottery Mound, New Mexico. M.S. Thesis. University of New Mexico, Albuquerque.

Brody, J. J., 1970, The Kiva Murals of Pottery Mound. Verhandlungen Des XXXVIII, Vol. 2, pp. 101–110. Paper Presented, Internationalen Amerikishen Kongress, Munich, Germany.

Brown, J. K., 1970, A Note on the Division of Labor by Sex. *American Anthropologist* 72:1073–1078.

Brown, J. A., and Vierra, R. K., 1983, What Happened in the Middle Archaic? Introduction to an Ecological Approach to Coster Site Archaeology. In *Archaic Hunters and Gatherers in The American Midwest*, edited by J. L. Phillips, J. A. Brown, pp. 165–195. Academic Press, New York.

Brown, K. M., 1991, Prehistoric Economics at Baker Cave: A Plan for Research. In *Papers on Lower Pecos Prehistory*, edited by S. A. Turpin, pp. 87–140. University of Texas Press, Austin.

Bryant, B. M. Jr., 1974, Prehistoric Diet in Southwest Texas. *American Antiquity* 39:407–420.

Bryant, B. M. Jr., 1986a, Pollen: Natures Tiny Capsules of Information. In *Ancient Texans*, edited by G. Zappler, pp. 50–57. Texas Monthly Press, Austin.

Bryant, B. M. Jr., 1986b, Prehistoric Diet: A Case for Coprolite Analysis. In *Ancient Texans*, edited by G. Zappler, pp. 1322–135. Texas Monthly Press, Austin.

Buikstra, J. E., and Ubelaker, D., 1994, *Standards for Data Collection from Human Skeletal Remains*. Arkansas Archaeological Survey Research Series, 44.

Buikstra, J. E., 1992, Diet and Disease in Prehistory. In *Disease and Demography in the Americas*, edited by J. Verano and D. Ubelaker, pp. 87–101. Smithsonian Institution Press, Washington, D.C.

Carmichael, D. L., 1990, Patterns of Residential Mobility and Sedentism in the Jornada Mogollon Area. In *Perspectives on Southwestern Prehistory*, edited by P. E. Minnis and C. L. Redman, pp. 122–134. Westview Press, Boulder.

Carter, D. R., Orr, T. E., and Fyhrie, D. P., 1989, Relationships Between Loading History and Femoral Cancellous Bone Architecture. *Journal of Biomechanics* 22:231–244.

Chamay, A., and Tschantz, P., 1972, Mechanical Influences in Bone Remodeling Experimental Research on Wolff's Law. *Journal of Biomechanics* 5:173–80.

Collins, M. B., Labadie, J. H., 1999, Excavation, Rock Art Recordation, Surface Feature Documentation, and Survey at Amistad National Recreation Area. *Texas Archaeology* 43 Winter: 1, 3–7.

Cordell, L. S., 1984, *Prehistory of the Southwest*. Academic Press, New York.

Cordell, L. S., 1997, *Archaeology of the Southwest*. Academic Press, San Diego.

Cordell, L. S., n.d., *Pottery Mound: Values and Current Status*. Unpublished Report.

Crotty, H., 1979, The Kiva Murals and the Question of Mesoamerican Influence. Paper Presented, Symposium on New Directions in Native American Art, Albuquerque.

Currey, J. D., 1990, Physical Characteristics Affecting the Tensile Failure Properties of Compact Bone. *Journal of Biomechanics* 23:837–844.

Danielson, D. R., and Reinhard, K. J., 1998, Human Dental Microwear Caused by Calcium Oxalate Phytoliths in the Prehistoric Diet of the Lower Pecos Region, Texas. *American Journal of Physical Anthropology* 107:207–307.

Dering, J. P., 1979, Pollen and Plant Macrofossil Vegetation Recovered from Hinds Cave, Val Verde County, Texas. M.A. Thesis. Texas A&M University, College Station.

Dering, J. P., 1999, Earth Oven Plant Processing in Archaic Period Economies: An Example from a Semi-Arid Savannah in South-Central North America. *American Antiquity* 64:659–671.

Dibble, D. S., and Prewitt, E. R., 1967, *Survey and Test Excavations At Amistad Reservoir, 1964–1965. Survey Report 3*. University of Texas, Texas Archaeological Survey, Austin.

Draper, P., 1985, Two Views of Sex Differences in Socialization. In *Male-Female Differences: A Biocultural Perspective*, edited by R. L. Hall, pp. 5–25. Praeger Press, New York.

Draper, P., and Cashdan, E., 1988, Technological Change and Child Behavior among the Kung. *Ethnology* 27:339–365.

Eddy, F. W., 1958, A Sequence of Cultural and Alluvial Deposits in the Cienega Creek Basin, Southeastern Arizona. M.S. Thesis. University of Arizona, Tucson.

Eddy, F. W., and Cooley, M. E., 1983, Cultural and Environmental History of Cienega Valley, Southeastern Arizona. *Anthropological Papers 43*. University of Arizona Press, Tucson.

Ember, C. R., 1983, The Relative Decline in Women's Contribution to Agricultural Intensification. *American Anthropologist* 85:285–304.

Ember, M., and Ember, C., 1976, The Conditions Favoring Matrilocal Versus Patrilocal Residence. *American Anthropologist* 73:571–594.

Emslie, S. D., and Hargrave, L. L., 1978, An Ethnobiological Study of the Avifauna of Pottery Mound, New Mexico. Paper Presented, Society For American Archaeology, Tucson.

Emslie, S. D., 1981, Prehistoric Agricultural Ecosystems: Avifauna From Pottery Mound. *American Antiquity* 46:853–861.

Eschman, P., 1992, SLCOMM. Eschman Archaeological Services, Albuquerque.

Fish, S. K., Fish, P. R., and Matson, J. H., 1990, Sedentism and Settlement Mobility in the Tucson Basin Prior To AD1000. In *Perspectives on Southwestern Prehistory*, edited by P. E. Minnis and C. L. Redman, pp. 76–163. Westview Press, Boulder.

Fish, S. K., Fish, P. R., and Matson, J. H., 1992, Sedentism and Agriculture in the Northern Tucson Basin. In *The Marana Community in the Hohokam World.* Anthropological Papers 56, edited by S. K. Fish, P. R. Fish, and J. H. Matson, pp. 11–19. University of Arizona, Tucson.

Fish, S. K., Fish, P. R., 1991, Comparative Aspects of Paradigms For the Neolithic Transition in The Levant and the American Southwest. In *Perspective on the Past, Theoretical Biases in Mediterranean Hunter-Gatherer Research*, edited by G. A. Clark, pp. 396–410. University of Pennsylvania Press, Philadelphia.

Gaulin, S., and Hoffman, H., 1988, Functional Significance of Sex Differences in Spatial Ability. In *Human Reproductive Behavior: A Darwinian Perspective*, edited by L. Betzig, P. Turke, and Borgerhoff-Mulder, pp. 129–152. University Press, Cambridge.

Hard, R. J., 1990, Agricultural Dependence in the Mountain Mogollon. In *Perspectives on Southwestern Prehistory*, edited by P. E. Minnis and C. L. Redman, pp. 135–149. Westview Press, Boulder.

Hard, R. J., Mauldin, R. P., Raymond, G. R., 1996, Mano Size, Stable Carbon Isotope Ratios, and Macrobotanical Remains as Multiple Lines of Evidence of Maize Dependence in the American Southwest. *J Archaeol Methods and Theory* 3:253–318.

Hartnady, P., Rose, J. C., 1991, Abnormal Tooth Loss Patterns among Archaic Period Inhabitants of the Lower Pecos Region, Texas. In *Advances in Dental Anthropology*, edited by M. A. Kelly and C. S. Larson, pp. 267–278. Wiley-Liss, New York.

Hemmings, E. T., Robinson, M. D., and Rogers, R. N., 1968, *Field Report on Pantano Site (AZEE: 2:50)*. M.S. Thesis, on File), Arizona State Library. University of Arizona, Tucson.
Hester, T. R., 1983, Late Paleoindian Occupation Baker Cave, Southwestern Texas. *Bulletin of the Texas Archaeological Society* 53:101–119.
Hibben, F. C., 1955, Excavations at Pottery Mound, New Mexico. *American Antiquity* 21:179–180.
Hibben, F. C., 1960, Prehispanic Paintings at Pottery Mound. *Archaeology* 13:267–271.
Hibben, F. C., 1961, The Dating of Pottery Mound. In *Illustrated World History*, edited by G. Rainerd. George Rainbird, Ltd, London.
Hibben, F. C., 1966, A Possible Pyramidal Structure and Other Mexican Influences At Pottery Mound, New Mexico. *American Antiquity* 31:522–529.
Hibben, F. C., 1967, Mexican Features of Mural Paintings at Pottery Mound. *Archaeology* 20:84–87.
Hibben, F. C., 1975, *Kiva Art of the Anasazi at Pottery Mound*. KC Publications, Las Vegas.
Hibben, F. C., 1996, The Prehistoric Site of Pottery Mound. Presented Lecture, Maxwell Museum Lecture Series. University of New Mexico, Albuquerque.
Hibben, F. C., n.d., Excavation Procedures Used in the Salvage Project at Pottery Mound. Unpublished Field Notes.
Hilton, C. E., and Greaves, R. D., 1995, Mobility Patterns in Modern Human Foragers (Abstract). *American Journal of Physical Anthropology* Supplement 20:111.
Holt, B., Churchill, S., 2000, Behavioral Changes in European Upper Paleolithic Foragers: Evidence from Biomechanical Analysis of the Appendicular Skeleton. *American Journal of Physical Anthropology* Supplement 30:182.
Huckell, B. B., 1977, Excavations at the Hasqin Site: A Multi-Component Preceramic Site Near Ganado, Arizona. Report Submitted To the Arizona Department of Transportation.
Huckell, B. B., 1983, Additional Chronometric Data on Cienega Valley, Arizona. In *The Cultural and Environmental History of Cienega Valley, Arizona. Anthropological Papers No. 43*, edited by F. W. Eddy and M. E. Cooley, pp. 57–58. University of Arizona, Tucson.
Huckell, B. B., 1984a, The Archaic Occupation of the Rosemont Area, Northern Santa Rita Mountains, Southeastern Arizona. *Arizona State Museum Archaeological Series* 147(1). University of Arizona, Tucson.
Huckell, B. B., 1984b, The Paleoindian and Archaic Occupation of the Tucson Basin: An Overview. *Kiva* 49:133–145.
Huckell, B. B., and Haury, E. W., 1988, *Excavations At Two Late Archaic Sites in the Cienega Valley, Southeastern, Arizona*. University of Arizona, Tucson.
Huckell, B. B., 1988, Late Archaeology of the Tucson Basin: A Status Report. In *Recent Research on Tucson Basin Prehistory: Proceedings of the Second Tucson Basin Conference. Anthropological Papers No. 10*, edited by W. H. Doelle and P. R. Fish, pp. 57–80. Institute for American Research, Tucson.
Huckell, B. B., 1990, *Late Preceramic Farmer-Foragers in Southeastern Arizona: A Cultural and Ecological Consideration of the Spread of Agriculture in the Arid Southwestern United States*. Ph.D. dissertation, University of Arizona. University Microfilms, Ann Arbor.
Huckell, B. B., 1995, *Of Marshes and Maize: Preceramic Agricultural Settlements in the Cienega Valley, Southeastern Arizona*. University of Arizona Press, Tucson.
Huckell, L. B., 1995, Farming and Foraging in the Cienega Valley. In *Of Marshes and Maize: Preceramic Agricultural Settlements in the Cienega Valley, Southeastern Arizona*, edited by B. B. Huckell, pp. 74–97. University of Arizona Press, Tucson.
Huckell, L. B., 1996, Paleoethnobotany of Late Preceramic/Early Ceramic Sites Along the Santa Cruz River, Tucson, Arizona. Paper Presented, Society For American Archaeology, New Orleans.
Huebner, J. A., 1991, Cactus For Dinner, Again! An Isotopic Analysis of Late Archaic Diet in the Lower Pecos Region of Texas. In *Papers on Lower Pecos Prehistory*, edited by S. A. Turpin, pp. 175–190. University of Texas Press, Austin.
Huebner, J. A., 1995a, The Isotopic Composition and Ecology of Archaic Human Diet in the Eastern Chihuahuan Desert. Ph.D. dissertation, University of Texas. University Microfilms, Ann Arbor.
Huebner, J. A., 1995b, Stable Isotope Analysis of Bone and Soft Tissue From Four Mummies From the Eastern Chihuahuan Desert of Texas, Texas Archaeological Research Laboratory. Paper Presented, II World Congress on Mummy Studies, Cartegena, Columbia.
Hurtado, A. M., and Hill, K. R., 1990, Seasonality in a Foraging Society: Variation in Diet, Work Effort, Fertility, and the Sexual Division of Labor among the Hiwi of Venezuela. *Journal of Anthropological Research* 46:293–345.

Hurtado, A. M., Hill, K., Kaplan, H., and Hurtado, I., 1992, Trade-Offs Between Female Food Acquisition and Child Care Among Hiwi and Ache Foragers. *Human Nature* 3(3):185–216.

Kelly, R. L., 1983, Hunter-Gatherer Mobility Strategies. *Journal of Anthropol Research* 39:277–306.

Kelly, R. L., 1992, Mobility/Sedentism: Concepts, Archaeological Measures, and Effects. *Annual Review of Anthropology* 21:43–66.

Kelly, R. L., 1995, *The Foraging Spectrum: Diversity In Hunter-Gatherer Lifeways*. Washington, DC: Smithsonian Institution Press.

Lancaster, J., 1986, Groundstone. In *Short Term Sedentism in the American Southwest, the Mimbres Valley Salado*, edited by M. A. Nelson and S. Leblanc. University of New Mexico Press, Albuquerque.

Larsen, C. S., 1997, *Bioarchaeology: Interpreting Behavior from the Human Skeleton*. University Press, Cambridge.

Larsen, C. S., Ruff, C. B., Schoeninger, M. J., and Hutchinson, D. L., 1992, Population Decline and Extinction in La Florida. In *Disease and Demography in the Americas*, edited by J. W. Verano and D. H. Ubelaker, pp. 25–39. Smithsonian Institution Press, Washington.

Larsen, C. S., and Kelly, R., 1995, *Bioarchaeology of the Stillwater Marsh: Prehistoric Human Adaptation in the Western Great Basin*. American Museum of Natural History, New York.

Leviness, W. T., 1959, Pottery Mound Murals. *New Mexico Magazine* 37:22–23, 52.

Lovejoy, C. O., and Trinkaus, E., 1980, Strength and Robusticity of the Neandertal Tibia. *American Journal of Physical Anthropology* 53:465–470.

Lovejoy, C. O., Burstein, A. H., and Heiple, K. G., 1976, The Biomechanical Analysis of Bone Strength: A Method and Its Application to Platycnemia. *American Journal of Physical Anthropology* 44:489–506.

Lundelius, E. L. Jr., 1974, The Last 15,000 Years of Faunal Change in North America. *Museum Journal* 15:141–160.

Mabry, J. B., 1998, *Archaeological Investigations of Early Village Life in the Middle Santa Cruz Valley: Analysis and Synthesis*. Anthropological Papers No. 10, Institute For American Research, Tucson.

Mabry, J. B., and Clark, J. J., 1994, Early Village Life on the Santa Cruz River. *Archaeology in Tucson* 8(1):1–5. Center For Desert Archaeology, Tucson.

Mabry, J. B., Swartz, D. L., Wocherl, H., Clark, J. J., Archer, G. H., and Linderman, M. W., 1995, *Archaeological Investigations of Early Village Sites in the Middle Santa Cruz Valley: Descriptions of the Santa Cruz Bend, Square Hearth, Stone Pipe, and Canal Sites*. Anthropological Papers 13. Center For Desert Archaeology, Tucson.

Marmaduke, W. S., 1978, *Prehistoric Culture in Trans Pecos, Texas: An Ecological Explanation*. Ph.D. dissertation, University of Texas. University Microfilms, Ann Arbor.

Matson, R. G., 1991, *The Origins of Southwestern Agriculture*. University of Arizona Press, Tucson.

Minnis, P. E., 1985, *Domesticating Plants and People in the Greater American Southwest*. In *Prehistoric Food Production in North America*, edited by R. I. Ford, pp. 309–340. University of Michigan, Ann Arbor.

Minnis, P. E., 1992, Earliest Plant Cultivation in the Desert Borderlands of North America. In *The Origins of Agriculture: An International Perspective*, edited by C. W. Watson and P. J. Watson, pp. 121–141. Smithsonian Institution Press, Washington, D.C.

Minturn, P. D., 1998, Osteology of the Santa Cruz Bend Site: AZAA:12:746 (ASM). In *Archaeological Investigations of Early Village Sites in the Middle Santa Cruz Valley: Analysis and Synthesis, Anthropological Papers 19*, edited by J. B. Mabry, pp. 739–755. Center For Desert Archaeology, Tucson.

Minturn, P. D., and Lincoln-Babb, L., 1995, Bioarchaeology of the Donaldson Site and Los Ojitos. In *Of Marshes and Maize: Preceramic Agricultural Settlements in the Cienega Valley, Southeastern Arizona*, edited by B. B. Huckell, pp. 106–116. University of Arizona Press, Tucson.

Murdock, G. P., and Provost, C., 1973, Factors in the Division of Labor by Cultural Analysis. *Ethnology* 12:203–225.

Nagurka, M. L., and Hayes, W. C., 1980, An Interactive Graphics Package for Calculating Cross-Sectional Properties of Complex Shapes. *Journal of Biomechanics* 13:59–64.

Ogilvie, M. D., 2000, A *Biological Reconstruction of Mobility Patterns at the Foraging to Farming Transition in the American Southwest*. Ph.D. dissertation, University of New Mexico. University Microfilms, Ann Arbor.

Rafferty, G., 1985, The Archaeological Record on Sedentariness, Recognition, Development, and Implications. In *Advances in Archaeological Method and Theory*, Vol. 8, edited by M. B. Schiffer, pp. 113–156. Academic Press, New York.

Reinhard, K. J., Olive, B. W., and Steele, D. G., 1989, Bioarchaeology Synthesis. Study Unit 3, Southwestern Division Archaeological Overview, U.S. Army Corps of Engineers. In *From the Gulf To the Rio Grande: Human Adaptation in Central, South, and Lower Pecos, Texas*, edited by T. R. Hester, S. L. Black, D. G. Steele, B. W. Olive, A. A. Fox, K. J. Reinhard, and L. C. Bement, pp. 129–140. Arkansas Archaeological Survey, Fayetteville.

Riskind, D. H., 1970, Pollen Analysis of Human Coprolites of Parida Cave. In *Archaeological Investigations At Parida Cave, Val Verde County, Texas*, edited by R. K. Alexander, pp. 89–101. Papers of the Archaeological Salvage Project 19, Austin.

Ruff, C. B., 1987, Sexual Dimorphism in Human Lower Limb Bone Structure: Relationship to Subsistence Strategy and Sexual Division of Labor. *Journal of Human Evolution* 16:391–416.

Ruff, C. B., 1992, Biomechanical Analyses of Archaeological Human Skeletal Samples. In *Skeletal Biology of Past Peoples: Research Methods*, edited by S. R. Saunders and M. A. Katzenberg, pp. 37–58. Wiley-Liss, New York.

Ruff, C. B., 1994, Biomechanical Analysis of Northern and Southern Plains Femora: Behavioral Implications. In *Skeletal Biology in the Great Plains: Migration, Warfare, Health, and Subsistence*, edited by D. W. Owsley and R. L. Jantz, pp. 235–245. Smithsonian Press, Washington, D.C.

Ruff, C. B., 1997, Structural Analysis of Long Bones from La Florida: Interpreting Behavior. *American Journal of Physical Anthropology* Suppl. 24:201.

Ruff, C. B., and Hayes, W. C., 1983a. Cross-Sectional Geometry of Pecos Pueblo Femora and Tibiae-A Biomechanical Investigation. I: Method and General Patterns of Variation. *American Journal of Physical Anthropology.* 60:359–381.

Ruff, C. B., and Hayes, W. C., 1983b. Cross-Sectional Geometry of Pecos Pueblo Femora and Tibiae-A Biomechanical Investigation. II: Sex, Age, and Side Differences. *American Journal of Physical Anthropology* 60:383–400.

Ruff, C. B., and Larsen, C. S., 1990, Postcranial Biomechanical Adaptations to Subsistence Strategy Changes on the Georgia Coast. In *The Archaeology of Mission Santa Catalina De Guale: Bicultural Interpretation of A Population in Transition*, edited by C. S. Larsen, pp. 94–120. American Museum of Natural History, New York.

Ruff, C. B., Larsen, C. S., and Hayes, W. C., 1984, Structural Changes in the Femur with the Transition to Agriculture on the Georgia Coast. *American Journal of Physical Anthropology* 64:125–136.

Ruff, C. B., and Leo, F. P., 1986, Use of Computed Tomography in Skeletal Structure Research. *Yearbook of Physical Anthropology* 29:181–196.

Ruff, C. B., Walker, A., Trinkaus, E., and Larsen, C. S., 1993, Postcranial Robusticity in *Homo*. I: Temporal Trends and Mechanical Interpretation. *American Journal of Physical Anthropology* 91:21–53.

Ruff, C. B., Walker, A., and Trinkaus, E., 1994, Postcranial Robusticity in Homo. III: Ontogeny. *American Journal of Physical Anthropology* 95:35–54.

Ruff, C. B., McHenry, H. M., and Thackeray, J. F., 1999, The "Robust" Australopithecine Hip: Cross-Sectional Morphology of the SK 82 and 97 Proximal Femora. *American Journal of Physical Anthropology*. 109:509–521.

Ruff, C. B., 2000a, Biomechanical Analyses of Archaeological Human Skeletal Material. In *Biological Anthropology of the Human Skeleton*, edited by M. A. Katzenberg and S. H. Saunders, pp. 71–102. Alan R. Liss, New York.

Ruff, C. B., 2000b, Skeletal Structure and Behavioral Patterns of Prehistoric Great Basin Populations. In *Understanding Prehistoric Lifeways in the Great Basin Wetlands: Bioarchaeological Reconstruction and Interpretation*, edited by B. E. Hemphill and C. S. Larsen, pp. 290–320. University of Utah Press, Salt Lake City.

Rubin, C. T., Mcleod, K. J., and Bain, S. D., 1990, Functional Strains and Cortical Bone Adaptation: Epigenetic Assurance of Skeletal Integrity. *Journal of Biomechanics* Supplement 1, 23:43–54.

Schorsch, R. L. G., 1962, *The Physical Anthropology of Pottery Mound: A Pueblo IV Site in West Central New Mexico*. M.A. thesis. University of New Mexico, Albuquerque.

Shafer, H. J., 1986a, The Lower Pecos Environment: Evolution of the Present Landscape. In *Ancient Texans*, edited by G. Zappler, pp. 34–49. Texas Monthly Press, Austin.

Shafer, H. J., 1986b, Lower Pecos Lifeways: Housing and Daily Rounds. In *Ancient Texans*, edited by G. Zappler, pp. 94–131. Texas Monthly Press, Austin.

Smith, B. D., 1992, *Eastern North American Horticulture: Rivers of Change.* Smithsonian Institution Press, Washington, DC.

Speth, J. D., 1990, The Study of Hunter-Gatherers in the American Southwest: New Insights From Ethnology. In *Perspectives on Southwestern Prehistory*, edited by P. E. Minnis and C. L. Redman, pp. 15–25. Westview Press, Boulder.

Steele, D. G., and Olive, B. W., 1989, Bioarchaeology of Region 3 Study Area. Study Unit 3, Southwestern Division Archaeological Overview, U.S. Army Corp of Engineers. In *From The Gulf To the Rio Grande: Human Adaptation in Central, South, and Lower Pecos Texas*, edited by T. R. Hester, S. L. Black, D. G.

Steele, B. W. Olive, A. A. Fox, K. J. Reinhard, and L. C. Bement, pp. 93–114. Arkansas Archaeological Survey, Fayetteville.

Stock, J. A., 1983, *The Prehistoric Diet of Hinds Cave (41VV 456), Val Verde, County, Texas: The Coprolite Evidence*. M.A. thesis. College Station: Texas A&M University.

Storey, D. A., and Bryant, B. M. Jr., 1966, *A Preliminary Study of the Paleoecology of the Amistad Reservoir Area. National Science Foundation Research Report G2667*. University of Texas, Austin.

Taylor, W. W., 1964, *Tethered Nomadism and Water Territoriality: A Hypothesis. Actase Memorias*. Sobritiro Del 35 Congresso Internacional De Americanistas, Mexico.

Thiel, J. H., 1996, Faunal Exploitation of Early Villagers in the Sonoran Desert. Paper Presented, Society For American Archaeology, New Orleans.

Timoshenko, S. P., and Gere, J. M., 1972, *Mechanics of Materials*. Van Nostrand Reinhold, New York.

Trinkaus, E., and Ruff, C. B., 1989, Diaphyseal Cross-Sectional Morphology and Biomechanics of the Fond-De Foret 1 Femur and the Spy 2 Femur and Tibia. *Bull Soc Roy Bel Anthrop Prehist* 100:33–42.

Trinkaus, E., Churchill, S., and Ruff, C. B., 1994, Postcranial Robusticity in Homo. II: Humeral Bilateral Asymmetry and Bone Plasticity. *American Journal of Physical Anthropology* 93:1–34.

Turpin, S. A., Hennberg, M., and Riskind, D. W., 1986, Late Archaic Mortuary Practices in the Lower Pecos River Region, Texas. *Plains Anthropologist* 31:295–315.

Turpin, S. A., 1988, Seminole Sink: Excavations of A Vertical Shaft Tomb, Val Verde County, Texas. *Plains Anthropol Memoir* 22 (33, Part 2).

Turpin, S. A., 1992, More About Mortuary Practices in the Lower Pecos River Region of Southwest Texas. *Plains Anthropologist* 37:7–17.

Turpin, S. A., 1995, The Lower Pecos River Region of Texas and Northern Mexico. In *Bulletin of the Texas Archaeological Society*, edited by T. K. Pertula and N. Reese. *Texas Archaeological Society* 66:541–560.

Van Devender, T. R., and Spaulding, W. G., 1979, Development of Vegetation and Climate in the Southwestern United States. *Science* 222:701–710.

Van Devender, T. R., Betancourt, J., and Wimberly, M., 1984, Biogeographic Implications of A Packrat Midden Sequence From the Sacramento Mountains, South Central New Mexico. *Quaternary Research* 22:344–360.

Vierra, B. J., 1990, Archaic Hunter-Gatherer Archaeology in Northwestern New Mexico. In *Perspectives on Southwestern Prehistory*, edited by P. E. Minnis and C. L. Redman, pp. 57–70. Westview Press, Boulder.

Vierra, B. J., 1998. *41MV120: A Stratified Late Archaic Site in Maverick County, Texas. Archaeology Studies Report 7, Archaeological Survey Report 251*. Center For Archaeological Research, the University of Texas At San Antonio, and Environmental Affairs Division, Texas Department of Transportation, Austin.

Vivian, P. B., 1961, *Kachina: the Study of Animism and Anthropomorphism Within the Ceremonial Wall Paintings of Pottery Mound and Jeddito*. M.A. thesis. Iowa State University, Ames.

Voll, C., 1967, *The Glaze Paint Ceramics of Pottery Mound*. M.A. thesis. University of New Mexico, Albuquerque.

Ward, C. G., 1992, *Shelby Brooks Cave: The Archaeology of a Dry Cave in the Texas Trans-Pecos*. MA thesis. University of Texas, Austin.

Williams-Dean, G., 1978, *Ethnobotany and Cultural Ecology of Prehistoric Man in South Texas*. Ph.D. dissertation, Texas A&M University. University Microfilms, Ann Arbor.

Wills, W. H., 1988a, *Early Prehistoric Agriculture in the American Southwest*. School of American Research, Santa Fe.

Wills, W. H., 1988b, Early Agriculture and Sedentism in the American Southwest: Evidence and Interpretations. *Journal of World Prehistory* 2:445–488.

Wills, W. H., 1992, Plant Cultivation and the Evolution of Risk Prone Economies in the Prehistoric American Southwest. In *Transitions To Agriculture in Prehistory*, edited by A. B. Gebauer and T. D. Price, pp. 153–176. Prehistory Press, Madison.

Wills, W. H., and Huckell, B. B., 1994, Economic Implications of Changing Land-Use Patterns in the Late Archaic. In *Themes in Southwest Prehistory*, edited by G. Gumerman, pp. . School of American Research Press, Santa Fe.

Wills, W. H., Crown, P., Dean, J., and Laudton, C., 1994, Complex Adaptive Systems and Southwest Prehistory. In *Understanding Complexity in the Prehistoric Southwest*, edited by G. Gumerman and M. Gell-Mann, pp. 297–339. Addison-Wesley, New York.

Chapter **12**

Uplifted Head, Free Hands, and the Evolution of Human Walking

HENRY M. MCHENRY

We enter the 21st Century with so much more information about our ancestors than was available to our intellectual predecessors in the last 150 years. This volume is a testament to the richness of human origins and the strength of the scientific study of what it means. A sign health and vigor in these studies is apparent in the strong disagreements in the interpretations of the details. The hominids that sloshed through the muddy ash of Laetoli 3.6 m.y. ago left footprints that appear essentially human to some eyes (e.g., Charteris et al., 1982; Clarke, 1979; Day and Wickens, 1980; Hay and Leakey, 1982; Leakey, 1978; Leakey, 1987; Leakey and Hay, 1979; Robbins, 1987; Tuttle, 1985; Tuttle, 1987; Tuttle et al., 1991a; Tuttle et al., 1991b; White, 1980; White and Suwa, 1987; Schmid, this volume) and ape-like to other eyes (e.g., Clarke, 1999; Clarke and Tobias, 1995; Berillon, Deloison, and Meldurm this volume). This difference of opinion can be seen as reassuring: We are not lock step and blind followers of received evolutionary wisdom but instead a bunch of feisty individualists trying to discover what is most likely to be true based on evidence.

Evidence did not constrain early evolutionists in their speculation about human evolution. Lamarck (1809) had this to say about the evolutionary origin of bipedalism:

> As a matter of fact, if some race of quadrumanous animals, especially one of the most perfect of them, were to lose, by force of circumstances or some other cause, the habit of climbing in trees and grasping branches with its feet in the same way as with its hands, in order to hold on to them; and if the individuals of this race were forced for a series of generations to use their feet only for walking, and to give up using their hands like feet; there is no doubt, according to the observations detailed in the preceding chapter, that these quadrumanous animals would at length be transformed into bimanous, and that the thumbs on their feet would cease to be separated from the other digits, when they only used their feet for walking.
>
> (Lamarck, , 1914 p. 170 translated by Hugh Elliot)

Darwin (1859) was famously restrained on the implications of evolution for humans in the *Origin of Species* with only a single line about us : "Light will be thrown on the

From Biped to Strider: The Emergence of Modern Human Walking, Running, and Resource Transport, edited by D. Jeffrey Meldrum and Charles E. Hilton. Kluwer Academic/Plenum Publishers, New York, 2004.

origin of man and his history" (Darwin, 1859 p. 488). His supporters quickly entered where he was cautious to tread. Huxley (1863) was quite specific about human's place in nature from an evolutionary point of view. Haeckel (1868) detailed just how apelike extremities were transformed into the human hand and foot and how this transformation happened long before speech and developed. When Darwin (1872) finally did speculate on human evolution, he too stressed the primacy of bipedalism that arose either because of changes in feeding or habitat:

> As the progenitors of man became more and more erect, with their hands and arms more and more modified for prehension and other purposes, with their feet and legs at the same time modified for firm support and progression, endless other changes of structure would have been necessary. The pelvis would have had to be made broader, the spine peculiarly curved and the head fixed in an altered position"
>
> (Darwin, , 1872 p. 137–138)

Nineteenth century speculation turned into 20th and 21st Century reality. As Huxley prophesized "In still older strata do the fossilized bones of an ape more anthropoid, or a Man more pithecoid, than yet known await the researchers of some unborn paleontologist?" (Huxley, 1863 p. 159). We can now celebrate the birth of the great fossil finders whose skill, intelligence, persistence, and good fortune have revealed our rich collection of ancient hominids.

We know less than we would like about earliest transitions in the human line. Begun (this volume) makes a convincing case that our last common ancestor with the animal world may have walked on its knuckles as do modern African great apes. He and others (e.g., Corruccini, 1978; Corruccini and McHenry, 2001; Dainton and Macho, 1999; Hunt, 1991; Hunt, 1992; Jenkins and Fleagle, 1975; Richmond et al., 2001; Richmond and Strait, 2000; Tuttle, 1969; Washburn, 1968) point out that the early hominids bear some imprint of knuckle-walking adaptations in their wrists. The alternative view is less attractive but possible. Given the morphological, biomechanical, and phylogenetic constraints on large-bodied members of Hominoidea, curling those long and arboreally adapted fingers into a knuckle-walking posture may have been an evolutionary route for several ape lineages. But Begun makes a convincing case that such a scenario is unlikely.

In the first half of the 20th Century few believed that there was reliable fossil evidence for an early stage in human evolution when brains were ape-like, but bodies were human-like. Dart (1925) described the first known australopithecine as being erect, but he had no limb bones to make his case convincing. Broom (1936) published his sketch of a Sterkfontein distal femur that certainly looked like it belonged to a biped, but skeptics prevailed (e.g., Kern and Straus, 1949). The influential author of *Early Forerunners of Man*, Le Gros Clark (1934), held what appeared to be the reasonable view that these ape-brained South African fossils were not bipedal hominids. That was until Broom and Robinson discovered Sts 14 at Sterkfontein in 1947 (Broom and Robinson, 1947).

The discovery of Sts 14 caused a profound change of view. Le Gros Clark visited South Africa in 1947 just as the pelvic bones were being extracted (Clark, 1967). They were to his trained eyes quite human and this view quickly became perceived wisdom—the ape-brained australopithecines were human and bipedal (Clark, 1947; 1948).

As this volume attests, interpretations of fossils do not go unchallenged. Le Gros Clark's (1947; 1948) view was supported by a majority of scholars (e.g., Keith, 1948; Robinson, 1972a; Washburn and Patterson, 1951) but certainly not all (e.g., Oxnard, 1973; Oxnard, 1975b; Zuckerman, 1950a, b; 1966; 1970; Zuckerman et al., 1973). The discovery of the OH 8 foot in 1960 lead many to the view that it was essentially human and not at all ape-like (e.g., Aiello and Dean, 1990; Aiello et al., 1998; Clark, 1967; Day, 1973; Day, 1978; Day et al., 1976; Day and Napier, 1964; Day and Wood, 1968; Harcourt-Smith et al., 2002; McHenry, 1972; Preuschoft, 1971; Robinson, 1972b; Susman and Stern, 1982; Tuttle, 1988) but some experts doubted that interpretation (Kidd et al., 1996; Lamy, 1986; Le Floch-Prigent and Deloison, 1985; Lewis, 1989; Oxnard, 1975a; Oxnard, 1979; Oxnard and Lisowski, 1978; Oxnard and Lisowski, 1980; Sarmiento and Marcus, 2000) and arguments persist to this day on the meaning of early hominid foot morphology (Kidd et al., 1996; Lamy, 1986; Le Floch-Prigent and Deloison, 1985; Lewis, 1989; Oxnard, 1975a; Oxnard, 1979; Oxnard and Lisowski, 1978; Oxnard and Lisowski, 1980; Sarmiento and Marcus, 2000). The original describers of *Australopithecus afarensis* emphasized its total commitment to terrestrial bipedality and loss of ape-like climbing abilities (Gebo, 1992; Johanson et al., 1982; Latimer, 1988; Latimer, 1991; Latimer and Lovejoy, 1989; Latimer and Lovejoy, 1990a; Latimer and Lovejoy, 1990b; Latimer et al., 1982; Latimer et al., 1987; Leakey, 1981; Lovejoy et al., 1982; McHenry, 1991; McHenry 1982; McHenry, 1984; McHenry, 1994; Tuttle et al., 1991b; e.g., White, 1980) but as many authors in this volume point out (e.g., Deloison, Berillon, Schmid) there is a strong alternative view proposed long ago by several scholars (e.g., Bacon, 1994; Christie, 1977; Clarke and Tobias, 1995; Deloison, 1984; Deloison, 1985; Deloison, 1991; Deloison, 1997; Lamy, 1983; Lamy, 1986; Le Floch-Prigent and Deloison, 1985; Lewis, 1989; Sarmiento and Marcus, 2000; Senut, 1980; Senut and Tardieu, 1985; Stern and Susman, 1983; Stern and Susman, 1991; Susman and Stern, 1991; Susman et al., 1984; Tuttle, 1981).

It strikes those of us who have only seen casts of the Laetoli footprints as odd that so many differences of opinion can arise about their interpretation. Clarke (1979) excavated much of the G-trail and his keen eyes saw human-like traits, but on further reflection and experimentation, he came to the view that they had ape-like characteristics (Clarke, 1999; Clarke and Tobias, 1995). White (1980) also excavated some of the prints and saw all the marks of humanness. Tuttle (Tuttle, 1985; 1987; Tuttle et al., 1991a, b) confirmed White's view about the Laetoli footprints, but did not agree that the foot bones from Hadar attributed to *A. afarensis* (A.L. 333–115) could be from the same species that made the Laetoli impressions. Deloison, Berillon and Medrum (this volume) find primitive features unlike modern humans in the Laetoli footprints. It is interesting to read Schimd's paper in this volume, because he has two important qualifications. He was one of the first to note (contra Johanson et al., 1982) that *A. afarenis* had striking ape-like qualities in its body (Schmid, 1983). His reconstruction of Lucy's body adorns many texts. He stated in his oral presentation preceding this volume (Schmid, 2000) that he expected to find overlooked primitiveness in the footprints when he was awarded the golden opportunity to examine the freshly re-excavated original footprints. As he reports herein, what this author considers one of the most important confirmations of the original describers conclusions, that the Laetoli footprints are remarkably human-like in all fundamental respects.

The fossil record reveals just enough to peak curiosity about the sequence of bipedal changes in human evolution. The most complete specimen of *A. afarensis*, A.L. 288–1 ("Lucy") has short legs and her contemporaries had long toes (A.L. 333–115). The forelimb elements attributed to *A. afarensis* vary enormously in size. The large and powerful hands presumably belong to the males of that species. McHenry (1996) suggested that such strong dimorphism may imply the action of sexual selection. Darwin (1872) showed how sexual selection and specifically male–male competition, could explain canine size dimorphism. In bipedal hominids free hands could hold weapons and these replaced canines in the role of threat and fighting. Carrier (this volume) develops this view more precisely and fully. He documents how male–male fighting could have led to the enlargements of forelimbs in *Australopithecus* and the innovations of tools could have resulted in forelimb reduction in *Homo*.

By 1.54 m.y. *Homo erectus* (= *ergaster*) had long legs and other specimens (e.g., KNM-ER 1808) that very likely belong to the same species had lengthened thighs as early as 1.8 m.y. ago (Ruff and Walker, 1993). The more human-like limb proportions may have been acquired as early as 1.95 m.y. as evidenced by the pelvic bone, KNM-ER 3228 (McHenry and Coffing, 2000), and perhaps by 2.5 m.y. in *A. garhi* (Asfaw et al., 1999). Kramer (this volume) explores the implications for the energetics and behavioral ecology of the transition between *Australopithecus* and *Homo* and particularly the differences between males and females in burden transport (see also Kramer and Eck, 2000). There is a striking gender difference in burden transport among modern Venezuelan foragres (Hilton and Greaves, this volume) where older women carry a much greater proportion of the food resources than do younger (reproductive) women and men. Gender differences in mobility are also apparent in these people and in earlier societies. Ogilvie (this volume) shows that cross-sectional properties of femora reveal that mobility levels remained high among males in early agricultural societies in the Southwest, but mobility dramatically declined in females with the advent of agriculture.

Dramatic changes characterize the origin of hominid bipedality and the transition from *Australopithecus* to early *Homo*, but is there evidence of changes in the evolution of the genus *Homo*? Gruss and Schmitt (this volume) explore the meaning of limb lengths in *H. erectus* (= *ergaster*). They make a strong case for kinematic and kinetic differences between this hominid and modern humans due to its relatively long tibia. Meldrum (this volume) emphasizes the non-modern features apparent to him and others (e.g., Berillon, this volume; Deloison, this volume) in the Laetoli footprints and the foot bones of *Australopithecus* and *H. habilis*. The human foot and footprint record is unfortunately relatively poor in middle Pleistocene, but enough is there to make a case that "flat flexible feet may have been a relatively stable adaptation" until the late Pleistocene (Meldrum, this volume).

Experts in the morphology and function of hominid postcranial fossils tread through a dangerous mine field. The record is imperfect, appreciation of normal variability is often limited, understanding of genetic and developmental pathways to morphological change is undeveloped (Lovejoy et al., 2002), and preconceptions are often blinding. Views are evolving. We can make the aspiration that those ideas that correspond best with the empirical evidence survive and we can all enjoy the unfolding of greater knowledge about our origins.

REFERENCES

Aiello, L., and Dean, C., 1990, *An Introduction to Human Evolutionary Anatomy*. Academic Press (Harcourt Brace Jovanovich), London.

Aiello, L., Wood, B., Key, C., and Wood, C., 1998, Laser Scanning and Paleoanthropology: an example from Olduvai Gorge, Tanzania. In *Primate Locomotion: Recent Advances*, edited by E. Strasser, J. Fleagle, A. Rosenberger and H. McHenry, pp. 223–236. Plenum Press, New York.

Asfaw, B., White, T., Lovejoy, O., Latimer, B., Simpson, S., and Suwa, G., 1999, *Australopithecus garhi*: A New Species of Early Hominid from Ethiopia. *Science* 284:629–635.

Bacon, A. M., 1994, Interpretetion Fonctionnelle des Proportions de la Trochlee Femorale en Relation avec L'aptitude a la Rotation Axiale du Genou Chez ies Primates Simiiformes Actuels. *Annales de Paleontologie* 80:194–210.

Broom, R., 1936, A New Fossil Anthropoid Skull from South Africa. *Nature* 138:486–488.
Broom, R., and Robinson, J. T., 1947, The Jaw of the Male Sterkfontein Ape-Man. *Nature* 160:153.
Charteris, J., Wall, J. C., and Nottrodt, J. W., 1982, Pliocene Hominid Gait: New Interpretations Based on Available Footprint Data from Laetoli. *American Journal of Physical Anthropology* 58:133–144.
Christie, P. W., 1977, Form and Function of the Afar Ankle. *American Journal of Physical Anthropology* 47:123(abst).
Clark, W. E. L., 1947, Observations on the Anatomy of the Fossil Australopithecinae. *Journal of Anatomy, London* 81:300–333.
Clark, W. E. L., 1948, The Significance of the Fossil "Ape-Men" (Australopithecinae) from South Africa. *Proc. Zool. Soc. Lond.* 117:607.
Clark, W. E. L., 1967, *Man-Apes or Ape-Men.* New York: Holt, Rinehart and Winston.
Clarke, R. J., 1979, Early Hominid Footprints from Tanzania. *South African Journal of Science* 75:148–149.
Clarke, R. J., 1999, Discovery of Complete Arm and Hand of the 3.3 Million-Year-Old *Australopithecus* Skeleton from Sterkfontein. *South African Journal of Science* 96:477–480.
Clarke, R. J., and Tobias, P. V., 1995, Sterkfontein Member 2 Foot Bones of the Oldest South African Hominid. *Science* 269:521–524.
Corruccini, R. S., 1978, Comparative Osteometrics of the Hominoid Wrist Joint, with Special Reference to Knuckle-Walking. *Journal of Human Evolution* 7:307–321.
Corruccini, R. S., and McHenry, H. M., 2001, Knuckle-Walkng Hominid Ancestors. *Journal of Human Evolution* 40:507–512.
Dainton, M., and Macho, G. A., 1999, Did Knuckle Walking Evolve Twice? *Journal of Human Evolution* 36:171–194.
Dart, R. A., 1925, *Australopithecus africanus*: The Man-Ape of South Africa. *Nature* 115:195–199.
Darwin, C., 1859, *On the Origin of Species by Means of Natural Selection.* John Murray, London.
Darwin, C., 1872, *The Descent of Man and Selection in Relation to Sex.* D. Appleton & Co, New York.
Day, M. H., 1973, Locomotor Features of the Lower Limb in Hominids. *Symp. Zool. Soc. Lond.* 33:29–51.
Day, M. H., 1978, Functional Interpretations of the Morphology of Postcranial Remains of Early African Hominids. In *Early Hominids of Africa*, edited by C. Jolly, pp. 311–345. St. Martin's Press, New York.
Day, M. H., Leakey, R. E. F., Walker, A. C., and Wood, B. A., 1976, New Hominids from East Turkana, Kenya. *American Journal of Physical Anthropology* 45:369–436.
Day, M. H., and Napier, J. R., 1964, Hominid Fossils from Bed I, Olduvai Gorge, Tanganyika: Fossil Footbones. *Nature* 201:969–970.
Day, M. H., and Wickens, E. H., 1980, Laetoli Pliocene Hominid Footprints and Bipedalism. *Nature* 286:385–387.
Day, M. H., and Wood, B. A., 1968, Functional Affinities of the Olduvai Hominid 8 Talus. *Man* 3:440–455.
De Lamarck, J. B., 1809, *Philosophie Zoologique.* 1.
Deloison, Y., 1984, Comparative Study of Calcaneums of Primates and *Pan-Australopithecus-Homo* Relationships. *Comptes-rendus Hebdomadaires des Seances de l'Academie des Sciences, Paris* 299:115–118.
Deloison, Y., 1985, Comparative Study of Calcanei of Primates and *Pan-Australopithecus-Homo* Relationship. In *Hominid Evolution: Past, Present and Future*, edited by P. V. Tobias, pp. 143–147. Alan R. Liss, New York.
Deloison, Y., 1991, Les Australopitheques Marchaient-Ils comme Nous? In *Origine(s) de la bipedie chez les hominides*, edited by Y. Coppens and B. Senut, pp. 177–186. Cahiers de Paleoanthropologie, CNRS, Paris.
Deloison, Y., 1997, Description d'un Astragale Fossile de Primate et Comparaison avec des Astragales de Chimpanzes, d'*Homo sapiens* et d'Hominides Fossiles: Australopitheques et *Homo habilis*. Human *Paleontology* 324:685–692.
Gebo, D. L., 1992, Plantigrady and Foot Adaptation in African Apes: Implications for Hominid Origins. *American Journal of Physical Anthropology* 89:29–58.
Haeckel, E., 1868, *Naturliche Schopfungsgeschichle.* Reimer, Berlin.
Harcourt-Smith, W., O'Higgens, P., and Aiello, L., 2002, From Lucy to Littlefoot: A Three Dimensional Analysis of Plio-Pleistocene Hominin Tarsal Remains. *American Journal of Physical Anthropology Annual Meeting Issue* 2002:82.
Hay, R. L., and Leakey, M. D., 1982, The Fossil Footprints of Laetoli. *Scientific American* 246:50–57.
Hunt, K. D., 1991, Mechanical Implications of Chimpanzee Positional Behavior. *American Journal of Physical Anthropology* 86:521–536.
Hunt, K. D., 1992, Positional Behavior of *Pan troglodytes* in the Mahale Mountains and Gombe Stream National Parks, Tanzania. *American Journal of Physical Anthropology* 87:83–105.
Huxley, T. H., 1863, *Evidence As to Man's Place in Nature.* Williams & Norgate, London & Edinburgh.

Jenkins, F. A., and Fleagle, J. G., 1975, Knuckle-Walking and the Functional Anatomy of the Wrists in Living Apes. In *Primate Functional Morphology and Evolution*, edited by R. H. Tuttle, pp. 213–227. Mouton Publishers, The Hague.

Johanson, D. C., Lovejoy, C. O., Kimbel, W. H., White, T. D., Ward, S. C., Bush, M. E., Latimer, B. M., and Coppens, Y., 1982, Morphology of the Pliocene Partial Hominid Skeleton (A.L. 288–1) from the Hadar Formation, Ethiopia. *American Journal of Physical Anthropology* 57:403–452.

Keith, A., 1948, *A New Theory of Human Evolution*. Watts and Co, London.

Kern, H. M., and Straus, W. L., 1949, The Femur of *Plesianthropus transvaalensis*. *American Journal of Physical Anthropology* 7:53–77.

Kidd, R. S., O'Higgins, P., and Oxnard, C. E., 1996, The OH8 foot: A Reappraisal of the Functional Morphology of the Hindfoot Utilizing a Multivariate Analysis. *Journal of Human Evolution* 31:269–291.

Kramer, P. A., and Eck, G. G., 2000, Locomotor Energetics and Leg Length in Hominid Bipedality. *Journal of Human Evolution* 38:651–666.

Lamarck, J. B., 1914, *Zoological Philosophy*. Macmillon, London.

Lamy, P., 1983, Le Systeme Podal de Certains Hominides Fossilles du Plio-Pleistocene d'Afrique de l'est: Etude Morphodynamique. *L'Anthropologie* 87:435–464.

Lamy, P., 1986, The Settlement of the Longitudinal Plantar Arch of Some African Plio-Pleistocene Hominids: A Morphological Study. *Journal of Human Evolution* 15:31–46.

Latimer, B. M., 1988, Functional Analysis of the Pliocene Hominid Ankle and Pedal Bones Recovered from the Hadar Formation, Ethiopia: 1974–1977 Collections. U.M.I. Dissertation Information Service, Ann Harbor.

Latimer, B. M., 1991, Locomotor Adaptations in *Australopithecus afarensis*: The Issue of Arboreality. In *Origine(s) de la bipedie chez les hominides*, edited by Y. Coppens and B. Senut, pp. 169–176. Cahiers de Paleoanthropologie, CNRS, Paris.

Latimer, B. M., and Lovejoy, C. O., 1989, The Calcaneus of *Australopithecus afarensis* and Its Implications for the Evolution of Bipedality. *American Journal of Physical Anthropology* 78:369–386.

Latimer, B. M., and Lovejoy, C. O., 1990a, Hallucal Tarsometatarsal Joint in *Australopithecus afarensis*. *American Journal of Physical Anthropology* 82:125–134.

Latimer, B. M., and Lovejoy, C. O., 1990b, Metatarsophalangeal Joints of *Australopithecus afarensis*. *American Journal of Physical Anthropology* 83:13–23.

Latimer, B. M., Lovejoy, C. O., Johanson, D. C., and Coppens, Y., 1982, Hominid Tarsal, Metatarsal, and Phalangeal Bones Recovered from the Hadar Formation: 1974–1977 Collections. *American Journal of Physical Anthropology* 57:701–720.

Latimer, B. M., Ohman, J. C., and Lovejoy, C. O., 1987, Talocrural Joint in African Hominoids: Implication for *Australopithecus afarensis*. *American Journal of Physical Anthropology* 74:155–175.

Le Floch-Prigent, P. P., and Deloison, Y., 1985, Anatomical Comparison Between Entire feet on Chimpanzee, Gorilla, and Man. In *Hominid Evolution: Past, Present and Future*, edited by P. V. Tobais, pp. 135–142. Alan R. Liss, New York.

Le Gros, C. W. E., 1934, *Early Forerunners of Man, The Morphological Study of the Evolutionary Origin of the Primates*. Baillière, Tindall and Cox, London.

Leakey, M. D., 1978, Pliocene Footprints at Laetolil, Northern Tanzania. *Antiquity* 52:133.

Leakey, M. D., 1981, Tracks and Tools. *Phil. Trans. R. Soc. Lond.* B 292:95–102.

Leakey, M. D., 1987, The Hominid Footprints. In *Laetoli A Pliocene Site in Northern Tanzania*, edited by M. D. Leakey and J. M. Harris, pp. 490–496. Clarendon Press, Oxford.

Leakey, M. D., and Hay, R. L., 1979, Pliocene Footprints in the Laetolil beds at Laetoli, Northern Tanzania. *Science* 278:317–323.

Lewis, O. J., 1989, *Functional Morphology of the Evolving Hand and Foot*. Oxford Press, Clarendon.

Lovejoy, C. O., Donald, C. D. C., and Yves, Y., 1982, Hominid Lower Limb Bones Recovered from the Hadar Formation: 1974–1977 Collections. *American Journal of Physical Anthropology* 57:679–700.

McHenry, H. M., 1972, *The Postcranial Anatomy of Early Pleistocene Hominids*. Ph.D. dissertation, Harvard University, Cambridge.

McHenry, H. M., 1982, The Pattern of Human Evolution: Studies on Bipedalism, Mastication and Encephalization. *Annual Review of Anthropology* 11:151–173.

McHenry, H. M., 1984, The Common Ancestor: A Study of The Post-Cranium of *Pan Paniscus*, *Australopithecus*, and Other Hominoids. In *The Pygmy Chimpanzee*, edited by R. L. Susman, pp. 201–230. Plenum Press, New York.

McHenry, H. M., 1994, Early Hominid Postcrania: Phylogeny And Function. In *Integrative Pathways to the Past: Paleoanthropological Papers in Honor of F. Clark Howell*, edited by R. S. Corruccini and R. L. Ciochon, pp. 251–268. Prentice-Hall, Engelwood Cliffs.

McHenry, H., 1991, First Steps? Analyses of the Postcranium of Early Hominids. In *Origine(s) de la bipedie chez les hominides*, edited by Y. Coppens and B. Senut, pp. 133–142. Cahiers de Paleoanthropologie, CNRS, Paris.

McHenry, H. M., and Coffing, K., 2000, *Australopithecus* to *Homo*: Transformations in Body and Mind. *Annual Review of Anthropology* 29:129–146.

Oxnard, C. E., 1973, *Form and Pattern in Human Evolution: Some Mathematical, Physical, and Engineering Approaches*. University of Chicago Press, Chicago.

Oxnard, C. E., 1975a, The Place of the Australopithecines in Human Evolution: Grounds for Doubt. *Nature* 258:389–395.

Oxnard, C. E., 1975b, *Uniqueness and Diversity in Human Evolution*. University of Chicago Press, Chicago.

Oxnard, C. E., 1979, Relationship of *Australopithecus* and *Homo*: Another View. *Journal of Human Evolution* 8:427–432.

Oxnard, C. E., and Lisowski, F. P., 1978, The Olduvai Foot. *Anatomy Research* 190:499–50.

Oxnard, C. E., and Lisowski, F. P., 1980, Functional Articulation of Some Hominoid Foot Bones: Implications for the Olduvai (Hominid 8) Foot. *American Journal of Physical Anthropology* 52:107–118.

Preuschoft, H., 1971, Body Posture and Mode of Locomotion in Early Pleistocene Hominids. *Folia Primatologica* 14:209–240.

Richmond, B. G., Begun, D. R., and Strait, D. S., 2001, Origin of Human Bipedalism: The Knuckle-Walking Hypothesis Revisited. *Yearbook of Physical Anthropology* 44:70–105.

Richmond, B. G., and Strait, D. S., 2000, Evidence that Humans Evolved from a Knuckle-Walking Ancestor. *Nature* 404:382–385.

Robbins, L. M., 1987, Hominid Footprints from Site G. In *Laetoli A Pliocene Site in Northern Tanzania*, edited by M. D. Leakey and J. M. Harris, pp. 497–502. Oxford: Clarendon Press.

Robinson, J. T., 1972a, The Bearing of East Rudolf Fossils on Early Hominid Systematics. *Nature* 240:239–240.

Robinson, J. T., 1972b, *Early Hominid Posture and Locomotion*. University of Chicago Press, Chicago.

Ruff, C. B., and Walker, A., 1993, Body Size and Body Shape. In *The Nariokotome Homo erectus Skeleton*, edited by A. Walker and R. Leakey, pp. 234–265. Harvard University Press, Cambridge.

Sarmiento, E. E., and Marcus, L. F., 2000, The Os Navicular of Humans, Great Apes, OH 8, Hadar, And *Oreopithecus*: Function, Phylogeny, and Multivariate Analyses. *American Museum Novitates* 3288:1–38.

Schmid, P., 1983, Eine Rekonstruktion des Skelettes von A.L. 288–1 (Hadar) und deren Konsequenzen. *Folia primatologica* 40:283–306.

Schmid, P., 2000, Functional Interpretation of the Laetoli Footprints. *American Journal of Physical Anthropology* Supplement 30:271.

Senut, B., 1980, New Data on the Humerus and Its Joints in Plio-Pleistocene Hominoids. *Collection Anthropologie* 4:87–93.

Senut, B., and Tardieu, C., 1985, Functional Aspects of Plio-Pleistocene Hominoid Limb Bones: Implications for Taxonomy and Phylogeny. In *Ancestors: The Hard Evidence*, edited by E. Delson, pp. 193–201. Alan R. Liss, Inc., New York.

Stern, J. T., and Susman, R. L., 1983, The Locomotor Anatomy of *Australopithecus afarensis*. *American Journal of Physical Anthropology* 60:279–318.

Stern, J. T., and Susman, R. L., 1991, "Total Morphological Pattern" Versus the "Magic Trait": Conflicting Approaches to the Study of Early Hominid Bipedalism. In *Origine(s) de la bipedie chez les hominides*, edited by Y. Coppens and B. Senut, pp. 99–112. Cahiers de Paleoanthropologie, CNRS, Paris.

Susman, R. L., and Stern, J. T., 1982, Functional Morphology of *Homo habilis*. *Science* 217:931–934.

Susman, R. L., and Stern, J. T., 1991, Locomotor Behavior of Early Hominids: Epistemology and Fossil Evidence. In *Origine(s) de la bipedie chez les hominides*, edited by Y. Coppens and B. Senut, pp. 121–132. Cahiers de Paleoanthropologie, CNRS, Paris.

Susman, R. L., Stern, J. T., and Jungers, W. L., 1984, Arboreality and Bipedality in the Hadar Hominids. *Folia Primatologica* 43:113–156.

Tuttle, R. H., 1969, Knuckle-Walking and the Problem of Human Origins. *Science* 166:953–961.

Tuttle, R. H., 1981, Evolution of Hominid Bipedalism and Prehensile Capabilities. *Phil. Trans. R. Soc. London* B 292:89–94.

Tuttle, R. H., 1985, Ape Footprints and Laetoli Impressions: A Response to SUNY Claims. In *Hominid Evolution: Past, Present and Future*, edited by P. V. Tobias, pp. 129–134. Alan R. Liss, New York.

Tuttle, R. H., 1987, Kinesiological Inferences and Evolutionary Implications from Laetoli Bipedal Trails G-1, G-2/3 and A. In *Laetoli: A Pliocene Site in Northern Tanzania*, edited by M. D. Leakey and J. M. Harris, pp. 503–523. Clarendon Press, Oxford.

Tuttle, R. H., 1988, *Comparative Anatomy and Evolution: The Foot,* pp. 12–24. Churchill Livingstone, Edinburgh.

Tuttle, R. H., Webb, D. M., and Baksh, M. 1991a, Laetoli Toes and *Australopithecus afarensis. Human Evolution* 6:193–200.

Tuttle, R. H., Webb, D. M., and Tuttle, N. I., 1991b, Laetoli Footprint Trails and the Evolution of Hominid Bipedalism. In *Origine(s) de la bipedie chez les hominides*, edited by Y. Coppens and B. Senut, pp. 187–198. Cahiers de Paleoanthropologie, CNRS, Paris.

Washburn, S. L., 1968, *The Study of Human Evolution.* Candon Lectures, Eugene.

Washburn, S. L., and Patterson, B., 1951, Evolutionary Importance of the South African Man-Apes. *Nature* 167:650–651.

White, T. D., 1980, Evolutionary Implications of Pliocene Hominid Footprints. *Science* 208:175–176.

White, T. D., and Suwa, G., 1987, Hominid Footprints at Laetoli: Facts and Interpretations. *American Journal of Physical Anthropology* 72:485–514.

Zuckerman, S. S., 1950a, South African Fossil Anthropoids. *Nature* 165:652.

Zuckerman, S. S., 1950b, South African Fossil Hominoids. *Nature* 166:158–159.

Zuckerman, S. S., 1966, Myth and Method in Anatomy. *Journal of the Royal College of Surgeons of Edinburgh* 11:27–114.

Zuckerman, S. S., 1970, *Beyond the Ivory Tower.* Taplinger Publishing Co., New York.

Zuckerman, S. S., Ashton, E. H., Flinn, R. M., and Oxnard, C. E., 1973, Some Locomotor Features of the Pelvic Girdle in Primates. *Symposium of the Zoological Society of London* 33:71–165.

Index